Springer Undergraduate Texts in Mathematics and Technology

More information about this series at http://www.springer.com/series/7438

Alko R. Meijer

Algebra for Cryptologists

Alko R. Meijer
Tokai, Cape Town
South Africa

ISSN 1867-5506 ISSN 1867-5514 (electronic)
Springer Undergraduate Texts in Mathematics and Technology
ISBN 978-3-319-30395-6 ISBN 978-3-319-30396-3 (eBook)
DOI 10.1007/978-3-319-30396-3

Library of Congress Control Number: 2016945761

Mathematics Subject Classification (2010): 94A60, 08-01, 11A, 11T

Printed on acid-free paper

This Springer imprint is published by Springer Nature
The registered company is Springer International Publishing AG Switzerland

Foreword

It is sometimes claimed that the world's population falls into two classes: those who believe that the population can be divided into two classes, and those who don't. We won't enter into that argument, but it is true that the reading population can be divided into those who read forewords and those who don't. Since you have read this far, you clearly belong to the first class, which gives me a chance to explain what this book is about.

Cryptology is a subject that lies in the intersection of three major fields of science: Computer Science, Electrical and Electronic Engineering and Mathematics—given in alphabetical order, to avoid fruitless discussions about their relative importance. Mathematics, in the form of ("abstract") algebra, number theory and discrete mathematics in general, provides much of its logical foundation, but computer scientists provide many of the opportunities for its use, while the electrical and electronic engineers supply the platforms on which implementation can take place.

The primary purpose of this book is to provide individuals with Electronic Engineering or Computer Science backgrounds, who find themselves entering the world of Cryptology, either as practitioners or as students, with some essential insights into the Algebra used. Here "cryptology" encompasses both the secret key/symmetric and public key/asymmetric aspects of the field, but the emphasis is perhaps more on the symmetric side than is usual in the textbook literature. Partial justification for this, if justification is needed, is given in Sect. 1.6.3, and the importance of symmetric cryptography in practice will, I hope, become apparent in the course of reading this book.

This should give an indication of the mathematical knowledge you, the reader, are assumed to have: firstly, that mysterious quality which used to be called "mathematical maturity", meaning you don't run away screaming in fright when you encounter a Σ sign. More constructively, you need to have some idea of what a mathematical proof is (e.g. proof by induction), and have a *very* basic understanding of propositional logic and Boolean algebra. You are assumed to be familiar with linear algebra (or at least with vectors and matrices). Familiarity with elementary probability theory would be useful, but the concepts most essential for our purposes are dealt with in the Appendix.

On the other hand, the reader is not expected to have had any previous exposure to ("abstract" or, as it was called in the middle decades of the last century, "modern") Algebra as such, and I believe that this book could, in fact, be used as an introduction to Algebra, where, unusually, the emphasis is on its applications, while the purely algebraic content itself is severely limited.

It is assumed that your "mathematical maturity" will allow for the development of the material at a fairly rapid pace and allow for explanations of how and where these concepts appear in the field. If the reader already has some, or even a great, knowledge of Algebra, I apologise for boring him or her but hope that the applications may nevertheless be interesting. My advice to such readers is to scan the relevant sections very quickly. The same applies to readers who, say, have previous acquaintance with Coding Theory or Information Theory: just scan the brief introductions to these fields.

Just three final comments (and thank you for reading this far): Firstly, many sections are followed by a short sequence of generally not very demanding exercises. Please do not skip them, even if you don't actually solve all the problems. Some of them are simple applications, a few others demand considerable thought. Give at least some thought to the latter kind. There are also expressions like "the diligent reader is invited to prove this" in the text, not explicitly labelled "Exercises". Such statements mean that you are welcome to accept the relevant statement as true, without actually constructing a proof yourself.

Secondly: While this book is aimed at those interested in Cryptology, I must stress that the book concentrates on the Mathematics involved: our descriptions blithely ignore implementation issues, including their security, and all matters concerned with the complexity of any algorithms that we deal with.

Finally, you will notice that there are many footnotes.[1] Most of these refer to original sources of definitions, theorems, etc., but some take the form of more or less parenthetical remarks, which are relevant to the topic discussed, but don't fit into the argument being developed.[2] If footnotes get you down, blame the LaTeX typesetting language, which makes inserting them far too easy.

James Thurber quoted with approval the schoolgirl who reviewed a book she had been told to read and review in the single sentence "This book told me more about penguins than I wanted to know." I hope that in reading this book you won't find yourself muttering the equivalent of "Perish these pestilential penguins!" to yourself. But there will be occasions where the subject matter is already familiar to you, in which case you should merely scan the relevant section. This may apply in particular to the early sections of Chap. 1; you are probably familiar with the ideas considered there, but the material is included in an attempt to make the book as self contained as possible. It is my pleasure to thank the reviewers of the manuscript, and in particular Dr Christine Swart of the University of Cape Town, for their encouragement and constructive criticism. Thanks are also due to various groups of aspiring cryptologists on whom I have tested much of the material in this book.

But I must accept responsibility for any remaining errors and other shortcomings.

---

[1] I once read the academic work of an author described as "laboriously hobbling along on a set of footnotes". I hope this description does not fit here.

[2] And some of the footnotes serve no purpose at all, like the two on this page.

I hope you will get as much enjoyment out of this book as I have out of my involvement with Cryptology.

Tokai, Cape Town, South Africa
December 2015

Alko R. Meijer

Contents

1 Prerequisites and Notation

In this introductory chapter we shall quickly review some mathematical concepts and in the process establish some notational conventions which we shall follow, or at least attempt to follow, in the coming chapters.

1.1 Sets

The concept of a *set* is fundamental in Mathematics, but undefined (like the term "point" in Euclidean Geometry). One can only explain its meaning by giving some synonyms for the word "set", such as "collection", "aggregate", etc. A set is thus a collection of things, which are called its *elements*.

One way of specifying a set is to list all its elements, e.g. $A=\{a, b, c, d\}$, using braces ($\{ \, , \}$).

Another way is to specify some property that the elements of the set share, e.g. "the set A consists of the first four letters of the alphabet". The notation $\{x|P(x)\}$ will always be used to denote the set of all "things" which satisfy the condition P. Thus, for example, $\{x|x \text{ is an even natural number}\}$ denotes the set whose elements are the integers 0, 2, 4, 6, ... In this notation the *empty set* ϕ may be defined as $\phi = \{x|x \neq x\}$.

The fact that some x is an element of a set A is denoted by $x \in A$. If x does not belong to A this is denoted by $x \notin A$.

Given two sets A and B their *intersection* $A \cap B$ is defined by

$$A \cap B = \{x|x \in A \text{ and } x \in B\}$$

and their *union* $A \cup B$ by

$$A \cup B = \{x|x \in A \text{ or } x \in B\}.$$

A.R. Meijer, *Algebra for Cryptologists*, Springer Undergraduate Texts in Mathematics and Technology,
DOI 10.1007/978-3-319-30396-3_1

Note that the "or" in this expression is to be interpreted in its *inclusive* sense: $A \cup B$ contains all elements which belong to A or to B *or to both.* We shall come to the *exclusive or* in a little while.

A set A is called a *subset* of a set B (denoted by $A \subseteq B$) if every element of A also belongs to B, i.e. if (and only if) the following[1] holds:

$$x \in A \Longrightarrow x \in B$$

and in such a case we write $A \subseteq B$. We might also express this as A being contained in B or B containing A.

If A is a subset of B but the two sets are not the same (i.e. every element of A is also in B, but there is at least one element in B which does not belong to A) we may write $A \subset B$. In such a case we may, if also $A \neq \phi$, refer to A as a *proper* subset of B.

The definition of subset given above has an interesting consequence. It is logically equivalent to

$$A \subseteq B \iff [x \notin B \Longrightarrow x \notin A].$$

But, for any set B and any $x \in B$, $x \notin \phi$ is true, so the empty set ϕ is actually a subset of any set whatsoever.[2]

Just a little less trivially the definition shows that, for any sets A and B

$$A \cap B \subseteq A \subseteq A \cup B.$$

To prove that two sets A and B are equal it is frequently easiest to prove that both inclusions $A \subseteq B$ and $B \subseteq A$ hold.

The reader is also assumed to be familiar with the *Distributive Laws* (and if he/she isn't they are fun and easy to prove): For any sets A, B and C

$$A \cup (B \cap C) = (A \cup B) \cap (A \cup C)$$
$$A \cap (B \cup C) = (A \cap B) \cup (A \cap C).$$

Two sets A and B are called *disjoint* if they have no element in common, or, in other words, if $A \cap B = \phi$.

We shall (mainly) deal with finite sets, i.e. sets containing only a finite number of elements (like, say, the set of all bitstrings of length 128, but unlike the set of all integers). We shall denote the number of elements of a set A (also called the *cardinality* of A) by $\#A$ or by $|A|$. Thus, for example, if $A = \{a, 5, 67, \alpha, x\}$, then $\#A = 5$. If A and B are disjoint finite sets, then $\#(A \cup B) = \#A + \#B$, but generally one can't say more than that $\#(A \cup B) \leq \#A + \#B$

[1]Here, and everywhere else, "$p \Longrightarrow q$", where p and q are propositions, means, of course, "If p then q". Similarly " $\iff$ " means "if and only if".

[2]From which it follows that there is only one empty set, since if there were two, each would be contained in the other, so they would be equal after all.

(and similarly that $\#(A \cap B) \leq \#A$ and $\#(A \cap B) \leq \#B$). If A is a finite set, and $B \subseteq A$, then the mere fact that $\#B = \#A$ shows that $B = A$. But in the infinite case it is quite possible to have $B \subset A$ even though they have the same (infinite) cardinality.

Finally, we shall denote by $A \backslash B$ the set of all elements of A which do not belong to B. This is sometimes called the (relative) *complement* of B in A. Thus

$$A \backslash B = \{x | x \in A \text{ and } x \notin B\}.$$

Clearly B and $A \backslash B$ are disjoint sets.

De Morgan's Laws state that for any sets A, B and C:

$$A \backslash (B \cup C) = (A \backslash B) \cap (A \backslash C),$$
$$A \backslash (B \cap C) = (A \backslash B) \cup (A \backslash C).$$

Again, proving these gives one a few minutes of innocent amusement.

The set $(A \backslash B) \cup (B \backslash A)$, sometimes denoted by $A \Delta B$, is frequently referred to as the *symmetric difference* of A and B. It clearly consists of all elements which belong to A or to B but not to both. In this way it corresponds to the *exclusive or.*

Sometimes one has to deal with families of sets all of which are subsets of one large set U, the nature of which is understood from the context of the discussion, and which may be referred to as the "universe" of the discussion. In such cases, rather than always using the notation $U \backslash A$ for the (relative) complement of a member of this family, we shall refer simply to the *complement of a set*, and use some notation like A^c or $\overline{A}$ to denote it: Thus

$$A^c = U \backslash A.$$

De Morgan's laws may then be written as

$$(B \cap C)^c = B^c \cup C^c,$$
$$(B \cup C)^c = B^c \cap C^c,$$

or, if a family $\{B_i\}_{i \in I}$ is involved,

$$\Big(\bigcap_{i \in I} B_i\Big)^c = \bigcup_{i \in I} B_i^c,$$
$$\Big(\bigcup_{i \in I} B_i\Big)^c = \bigcap_{i \in I} B_i^c.$$

In what follows we shall deal with lots of different sets, but sets that immediately spring to mind in cryptography are sets like the *plaintext space* and the *ciphertext space*, which are, respectively, the set of all messages that might be sent and the set of the encryptions of those messages. The encryption process itself then becomes a set of functions from the one set to the other. We shall clear up that vague and imprecise description in Sect. 1.6. (We refuse to get involved in any discussion of why the word "space" is used.)

Exercises

1. Show that if A and B are sets, the following conditions are equivalent:
   - $A \subseteq B$;
   - $A \cap B = A$;
   - $A \cup B = B$.
2. Show that a set with n elements has $\frac{n!}{(n-m)!m!}$ different subsets with m elements, for any m such that $0 \leq m \leq n$. (Note that the *binomial coefficient* $\frac{n!}{(n-m)!m!}$ will usually be written as $\binom{n}{m}$.)
3. Show, using the binomial theorem or otherwise, that
$$\sum_{i=0}^{n} \binom{n}{i} = 2^n.$$
4. Show that if A is a set with $\#A = n$, then A has 2^n different subsets.

1.2 Products of Sets

The (Cartesian) product of two sets A and B is the set of all ordered pairs (a, b) with the first entry from A and the second from B:

$$A \times B = \{(a, b) | a \in A \text{ and } b \in B\}.$$

Clearly $\#(A \times B) = \#A \times \#B$. By extension we can form the product of any number of sets by forming ordered n-tuples instead of pairs:

$$\prod_{i=0}^{n-1} A_i = \{(a_0, a_1, \ldots a_{n-1}) | a_i \in A_i \text{ for all } i \in \{0, 1, \ldots, n-1\}\}.$$

We shall frequently consider situations in which all the sets A_i are identical. In such a case we would use the notation A^n for the set

$$\prod_{i=0}^{n-1} A = \{(a_0, a_1, \ldots, a_{n-1}) | a_i \in A \text{ for all } i \in \{0, 1, \ldots, n-1\}\}.$$

A warning may be appropriate here: When we come to discuss groups and other algebraic structures, we may use the same notation (A^n) with a different meaning; the interpretation of the notation will, we hope, be clear from the context.

It is clear that the cardinality of the set $\prod_{i=0}^{n-1} A$ is $(\#A)^n$.

An example of the use of such a construction is in the well-known concept of *vector spaces*: let $\mathbb{R}$ denote the set of real numbers. Then we get a three-dimensional real vector space by considering the set $\mathbb{R}^3$ and defining addition of vectors and multiplication of a vector by a scalar $\alpha \in \mathbb{R}$ by

$$(x_0, x_1, x_2) + (y_0, y_1, y_2) = (x_0 + y_0, x_1 + y_1, x_2 + y_2)$$

and

$$\alpha(x_0, x_1, x_2) = (\alpha x_0, \alpha x_1, \alpha x_2)$$

respectively.

In our subsequent discussions we shall frequently look at products of the form $\prod_{i=0}^{n-1} A$ where A is the set consisting only of 0 and 1: $A = \{0, 1\}$. In this case A^n will be the set of all binary strings of length n, and, in fact, the set of all such strings can be considered as the union of all such sets, i.e. as the set

$$\bigcup_{i=0}^{\infty} A^i,$$

which is frequently denoted by $\{0, 1\}^*$.

Exercises

1. Let A, B, C be sets. Prove that $A \times (B \cup C) = (A \times B) \cup (A \times C)$.
2. Let $A = \{1, 2, \ldots, n\}$ and $B_k = \{x | x \in A, x \geq k\}$. What are

- $\#(A \times B_k)$, and
- $\#(\bigcup_{k=1}^{n} (A \times B_k))$?

1.3 Relations

In common parlance, the meaning of the word "relation" (other than when applied to family relationships!) is intuitively clear. Nobody will deny that "Dogs are more intelligent than cats" describes a relationship between the intelligences of a couple of animal species. When a proper definition, acceptable to the mathematical mind, is required, a considerable amount of abstraction is necessary:

A relation on a set A is a subset $R \subseteq A \times A$. If $(a_0, a_1) \in R$, we may write $a_0 R a_1$. As an example, consider the set $\mathbb{Z}$ of the integers: it would be correct but rather unusual (not to say, pedantic) to write $(-1, 1) \in \leq$ rather than $-1 \leq 1$.

A relation $\sim$ on a set A is called an *equivalence relation* (on A) if it satisfies the following three properties:

1. **Reflexivity**[3] :$a \sim a \; \forall a \in A$;
2. **Symmetry:** if $a \sim b$ then $b \sim a$;
3. **Transitivity:** if $a \sim b$ and $b \sim c$ then $a \sim c$.

Examples Consider the set of all integers:

1. The relation $<$ is transitive, but is neither reflexive nor symmetric;
2. The relation $\leq$ is reflexive and transitive, but not symmetric;
3. The relation $=$ is an equivalence relation, even if it is a rather dull one.

To get a less trivial example of an equivalence relation on the set of integers, we need the following definitions. These will prove to be very important for our future deliberations in both symmetric and asymmetric cryptology.

Definitions

1. Let m and n be integers. We call m *divisible* by n if there exists an integer q such that $m = q \cdot n$. We denote this relation by $n|m$ and its negation by $n \not| m$.
2. Let m_1, m_2 and n be integers. We call m_1 and m_2 *congruent modulo n* if $m_1 - m_2$ is divisible by n, and denote this by $m_1 \equiv m_2 \bmod n$ (and its negation by $m_1 \not\equiv m_2 \bmod n$).

Theorem *Let n be any nonzero integer. Then congruence modulo n is an equivalence relation on the set of integers.*

Proof

Reflexivity: Let a be any integer. Then $a - a = 0 = 0 \cdot n$, so that $n|(a-a)$, i.e. $a \equiv a \bmod n$.
Symmetry: Let $a \equiv b \bmod n$. Then $a - b = q \cdot n$ for some integer q. Hence $b - a = (-q) \cdot n$, so that $b \equiv a \bmod n$.
Transitivity: Let $a \equiv b \bmod n$ and $b \equiv c \bmod n$. This means that there exist integers q_1 and q_2 such that $a - b = q_1 \cdot n$ and $b - c = q_2 \cdot n$. But then $a - c = (a - b) + (b - c) = q_1 \cdot n + q_2 \cdot n = (q_1 + q_2) \cdot n$, so that $a \equiv c \bmod n$.

From this example, let us return to the general principles:

Definition Let $A \neq \phi$ be a set, and let $\{A_i\}_{i \in I}$ be a family of subsets of A such that

$$A_i \neq \phi \;\; \forall i \in I,$$

$$\bigcup_{i \in I} A_i = A,$$

$$A_i \cap A_j = \phi \;\; \forall i, j \in I, \text{ with } i \neq j.$$

Thus every element of A belongs to exactly one of the subsets. Then $\{A_i\}_{i \in I}$ is called a *partition* of A.

[3] Here and elsewhere we use the symbol $\forall$ to denote the "universal quantifier" from mathematical logic, meaning "for all". We shall also occasionally use "$\exists$" as the "existential quantifier", meaning "there exists".

We now show that there is a one-to-one correspondence between equivalence relation and partitions:

For let $\sim$ be an equivalence relation on a set A. For any $a \in A$ define the set $N_a = \{x | x \in A$ and $x \sim a\}$. Denote the family of all such (distinct) sets by $A/\sim$. Then $A/\sim$ is a partition of A:

1. Since $a \in N_a$, $N_a \neq \phi$, i.e. every element of $A/\sim$ is a nonempty set.
2. $\bigcup_{a \in A} N_a = A = \bigcup_{N_a \in A/\sim} N_a$.
3. $N_a = N_b$ if and only if $a \sim b$. Hence, for any $N_a, N_b \in A/\sim$ either $N_a = N_b$ or $N_a \cap N_b = \phi$.

Conversely, if one has a partition $\{A_i\}_{i \in I}$ of the set A, then we can define a relation $\sim$ on A by

$$a \sim b \iff \text{there exists an } i \text{ such that } a, b \in A_i,$$

i.e. $a \sim b$ if and only if a and b lie in the same subset belonging to the partition. It is easy to see that $\sim$ is an equivalence relation (and that the partition induced by this equivalence relation is precisely the given partition).

The elements of the family $A/\sim$ of subsets of A are called *equivalence classes*. What we have therefore shown is that any equivalence relation on a set determines a family of equivalence classes, and that, conversely, if one has a partition then the members of that partition are the equivalence classes for some equivalence relation.

In the next chapter we shall discuss some properties of the arithmetic of the integers. This will enable us to describe the equivalence classes which arise from the congruence modulo n relation. The ability to perform operations on the set of those equivalence classes will be shown in Chap. 4 to be fundamental in public key encryption and, in particular, in the creation of "unforgeable" digital signatures. Similar operations on other sets will prove essential in secret key encryption as well, as we shall describe in Chaps. 6 and 7. We will explain the terms "public key encryption" and its "secret key" counterpart in Sect. 1.6.

Exercises

1. Let $\mathbb{Z}$ denote the set of integers, and for any $a, b \in \mathbb{Z}$ define the relation $\sim$ by

$$a \sim b \iff a^2 + a = b^2 + b.$$

Show that $\sim$ is an equivalence relation, and describe the equivalence classes.

2. Let A be the set of all functions of the form $\sum_{1=0}^{n} a_i x^i$ for some $n \geq 0$, with the $a_i \in \mathbb{Z}$, or, in other words A is the set of all polynomials in x with integral coefficients. Let $\sim$ be the relation defined on A by

$$f(x) \sim g(x) \iff f(x) - g(x) \text{ is a multiple of } x.$$

Show that $\sim$ is an equivalence relation and describe a simple representation of the equivalence classes.

3. What is wrong with the following argument purporting to show that the requirement for reflexivity in the definition of equivalence relation is superfluous?
   Let a be any element of the set. If $a \sim b$, then by symmetry $b \sim a$, so that by transitivity $a \sim a$. Hence the relation $\sim$ is reflexive.

1.4 Functions

A *function* f on a set A to a set B assigns to every element $a \in A$ an element of B, which is (usually) denoted by $f(a)$. We may write this as

$$f : A \longrightarrow B$$
$$a \mapsto f(a) \qquad a \in A$$

To describe a function, its *domain* (A in the above case), its *codomain* (B in this case) and its effect on each of the elements of the domain must be specified. The *range* of the function f is the set of all values which it can assume, i.e. the set $\{b|b = f(a) \text{ for some } a \in A\}$, which is a subset of B. If the range of the function is all of its codomain, the function is called *surjective* or, in plainer language, *onto*. If the function satisfies the condition

$$f(a_1) = f(a_2) \Longrightarrow a_1 = a_2$$

then it is called *injective* or *one-to-one*.

A function which is both injective and surjective is called *bijective* or (you guessed it!) *one-to-one onto*. Such functions allow for the definition of an *inverse function*: If $f : A \longrightarrow B$ is one-to-one onto, then we define its inverse $f^{-1} : B \longrightarrow A$ by

$$f^{-1}(b) = a \text{ if } f(a) = b.$$

On the other hand, if a function f has an inverse, then, clearly, $f : A \to f(A)$ must be one-to-one. Jumping ahead a little, it seems reasonable that an encryption function should have an inverse; there is little point of sending a message if the recipient is unable to read it.[4]

A function f for which the domain and the codomain are the same set, i.e. $f : A \longrightarrow A$, may be called a *unary operation* on A. If A is a finite set, and such a function is one-to-one onto (but see question 5 below) then it is called a *permutation* of A. A permutation on A may be thought of as a reshuffling of the elements of A.

Next consider two functions $f : A \longrightarrow B$ and $g : B \longrightarrow C$. We define their *composition* $g \circ f : A \longrightarrow C$ (sometimes simply written as gf, if there is no risk of confusing the composition operation with multiplication of some kind) by

$$(g \circ f)(a) = g(f(a)).$$

[4] There is an accepted encryption scheme, not very commonly in use, where the equation $f(X) = B$ has more than one solution for X. The correct value, i.e. which of those solutions was the originally sent message, then has to be determined by other means. We'll get there in Sect. 10.9.

Note that $h \circ (g \circ f) = (h \circ g) \circ f$ if either of these is defined, so that we may as well leave out the brackets.

Exercises

1. If A is a finite set, $\#A = n$, and S is the set of all permutations of A, what is $\#S$?
2. If $A = \{0, 1\}$, describe all functions $f : A \longrightarrow A$ and classify them as injective, surjective, bijective or none of these.
3. With A as in question 2, what is $\#\{f | f : A^n \longrightarrow A\}$? And what is $\#\{f | f : A^n \longrightarrow A^n\}$? How many of the functions in the second set are bijective?
4. If f is one-to-one onto and $f \circ g$ is defined, show that g is one-to-one if and only if $f \circ g$ is, and that g is onto if and only if $f \circ g$ is.
5. Let A and B be finite sets with $\#A = \#B$ and let $f : A \longrightarrow B$ be a function. Show that the following are equivalent:
   - f is surjective;
   - f is injective;
   - f is bijective.
6. Let U be a set, and define for every $A \subseteq U$ the function $\chi_A : U \longrightarrow \{0, 1\}$ by $\chi_A(u) = 1$ if $u \in A$, $\chi_A(u) = 0$ if $u \notin A$. Express $\chi_{A \cup B}$, and $\chi_{A \cap B}$ in terms of χ_A and χ_B.

1.5 Binary Operations

In contrast to unary operations which have a single input, a binary operation has two variables. More precisely, a *binary operation* on a set A is a function $* : A \times A \longrightarrow A$. Instead of writing $*(a_1, a_2)$ for the value of the function, we shall denote it by $a_1 * a_2$.

At the risk of stating the obvious, we just remark here that the adjective "binary" refers to the fact that the operation "operates" on *two* elements of the set; it has nothing to do with the representation of integers (or of numbers in general) in terms of powers of 2 (as opposed to the more pedestrian decimal representation in terms of powers of 10).

The reader will be able to work out what a ternary operation would be, but we won't be needing such things.

Examples

1. On the set $\mathbb{Z}$ of integers, we have, for example, the binary operations "+", which maps a pair of integers onto their sum, "·", which maps a pair of integers onto their product, "−" which maps an ordered pair of integers onto the difference between the first and the second, etc.
2. Let U be a set and consider the set $\mathcal{P}(U) = \{S | S \subseteq U\}$. The symmetric difference is a binary operator on $\mathcal{P}(U)$.
3. Let A be a set and consider the set of functions $\mathcal{P} = \{f | f : A \longrightarrow A\}$. Composition of functions, i.e. the function $c : \mathcal{P} \times \mathcal{P} \longrightarrow \mathcal{P} : (f, g) \mapsto g \circ f$ is a binary operator on $\mathcal{P}$.
4. In particular, let A be a finite set and consider the set $\pi(A)$ of all permutations of A. The composition of permutations is a binary operation on $\pi(A)$.

A binary operation $*$ on a set A is called *associative* if

$$a * (b * c) = (a * b) * c \quad \forall a, b, c \in A.$$

If the operation is associative, we need therefore not use any brackets in such expressions, since the meaning of $a * b * c$ is unambiguous. Of the above examples, the operation "−" on the set of integers is the only one which is not associative. (Why isn't it?) Proving that the symmetric difference operator is associative is left as an exercise.

A binary operation $*$ on a set A is called *commutative* if

$$a * b = b * a \quad \forall a, b \in A.$$

A minor note on terminology is required here. Suppose we have a binary operation $*$ defined on a set B and we have a subset $A \subset B$. Now it is obviously meaningful to consider elements of the form $a_1 * a_2$ with $a_1, a_2 \in A$. But, in general, that does not give us a binary operation on A, because there is no guarantee that elements of this form belong to A – all we know is that they are well-defined elements of B. But if, for all $a_1, a_2 \in A$ we have that $a_1 * a_2 \in A$, then the "induced" binary operation on A makes sense. In such a case we say that A is *closed* under the operation $*$.

Exercises

1. Prove the above assertion that the symmetric difference operator is associative. Is it commutative?
2. Check all the other examples above for associativity and commutativity.
3. How many different binary operations can be defined on a set with three elements? How many of these are commutative?

1.6 Cryptography

This is a book mainly about Algebra, but with the intention of applying Algebra to Cryptography. As a consequence, we shall be dealing with some topics not usually covered in academic courses on Algebra, whereas, on the other hand, we shall be very selective in our choice of material and totally ignore most of what any algebraist would consider important, or even vital, topics in his or her chosen field. On the other hand, this book is, strictly speaking, not about Cryptography either, but, in order to justify to the reader why we selected the topics we did, and to show how these ideas are applied, we need to refer to Cryptography. We therefore give, in this part of this introductory chapter, a very brief overview of Cryptography, mainly in order to establish some common concepts, as well as our terminology.

1.6.1 Encryption Mechanisms

The main purpose of Cryptography (certainly at this very fundamental level) is the protection of the confidentiality or secrecy of messages from one entity (A) to another (B). Let $\mathcal{M}$ be the set of messages from which A might select one or more to send to B. $\mathcal{M}$ is called the *plaintext* space. Let $\mathcal{C}$ be another set of the same or greater cardinality, called the *ciphertext* space. An encryption mechanism is a family of functions $\{E_K()|K \in \mathcal{K}\}$ from $\mathcal{M}$ to $\mathcal{C}$, parametrized by a set $\mathcal{K}$, called the *key space*, such that

1. For every $K \in \mathcal{K}$ there exists a $\overline{K} \in \mathcal{K}$ and a function $D_{\overline{K}} : \mathcal{C} \longrightarrow \mathcal{M}$ such that $D_{\overline{K}} \circ E_K$ is the identity function on $\mathcal{M}$, i.e. $D_{\overline{K}} \circ E_K(m) = m$ for all $m \in \mathcal{M}$.
2. In the absence of knowledge of $\overline{K}$, $C = E_K(M)$ yields no information about M.

If $C = E_K(M)$, C is called the encryption of M under the key K, and the first requirement states that, given the key $\overline{K}$, it is possible to retrieve the plaintext from the ciphertext, but the second requirement states that anyone who does not know $\overline{K}$, cannot do so, and ideally should not be able to obtain any *information* about M. (The meaning of the word *information* will be made precise at a later stage; in any case, if, as is usually the case, the elements of $\mathcal{M}$ and $\mathcal{C}$ consist of binary strings, it is unlikely that this aim will be achieved completely, since at the very least there will be some indication of length of M when C is intercepted. An interceptor would therefore normally have no difficulty in distinguishing between two possible messages, however well encrypted: "Yes!" and "I hope that we'll always stay good friends".) When $\overline{K}$ always equals K or is easily obtainable from K, the mechanism is called *symmetric* (or secret key); if, on the other hand, it is not feasible to obtain $\overline{K}$ from K, the system is *asymmetric* (or public key).

The process of turning M into $C = E_K(M)$ is called encryption; the inverse mapping $D_{\overline{K}}$ is called the decryption function.

We shall usually refer to an encryption mechanism as a *cipher*. Symmetric ciphers come in two main varieties: block ciphers and stream ciphers. We shall have much more to say about these in later chapters.

1.6.2 Confusion and Diffusion

Claude E. Shannon, the father of Information Theory, in his seminal paper *Communication theory of secrecy systems*[5] identified two important features that an encryption function should have:

- *Confusion:* Confusion is intended to make the relationship between the plaintext and/or the key on the one hand, and the ciphertext on the other, as complex as possible, or as stated by J.L. Massey: "The ciphertext statistics should depend on the plaintext statistics in a manner too complicated to be exploited by the cryptanalyst." From this point of view

[5]Bell System Technical J. **28** (1949), 656–715.

a simple substitution cipher, which is essentially just a permutation π of the plaintext alphabet {a,b,c, ..., z}, is unsatisfactory: the frequency distribution of, say, English is inherited by the ciphertexts—the overwhelming frequency of the letter e in English is now reflected in the same frequency of $\pi(\mathtt{e})$ in the ciphertext.
- *Diffusion:* In order to avoid this kind of weakness, the second attribute of diffusion is required. Each symbol of the key and/or of the plaintext should affect as many ciphertext symbols as possible.[6]

1.6.3 Symmetric and Asymmetric Encryption

The 1970s were exciting times in cryptography.

On the one hand there was the invention (or discovery? Let's not debate that philosophical point) of public key cryptography following Diffie and Hellman's famous paper *New directions in cryptography* in 1976, followed shortly thereafter by Rivest, Shamir and Adleman's implementation of those ideas in their *A method for obtaining digital signatures and public-key cryptosystems* in 1978 (after having been discussed by Martin Gardner in *Scientific American* in its August 1977 issue).

On the other hand, the 1970s also saw the first ever standardization of a secret key cipher: Federal Information Processing Standard 41, defining the Data Encryption Standard (DES). Until then ciphers in use had never been made public. This public exposure of a cipher on the one hand, and the secrecy about the principles on which its design was based on the other, led to a major surge in interest in secret key encryption and the corresponding cryptanalysis.

Nevertheless, the asymmetric ciphers continue to appear as the "glamorous" side of cryptography, with the symmetric side seen as the poor workhorse of communication security, in spite of subsequent exposure through the project for finding a successor for DES, the Advanced Encryption Standard "competition". There is a fairly common perception that symmetric techniques became "obsolete after the invention of public-key cryptography in the mid 1970s" [7] and this perception is reinforced by some of the textbook literature.

> "However, symmetric techniques are still widely used because they are the only ones that can achieve some major functionalities as high-speed or low-cost encryption, fast authentication, and efficient hashing. Today, we find symmetric algorithms in GSM mobile phones, in credit cards, in WLAN connections, and symmetric cryptology is a very active research area.
> There is a strong need for further research in this area. On the one hand, new industrial needs are arising with the development of new application environments. For instance, the demand for low-cost primitives dedicated to low-power devices is pressing."[8]

[6] In so-called *additive* ciphers, in which each plaintext symbol is changed into a ciphertext symbol through the "addition", in some sense, of key-derived material, no diffusion between symbols can take place, and the emphasis is on spreading the effect of a key bit change as widely as possible. Stream ciphers, discussed in Chap. 7, are invariably of this kind.

[7] D.STVL.9, Ongoing Research Areas in Symmetric Cryptography, ECRYPT European Network of Excellence in Cryptology, 2008; Executive Summary.

[8] *op.cit.*

To the listed uses of symmetric cryptography we now add lightweight cryptography, for use in RFID tags, sensors, contactless smart cards, etc., currently a very active field. The practising or aspiring cryptologist reading this book is therefore quite likely to find himself/herself working with symmetric/secret key algorithms, which is why in this book we devote a considerable amount of our time to algebraic topics relevant thereto.

The main reason for the ubiquity of symmetric techniques in use is simple: it is just a matter of speed. It is difficult to compare the speeds of symmetric and asymmetric ciphers directly, because choice of cipher depends on the purpose for which it is used and the platform on which it is implemented, but as a very rough guide it is probably fair to say that a symmetric cipher will process data at about 1000 times the rate of an asymmetric cipher.

1.7 Notational Conventions

A few conventions briefly referred to in this section may be well known to the reader, but are included for the sake of completeness.

1.7.1 Floor and Ceiling

The following two functions are defined on the set $\mathbb{R}$ of all real numbers:

Definitions If x is any real number $\lfloor x \rfloor$ is defined to be the largest integer not greater than x. Thus, for example, $\lfloor \frac{7}{3} \rfloor = 2$, $\lfloor \frac{-130}{7} \rfloor = -19$, $\lfloor \frac{1001}{11} \rfloor = 91$.

$\lfloor\ \rfloor$ is frequently called the "floor" function.

Similarly the "ceiling" function $\lceil\ \rceil$ is defined as follows: for any real number x, $\lceil x \rceil$ is the smallest integer not less than x.

Clearly $\lfloor x \rfloor = \lceil x \rceil$ if and only if x is an integer. Some of the properties of these two functions are left as exercises.

Exercises Prove that for any real number x:

1. $x - 1 < \lfloor x \rfloor \le x$.
2. $\lfloor x \rfloor \le x < \lfloor x \rfloor + 1$.
3. $x \le \lceil x \rceil < x + 1$.
4. $\lceil x \rceil - 1 < x \le \lceil x \rceil$.
5. $\lceil x \rceil - \lfloor x \rfloor = 1$ if $x \notin \mathbb{Z}$.
6. $\lceil x \rceil = \lfloor x + 1 \rfloor$ if $x \notin \mathbb{Z}$.

1.7.2 Fractional Part

Rather less common is the notation $\{a\}$ for the fractional part of the number a. Thus

$$\{a\} = a - \lfloor a \rfloor.$$

For example, $\{\frac{7}{3}\} = \frac{1}{3}, \{-\frac{190}{7}\} = \frac{3}{7}$. We shall only need this concept when we get to the final chapter of this book!

1.7.3 Exclusive Or

We denote by "$\oplus$" the exclusive or (XOR) operation on the set $\{0, 1\}$, defined as follows:

$\oplus$	0	1
0	0	1
1	1	0

We shall in later chapters discover that this represents the group operation in the smallest non-trivial group, or, from the point of view of fields, the basis of the addition operation in fields of characteristic 2.

The XOR operation is very easily implemented in hardware and, applied bitwise to bytes or longer bitstrings, is therefore part of almost every secret key encryption function.

In the meantime we leave as an exercise the proof that, if the *characteristic function* χ_S of a set S is as defined in Exercise 6 of Sect. 1.4, then, for any sets A and B $\chi_{A\Delta B} = \chi_A \oplus \chi_B$. (At considerable risk of stating the obvious, this corresponds, of course, to the fact that $A\Delta B$ consists of all elements that belong to A *or* to B, *but not to both.*)

1.7.4 Matrix Multiplication

We assume that the reader is familiar with matrices. We draw attention to the idea of *block matrices*.

Any matrix (larger than 1×1) can be thought of as a matrix whose entries are themselves matrices. This is best explained by an example: The matrix

$$A = \begin{pmatrix} 1 & 2 & 3 & 4 \\ 5 & 6 & 7 & 8 \\ 9 & 10 & 11 & 12 \end{pmatrix}$$

could be written (and there might be good reasons for doing so, though off-hand it's hard to think of any) as

$$A = \begin{pmatrix} X & Y \\ Z & W \end{pmatrix},$$

where $X = \begin{pmatrix} 1 & 2 \\ 5 & 6 \end{pmatrix}$, $Y = (3\ 4)$, $Z = (9\ 10)$ and $W = \begin{pmatrix} 7 & 8 \\ 11 & 12 \end{pmatrix}$.

Note that the addition and multiplication of block matrices correspond to ordinary matrix addition and multiplication. For example

$$\begin{pmatrix} A & B \\ C & D \end{pmatrix} \begin{pmatrix} K & L \\ M & N \end{pmatrix} = \begin{pmatrix} AK + BM & AL + BN \\ CK + DM & CL + DN \end{pmatrix}$$

provided, naturally, that the dimensional requirements are satisfied, i.e. number of columns of A = number of rows of K, etc. in order for the block multiplications to make sense.

2

Basic Properties of the Integers

In this chapter we consider some of the elementary properties of the integers. Many of them are *really* elementary and known to every school child. The algebraist's approach to such things as greatest common divisors (alias "highest common factors") and similar things may, however, be experienced as refreshingly different. We shall use, as before, the notation $\mathbb{Z}$ for the set of all integers, i.e. $\mathbb{Z} = \{\dots, -2, -1, 0, 1, 2, 3, \dots\}$.

Much of modern Cryptology is based on arithmetic (or Number Theory, which at the level we aim at, would be a more appropriate appellation) as we shall show in later chapters, so we promise that you will eventually see this return to these concepts as justifiable.

2.1 Divisibility

As before, we have the following

Definitions Let $a, b \in \mathbb{Z}$ with $a \neq 0$. Then a is called a *divisor* or *factor* of b if there exists a $q \in \mathbb{Z}$ such that $b = q \cdot a$, and b is then said to be *divisible by* a, or a *multiple* of a. We denote this by $a|b$. Thus, by definition,

$$a|b \Longrightarrow \exists q \in \mathbb{Z} \text{ such that } b = qa.$$

q is called the *quotient*.

Theorem *For all* $a, b, c \in \mathbb{Z}$, *with* $a \neq 0$

- $a|a,\ 1|a;$
- $a|b \Longrightarrow a|-b;$
- $a|b$ *and* $a|c \Longrightarrow a|(b + c);$
- $a|b$ *and* $b|c \Longrightarrow a|c;$
- $a|b$ *and* $b|a \Longrightarrow a = \pm b.$

A.R. Meijer, *Algebra for Cryptologists*, Springer Undergraduate Texts in Mathematics and Technology,
DOI 10.1007/978-3-319-30396-3_2

Proof The proofs are trivial with the possible exception of that for the last statement: From the given statements we conclude that there exist integers q_1, q_2 such that $b = q_1 a$ and $a = q_2 b$, so that $a = q_2 q_1 a$, so that $q_1 q_2 = 1$. The only possibilities are that $q_1 = q_2 = 1$, in which case $a = b$, or that $q_1 = q_2 = -1$, in which case $a = -b$.

Any integer a is trivially divisible by ± 1 and by $\pm a$. If an integer p ($\neq \pm 1$) has only these trivial divisors it is called *prime*, otherwise it is called *composite*. The first (positive) primes are 2, 3, 5, 7, 11, 13, ... The numbers ± 1 are (by definition) not considered prime. 2 is the only even prime. (This is in itself not a profound statement: one might as well make the observation that 8923 has the remarkable property of being the only prime which is divisible by 8923. Nevertheless, the prime 2 seems, in many cases, to call for treatment different from that of odd primes.)

The **Fundamental Theorem of Arithmetic** states that every integer can be written as the product of primes in an essentially unique way:

Every nonzero integer n can be expressed as a product of the form

$$n = \pm p_1^{e_1} p_2^{e_2} \dots p_r^{e_r},$$

where the p_i are r distinct positive primes and the e_i are integers with $e_i > 0$. This representation is unique up to the order in which the factors are written.[1]

Note that if $n = \pm 1$, then $r = 0$, and the product of no terms at all is, by convention, equal to 1.

We shall not prove the Fundamental Theorem, assuming that everyone is aware of it (and believes that it is true). Those who would like to see a proof are referred to Victor Shoup's book *A Computational Introduction to Number Theory and Algebra*, Cambridge University Press, 2005, available online at www.shoup.net/ntb/ntb.v1.pdf.

We shall also make a lot of use of the following theorem:

Division with Remainder Property

For all $a, b \in \mathbb{Z}$ with $b > 0$ there exist unique $q, r \in \mathbb{Z}$ such that $a = qb + r$ and $0 \leq r < b$.

Clearly $q = \lfloor \frac{a}{b} \rfloor$ and $r = a - b \cdot \lfloor \frac{a}{b} \rfloor \geq 0$.

Proof Consider the set $\{a - zb | z \in \mathbb{Z} \text{ and } a - zb \geq 0\}$. This set is clearly nonempty and therefore contains a smallest element.[2] Let r be this smallest element, so $r \geq 0$ and $r = a - qb$ for some $q \in \mathbb{Z}$. Also $r < b$, since otherwise we could write $a = (q + 1)b + (r - b)$, and $r - b \geq 0$, so that $r - b$ belongs to the set under discussion, contradicting the choice of r as the smallest element in that set.

It remains to prove that q and r are uniquely determined. Suppose that we have $a = qb + r$ and $a = q'b + r'$, with $0 \leq r, r' < b$. Then, subtracting and rearranging, we have $r - r' = (q' - q)b$. In this equation the left-hand side has absolute value less than $|b|$, whereas if

[1] If we, foolishly, decided that 1 should be considered to be a prime, the uniqueness of this decomposition into primes would no longer hold! This would, at the very least, be inconvenient for mathematicians.

[2] The principle that any nonempty set of positive integers contains a smallest element is equivalent to the principle of mathematical induction, as the reader is invited to prove. Here "equivalent" means that assuming either one of these principles, the other one can be proved as a theorem.

$q \neq q'$ the right-hand side has absolute value at least $|b|$. Hence $q = q'$, from which $r = r'$ follows.

Exercises

1. Let a, b, c, d be integers such that $c|a$ and $d|b$. Prove that $cd|ab$.
2. Show that any product of four consecutive integers is divisible by 24.
3. Prove that $4 \nmid n^2 + 2$ for any integer n.
4. Prove by induction that $5|n^5 - n$ for every positive integer n.
5. Let n be a positive composite number. Show that there exists at least one prime divisor p of n satisfying $p \leq \sqrt{n}$.
6. Establish a one-to-one correspondence between the divisors of a positive integer n which are less than $\sqrt{n}$ and those that are greater than $\sqrt{n}$.
7. Prove the "Division with remainder property" in the following form: For all $a, b \in \mathbb{Z}$ with $b > 0$ there exist unique $q, r \in \mathbb{Z}$ such that $a = qb + r$ and $-b/2 < r \leq b/2$.

2.2 Ideals and Greatest Common Divisors

We introduce the concept of *ideals* in $\mathbb{Z}$.

Definition A nonempty set $I \subseteq \mathbb{Z}$ is called an *ideal* in $\mathbb{Z}$ if the following two conditions hold:

$$\forall a, b \in I, \quad a + b \in I,$$

$$\forall a \in I \text{ and } \forall z \in \mathbb{Z}, \quad za \in I.$$

In other words, the set I is closed under addition, as well as under multiplication by *any* integer (and *not* just under multiplication by numbers belonging to I).

Two trivial examples of ideals in $\mathbb{Z}$ are, firstly, the set $\{0\}$ consisting only of the integer 0, and, secondly, the set $\mathbb{Z}$ itself.

Note that if I is an ideal in $\mathbb{Z}$, and $a, b \in I$, then also $a - b \in I$, since, by the second condition $-b = (-1)b \in I$, and therefore, by the first condition $a + (-b) = a - b \in I$. Consequently, if I is any ideal in $\mathbb{Z}$, then $0 \in I$.

We introduce the following notation:

$$a\mathbb{Z} = \{x | x = az \text{ for some } z \in \mathbb{Z}\}.$$

Thus $a\mathbb{Z}$ consists precisely of all multiples of a. We leave as an exercise the easy proof that the set $a\mathbb{Z}$ is an ideal. It is called the *principal ideal generated by a*.

What makes the integers interesting[3] is that the principal ideals are in fact the only ideals in $\mathbb{Z}$. Even the two trivial ideals are of this kind: $\{0\} = 0\mathbb{Z}$ and $\mathbb{Z} = 1\mathbb{Z}$.

Theorem *Every ideal in* $\mathbb{Z}$ *is principal.*

Proof Let I be an ideal in $\mathbb{Z}$. If $I = \{0\}$, we are through, so assume that $I \neq \{0\}$. Then I contains a positive integer, since (as we noted) if $x \in I$ then also $-x \in I$. Let a be the least positive element of I. We prove that $I = a\mathbb{Z}$.

For let b be any element of I. By the theorem above on division with remainder, there exist integers q and r, with $0 \leq r < a$ such that $b = qa + r$, i.e.

$$r = b - qa.$$

But $qa \in I$ (by the second condition on ideals) and since $b \in I$, we have that $r \in I$, by the first condition. But a was the least positive element of I, so the only way this can happen is for r to be 0. Hence, $b = qa \in a\mathbb{Z}$. We have therefore proved that

$$I \subseteq a\mathbb{Z}.$$

The reverse inclusion is obvious, since any multiple of an element of an ideal must also belong to the ideal, so that

$$a\mathbb{Z} \subseteq I,$$

which completes the proof.

Notation A principal ideal $a\mathbb{Z}$ of $\mathbb{Z}$ will also sometimes be denoted by $< a >$.

Theorem *Let* $a, b \in \mathbb{Z}$. *Then* $a\mathbb{Z} \subseteq b\mathbb{Z}$ *if and only if* $b|a$.

Example

$$6\mathbb{Z} = \{\dots, -12, -6, 0, 6, 12, 18, \dots\},$$
$$3\mathbb{Z} = \{\dots, -12, -9, -6, -3, 0, 3, 6, 9, 12, \dots\},$$

so $6\mathbb{Z} \subset 3\mathbb{Z}$, corresponding to the fact that $3|6$.

We leave the (easy) proof of the above theorem as an exercise.

What happens if we have two ideals, neither of which is contained in the other? Let I and J be two such ideals, so $I \not\subseteq J$ and $J \not\subseteq I$. Any ideal containing both I and J must contain all sums of elements $x \in I$ and $y \in J$. But it is straightforward to verify that the set

$$K = \{x + y | x \in I, y \in J\}$$

[3]Or boring, depending on one's taste. Rings for which the following condition holds, the "principal ideal rings", are certainly easier to work with than a lot of other rings. We shall in Chap. 5 have occasion to consider an important family of other principal ideal rings which will turn out to be extremely useful in the construction of different kinds of symmetric key ciphers.

is itself already an ideal, so it must be the smallest ideal that contains both I and J. In fact, if $I = < a >$ and $J = < b >$, then K consists of all "linear combinations" of a and b:

$$K = \{ra + sb | r, s \in \mathbb{Z}\}.$$

We usually call the ideal K the *sum* of the ideals I and J: $K = I + J$.

But now something interesting can be noted. All ideals are principal, so $K = < d >$, for some integer d:

$$< a > + < b > = < d > .$$

Now $< a > \subseteq < d >$, so $d|a$, and similarly $d|b$, so that d is a *common divisor* of a and b. Suppose that d' is another divisor of both a and b. Then $< a > \subseteq < d' >$ and $< b > \subseteq < d' >$, so that $K \subseteq < d' >$, or $< d > \subseteq < d' >$, i.e. $d'|d$. Hence d is not only a common divisor of a and b, but *any common divisor of a and b is a divisor of d*. Thus d may be regarded as the *greatest common divisor* of a and b.

Definition If $a, b \in \mathbb{Z}$ then a greatest common divisor d of a and b is an integer such that

- $d|a$ and $d|b$;
- if $c|a$ and $c|b$, then $c|d$.

What we have therefore proved is that any two integers have a greatest common divisor. It is easy to see, and left to the reader as an exercise, that this result can easily be extended: any finite set of integers has a greatest common divisor. Note also that, following from the first theorem in this chapter, any two greatest common divisors differ at most simply by their sign: If d and d' are greatest common divisors of some (finite) set of integers, then $d = \pm d'$. When we speak of *the* greatest common divisor, we shall simply assume henceforth that we mean the positive one.

Notation Given two integers a and b we shall denote their greatest common divisor by $gcd(a, b)$.

We have, in the process, also proved an important property of the greatest common divisor of two integers:

If $d = gcd(a, b)$ then there exist integers x, y such that $xa + yb = d$.

This result is known as *Bezout's identity*.[4] It is, of course, immediate from the fact that $< a > + < b > = < d >$.

For example, the greatest common divisor of 56 and 36 is 4, which we can write as $4 = 2 \cdot 56 + (-3) \cdot 36$.

Definition Two integers a and b are called *relatively prime* if $gcd(a, b) = 1$.

---

[4]Incorrectly, strictly speaking. In the form in which we have just given it, as a theorem in Number Theory, the result is due to Claude Gaspard Bachet de Méziriac, like Bezout a Frenchman, who lived a century before Bezout.

Theorem *If a, b, c are integers such that $a|bc$ and $gcd(a, b) = 1$, then $a|c$.*

Proof By the previous observation, there exist integers x, y such that

$$xa + yb = 1.$$

Multiplying this equation by c we get

$$xac + ybc = c.$$

Now a is a divisor of each of the terms on the left-hand side, so a must be a divisor of c.

This result has the immediate

Corollary *Let $p > 0$ be a prime. If $p|ab$, then $p|a$ or $p|b$.*

Proof Suppose $p \not| a$. Since p is prime, its only divisors are ± 1 and $\pm p$, so that $gcd(p, b) = 1$ or $gcd(p, b) = p$. By the result just proved, the first case cannot hold. Hence $p|b$.

It is obvious that this result can be generalised: If p is a prime factor of $\prod_{i \in S} a_i$, where $\{a_i\}_{i \in S}$ is some finite set of integers, then $p|a_j$ for some $j \in S$.

Finally, we mention the following fact, which will occasionally prove useful:

If $gcd(a, b) = d$, then $gcd(a/d, b/d) = 1$. This also follows from the fact that $xa + yb = d$ for some integers x, y, and therefore

$$x\frac{a}{d} + y\frac{b}{d} = 1.$$

This means that any common divisor of a/d and b/d must be a divisor of 1, and therefore must be ± 1.

Exercises

1. Prove that the intersection $I \cap J$ of two ideals is itself an ideal. If $I = < a >$ and $J = < b >$, how can you describe $I \cap J$ in terms of a and b?
2. Define, for integers a, b, a *least common multiple* of a and b as an integer m such that

   1. $a|m$ and $b|m$

   and

   2. $\forall\, k \in \mathbb{Z}\ (a|k \text{ and } b|k \Longrightarrow m|k)$.

   Use the result from the previous exercise to show that any two integers have a least common multiple.
3. An ideal M in $\mathbb{Z}$ is called *maximal* if there is no ideal J such that $M \subset J \subset \mathbb{Z}$. Prove that an ideal M is maximal if and only if $M = < p >$ for some prime number p.
4. Prove that $\gcd(n, n + 2) = 1$ or 2 for every $n \in \mathbb{Z}$.
5. Prove that $a|bc$ if and only if $\frac{a}{\gcd(a,b)}|c$.

2.3 The Euclidean Algorithm

The "Division with Remainder" property can be used to prove the uniqueness of the factorization property of the integers, i.e. the Fundamental Theorem of Arithmetic, as shown in e.g. the book by Victor Shoup to which we referred in Sect. 2.1. Instead, however, we shall take uniqueness of factorisation as a fact, and doing so, it is easy to see that if we take two integers n and m with factorization into primes as

$$n = \prod_{i=0}^{r} p_i^{e_i},$$
$$m = \prod_{j=0}^{s} q_j^{f_j},$$

then $m|n$ if and only if every $q_j = p_i$ for some i and the corresponding $f_j \leq e_i$.

This observation allows one to express the least common multiple of two integers, once their factorisations into primes are known. For let

$$a = \prod_{i=0}^{r} p_i^{e_i},$$
$$b = \prod_{i=0}^{r} p_i^{f_i},$$

where the set of primes has been made the same for the two factorisations, by inserting exponents 0 where necessary, then

$$\gcd(a, b) = \prod_{i=0}^{r} p_i^{\min\{e_i, f_i\}}.$$

Also, for the least common multiple one has

$$\operatorname{lcm}(a, b) = \prod_{i=0}^{r} p_i^{\max\{e_i, f_i\}}.$$

This may be called the "high school method" of determining greatest common divisors and least common multiples. Note, incidentally, that this also shows that

$$\operatorname{lcm}(a, b) = \frac{ab}{\gcd(a, b)}.$$

Example

$$32170781418 = 2^1 \times 3^1 \times 19^4 \times 41143,$$
$$1323248675872933 = 2^0 \times 3^0 \times 19^1 \times 41143^3,$$

so that their gcd is

$$2^0 \times 3^0 \times 19^1 \times 41143^1 = 781717.$$

The disadvantage of this method is clearly that one needs to be able to factorise the integers in order to apply it, and factorisation is, in general, a non-trivial problem. It is therefore good to know that there exists an efficient algorithm which does not require this prior knowledge. This is called the "Euclidean Algorithm" and appears already in the famous book of Euclid of Alexandria (ca. 350 BCE). The efficiency of this method actually allows one to reverse the process: instead of using factorisation in order to find greatest common divisors, the technique of finding greatest common divisors is a standard component of methods for finding factors.

Let a and b be the two integers whose greatest common divisor we wish to determine, and assume without loss of generality that $0 < b < a$. By the division with remainder theorem we can write

$$\begin{aligned} a &= q_0 b + r_0 \qquad & 0 \le r_0 < b \\ b &= q_1 r_0 + r_1 \qquad & 0 \le r_1 < r_0 \\ r_0 &= q_2 r_1 + r_2 \qquad & 0 \le r_2 < r_1 \\ r_1 &= q_3 r_2 + r_3 \qquad & 0 \le r_3 < r_2 \end{aligned}$$

etc. This process cannot continue indefinitely, because the r_i form a strictly decreasing sequence of non-negative integers. The set $\{r_0, r_1, r_2, \ldots\}$ must contain a smallest element, and this smallest element must be 0, since otherwise we could use division with remainder once more to get a still smaller remainder. Thus we must eventually reach an n such that

$$\begin{aligned} r_{n-2} &= q_n r_{n-1} + r_n, \\ r_{n-1} &= q_{n+1} r_n + 0. \end{aligned}$$

We now claim that $r_n = \gcd(a, b)$. Indeed, $r_n | r_{n-1}$, so that, by the second to last equation, we find that $r_n | r_{n-2}$. Moving up one equation, we see that $r_n | r_{n-3}$. Continuing in this way, we eventually arrive at $r_n | r_2$, $r_n | r_1$, $r_n | r_0$ and finally at $r_n | b$ and $r_n | a$. Thus r_n is indeed a common divisor of a and b.

On the other hand, suppose that c is a common divisor of a and b, i.e. that $c|a$ and $c|b$. This time we move downwards through the equations. The first equation shows that $c|r_0$. Using this, we see from the second equation that $c|r_1$. Continuing in this way, we eventually arrive at $c|r_n$. Thus any common divisor of a and b is a divisor of r_n.

Example To find gcd(6035,1921) we note that

$$6035 = 3 \times 1921 + 272,$$
$$1921 = 7 \times 272 + 17,$$
$$272 = 16 \times 17 + 0,$$

and we conclude that the greatest common divisor is 17.

Working backwards through this example, we also note how we can express 17 as a linear combination of 6035 and 1921:

$$\begin{aligned} 17 &= 1921 - 7 \times 272 \\ &= 1921 - 7 \times (6035 - 3 \times 1921) \\ &= 22 \times 1921 - 7 \times 6035. \end{aligned}$$

Since writing the gcd of two integers as a linear combination of those two integers will turn out to be an exceedingly useful trick, we formalise the above in a general way.

Let, as before, a and b be positive integers, with $b < a$ and let

$$\begin{aligned} a &= q_0 b + r_0, \\ b &= q_1 r_0 + r_1, \\ r_0 &= q_2 r_1 + r_2, \\ r_1 &= q_3 r_2 + r_3, \\ &\cdots\cdots\cdots\cdots \\ r_{k-1} &= q_{k+1} r_k + r_{k+1}, \\ &\cdots\cdots\cdots\cdots \\ r_{n-1} &= q_{n+1} r_n. \end{aligned}$$

We shall define two sequences $\{P_i\}$ and $\{Q_i\}$ such that, for every $i = 0, 1, \ldots, n$

$$r_i = (-1)^i Q_i a + (-1)^{i+1} P_i b.$$

If, for consistency, we put $r_{-2} = a$ and $r_{-1} = b$, then this equation is satisfied for $i = -2$ and $i = -1$ if we choose $Q_{-2} = 1, Q_{-1} = 0, P_{-2} = 0$ and $P_{-1} = 1$. Now define, for $i \geq 0$, P_i and Q_i recursively by

$$P_i = q_i P_{i-1} + P_{i-2},$$
$$Q_i = q_i Q_{i-1} + Q_{i-2}.$$

An inductive argument then shows that the above equation holds for all i:

$$
\begin{aligned}
r_{k+1} &= r_{k-1} - q_k r_k \\
&= (-1)^{k-1} Q_{k-1} a + (-1)^k P_{k-1} b \\
&\quad - q_k [(-1)^k Q_k a + (-1)^{k+1} P_k b] \\
&= (-1)^{k-1} [Q_{k-1} + q_k Q_k] a + (-1)^k [P_{k-1} + q_k P_k] b \\
&= (-1)^{k+1} Q_{k+1} a + (-1)^{k+2} P_{k+1} b.
\end{aligned}
$$

Note, in particular, that since $r_{n+1} = 0$,

$$\gcd(a, b) = r_n = (-1)^n Q_n a + (-1)^{n+1} P_n b,$$

as we were hoping to get, and that, since $r_{n+1} = 0$

$$\frac{P_{n+1}}{Q_{n+1}} = \frac{a}{b}.$$

It is not hard to show (although we shall refrain from doing so) that the pairs P_i, Q_i consist of relatively prime integers. This implies that in the equation $\frac{P_{n+1}}{Q_{n+1}} = \frac{a}{b}$, the left-hand side cannot be simplified by cancelling out any common factors, or, in other words, the left-hand side represents the fraction $\frac{a}{b}$ in its lowest terms.

Example Let $a = 489$ and $b = 177$. We obtain

$$
\begin{aligned}
489 &= 2 \times 177 + 135, \\
177 &= 1 \times 135 + 42, \\
135 &= 3 \times 42 + 9, \\
42 &= 4 \times 9 + 6, \\
9 &= 1 \times 6 + 3, \\
6 &= 2 \times 3 + 0,
\end{aligned}
$$

so our quotients are as follows

i	0	1	2	3	4	:	5
q_i	2	1	3	4	1	:	2

Using the initial values and the recursion relations, we can complete the table as follows:

i	−2	−1	0	1	2	3	4	5
q_i			2	1	3	4	1	2
P_i	0	1	2	3	11	47	58	163
Q_i	1	0	1	1	4	17	21	59

Calculation confirms that $(-1)^4 Q_4 a + (-1)^5 P_4 b = 21 \times 489 - 58 \times 177 = 3 = r_4$ is indeed the greatest common divisor of 489 and 177.

It should be clear that any implementation of the Euclidean algorithm can, with very little extra effort, be extended to keep track of the P_i and Q_i at the same time and thus output at the end not only the greatest common divisor but also two integers x and y such that $xa + yb = \gcd(a, b)$. Such an implementation is commonly referred to as an implementation of the *Extended Euclidean Algorithm*.

One final comment may be made about the Euclidean algorithm: it is amazingly efficient. Its complexity is linear in the logarithm of its inputs, so finding the gcd of two 100 digit integers will take only twice as many steps as finding the gcd of two 10 digit integers.

The connections between this algorithm and the theory of continued fraction approximations to real numbers are not of any interest to cryptologists; with the possible exception of a (now outdated) technique for factoring,[5] its relevance to cryptology is minimal in any case. Factoring itself is relevant, and probably crucial, in breaking the RSA algorithm, as we shall see in Sect. 4.1.

Exercises

1. Find the gcd d of $a = 233968$ and $b = 282015$ and integers x, y such that $ax + by = d$.
2. Show that there are infinitely many pairs of integers (x, y) such that $x + y = 133$ and $\gcd(x, y) = 7$.

2.3.1 Stein's gcd Algorithm

Stein's version of the Euclidean algorithm is an adaptation convenient for machine implementation. Generally, division is a computationally demanding operation, whereas division by 2 is a simple right shift. Also, determining whether a given integer is even or odd, is a simple check on the least significant bit. Stein's implementation therefore makes no use of division except when it can be done by shifting.

Stein's algorithm depends on the following three observations, each of which allows one to reduce the size of at least one of the inputs by half:

1. If a and b are both even and not both zero, then $\gcd(a, b) = 2 \cdot \gcd(a/2, b/2)$.
2. If a is even and b is odd, then $\gcd(a, b) = \gcd(a/2, b)$.
3. If a and b are both odd, then $\gcd(a, b) = \gcd(\frac{a-b}{2}, b)$.

Stein's algorithm (also known as the Binary Algorithm) for finding greatest common divisors is faster than the standard method, because it does not require costly (in terms of time) integer divisions, even though it requires more iterations. The constants that are

[5]Lehmer, D.H.; Powers, R.E.: On Factoring Large Numbers; Bulletin of the American Mathematical Society **37** (10) (1931), pp. 770–776.

required in Bezout's identity, thereby obtaining an "extended" version of Stein's algorithm, can be found in a manner analogous to how it is done in the Euclidean algorithm.[6]

Still other algorithms are known, such as the Schönhage–Strassen algorithm, which are faster than either Euclid's or Stein's, but their efficiency only becomes apparent when the inputs are of the order of $2^{2^{15}}$ or more.

Please note that those Bezout constants will turn out to be quite important when we start dealing with $\mathbb{Z}_p$ (p prime) as a *field*. And to do that we clearly need to find out first of all what $\mathbb{Z}_p$ is supposed to mean.

2.4 Congruences

We recall the following from our introductory chapter:

Definition Let n be any nonzero integer. Two integers a and b are said to be *congruent modulo n* if $n|a - b$. We denote this by $a \equiv b \mod n$.

The relation $a \equiv b \bmod n$ is called a *congruence relation* or simply a *congruence*, and the integer n appearing in it is called the *modulus* of that congruence.

From the "division with remainder" property, the next theorem follows immediately:

Theorem *For any integers a and n with* $n \neq 0$, *there exists a unique integer* $b \in \{0, 1, \ldots, n-1\}$ *such that* $a \equiv b \mod n$. (We shall, possibly with an abuse of notation, denote this integer by $a \bmod n$. The reader is, however, warned that not all computer languages interpret $a \bmod n$ this way. The C language, for one, does not, and may return a negative value for $a \bmod n$.)

The following theorem, which we have already seen in Chap. 1, shows that a congruence is in fact an equivalence relation:

Theorem *Let* a, b, c *and* $n \neq 0$ *be integers. Then*

- $a \equiv a \bmod n$;
- *if* $a \equiv b \bmod n$ *then* $b \equiv a \bmod n$;
- *If* $a \equiv b \bmod n$ *and* $b \equiv c \bmod n$, *then* $a \equiv c \bmod n$.

Our next theorem shows that the arithmetical operations of addition (and subtraction) and multiplication on $\mathbb{Z}$ "carry over" to operations on the resulting set of equivalence classes. These equivalence classes are also known as *congruence* or *residue classes*, and we shall frequently use this terminology.

---

[6]See, for example, Joux, A.: *Algorithmic Cryptanalysis*, CRC Press, Boca Raton, 2009, pp. 30–32. Pseudocode for the extended Stein algorithm can be found in Menezes, A.J., van Oorschot, P.C. and Vanstone, S.A.: *Handbook of Applied Cryptography*, CRC Press, 1997, p. 606.

Theorem *Let a, b, a', b' and $n \neq 0$ be integers such that $a \equiv a' \bmod n$ and $b \equiv b' \bmod n$. Then*

$$a + b \equiv a' + b' \bmod n,$$
$$a \cdot b \equiv a' \cdot b' \bmod n.$$

Proof $n|a - a'$ and $n|b - b'$, so there exist integers q_1, q_2 such that $a - a' = q_1 n$ and $b - b' = q_2 n$. But then $(a+b) - (a'+b') = (q_1 + q_2)n$ and $ab - a'b' = ab - ab' + ab' - a'b' = a(b - b') + (a - a')b' = aq_2 n + q_1 nb' = (aq_2 + q_1 b')n$.

We can therefore define operations of "addition" and "multiplication" of the (finite) set of *equivalence classes*: Denote, temporarily, the equivalence class containing the integer x by $\overline{x}$, and keep the modulus n fixed. Then the theorem states that

$$\overline{a} + \overline{b} = \overline{a + b}$$

and

$$\overline{a} \cdot \overline{b} = \overline{a \cdot b}.$$

Examples In the following examples we shall, by a common abuse of notation, denote the equivalence class containing a by a, rather than by the more correct $\overline{a}$.

1. Let $n = 6$. Then the set of equivalence classes is $\{0, 1, 2, 3, 4, 5\}$ and the tables for the addition and the multiplication of these classes are

+	0	1	2	3	4	5
0	0	1	2	3	4	5
1	1	2	3	4	5	0
2	2	3	4	5	0	1
3	3	4	5	0	1	2
4	4	5	0	1	2	3
5	5	0	1	2	3	4

and

·	0	1	2	3	4	5
0	0	0	0	0	0	0
1	0	1	2	3	4	5
2	0	2	4	0	2	4
3	0	3	0	3	0	3
4	0	4	2	0	4	2
5	0	5	4	3	2	1

2. Let $n = 5$. Then the set of equivalence classes is $\{0, 1, 2, 3, 4\}$, and the tables are

+	0	1	2	3	4
0	0	1	2	3	4
1	1	2	3	4	0
2	2	3	4	0	1
3	3	4	0	1	2
4	4	0	1	2	3

and

·	0	1	2	3	4
0	0	0	0	0	0
1	0	1	2	3	4
2	0	2	4	1	3
3	0	3	1	4	2
4	0	4	3	2	1

3. Now let $n = 2$. In this case there are only two equivalence classes, viz. $\overline{0}$ (which is the class of all even integers) and $\overline{1}$ (the class of all odd integers), and the tables are as follows:

+	0	1
0	0	1
1	1	0

·	0	1
0	0	0
1	0	1

Note that the first table corresponds to the exclusive or (XOR, $\oplus$) operation. The second corresponds to the logical AND. (Alternatively, one might interpret these tables as "odd + even yields odd", "odd times odd yields odd", etc.)

Notation If n is a nonzero integer, then we shall refer to the set of congruence classes modulo n with the above definitions for addition and multiplication of such classes as the *ring of integers modulo n*, and denote this ring by $\mathbb{Z}_n$. We leave to the next chapter the answer to the question which immediately arises from this terminology: What is a ring? (Or, to avoid facetious replies: "What does an algebraist mean when he or she uses the word 'ring' in his or her professional capacity?")

2.5 Fermat's Factoring Method

The problem of finding the (prime or otherwise) factors of composite integers has been studied for centuries, but has in recent decades achieved greater attention because of its importance in the RSA public key system: it is generally assumed that a system based on RSA can be broken only by factoring the (publicly known) modulus used in the particular implementation.[7] One way of attempting to find a factorisation of an integer n would, of course, be to try the prime numbers less than $\sqrt{n}$, one by one, until one is found that divides neatly (i.e. without remainder) into n. This works fine if n is a small integer, but such an "exhaustive search" becomes totally impracticable when n has been selected to have only large prime factors.

More sophisticated methods of factoring an integer n usually depend on finding two integers a and b (with $a \not\equiv \pm b \bmod n$) such that

$$a^2 \equiv b^2 \bmod n$$

or, equivalently,

$$a^2 - b^2 \equiv 0 \bmod n,$$

$$(a - b)(a + b) \equiv 0 \bmod n.$$

Once this has been achieved, the problem is, for all practical purposes, solved, for this implies that $a - b$ and $a + b$ must have some divisor(s) in common with n, so that $\gcd(a - b, n)$ will be a divisor of n. We have the Euclidean algorithm available for finding that gcd efficiently.

Fermat's method of factoring represents an early, simple, version of this idea. Instead of trying to find a congruence

$$a^2 - b^2 = qn$$

one tries to find an equation

$$a^2 - b^2 = n$$

or

$$a^2 - n = b^2,$$

[7]But this has not been proved! (Or not yet?) It also does not mean that an *implementation* of RSA, or a protocol using RSA for encryption or signing is necessarily secure. An early example of what can go wrong is given by Bleichenbacher's successful attack on version 1 of RSA Data Security's standard PKCS 1 [Bleichenbacher, D.: Chosen ciphertext attacks against protocols based on the RSA Encryption Standard PKCS # 1; Proc. Crypto '98, LNCS 1462, Springer-Verlag, 1998, pp. 1–12].

RSA, by the way, is named after its inventors: **R**abin, **S**hamir and **A**dleman. We shall discuss RSA in some detail in Chap. 4.

whence the obtained factors of n are simply $a-b$ and $a+b$. Finding such an equation is done by, starting at $x = \lceil\sqrt{n}\rceil$, increasing x by 1 in each iteration, until $x^2 - n$ is a perfect square.

Example Factoring 3973 by this method, we start at $x = \lceil\sqrt{3973}\rceil = 64$, and obtain successively

$$
\begin{aligned}
64^2 - 3973 &= 123 \text{ which is not a perfect square,}\\
65^2 - 3973 &= 252 \text{ which is not a perfect square,}\\
66^2 - 3973 &= 383 \text{ which is not a perfect square,}\\
67^2 - 3973 &= 516 \text{ which is not a perfect square,}\\
68^2 - 3973 &= 651 \text{ which is not a perfect square,}\\
69^2 - 3973 &= 788 \text{ which is not a perfect square,}\\
70^2 - 3973 &= 927 \text{ which is not a perfect square,}\\
71^2 - 3973 &= 1068 \text{ which is not a perfect square,}\\
72^2 - 3973 &= 1211 \text{ which is not a perfect square,}\\
73^2 - 3973 &= 1356 \text{ which is not a perfect square,}\\
74^2 - 3973 &= 1503 \text{ which is not a perfect square,}\\
75^2 - 3973 &= 1652 \text{ which is not a perfect square,}\\
76^2 - 3973 &= 1803 \text{ which is not a perfect square,}\\
77^2 - 3973 &= 1956 \text{ which is not a perfect square,}\\
78^2 - 3973 &= 2111 \text{ which is not a perfect square,}\\
79^2 - 3973 &= 2268 \text{ which is not a perfect square,}\\
80^2 - 3973 &= 2427 \text{ which is not a perfect square,}\\
81^2 - 3973 &= 2588 \text{ which is not a perfect square,}\\
82^2 - 3973 &= 2751 \text{ which is not a perfect square,}\\
83^2 - 3973 &= 2916 = 54^2 \text{ which IS a perfect square.}
\end{aligned}
$$

From the last step we get that $3973 = 83^2 - 54^2 = (83 - 54)(83 + 54) = 29 \times 137$.

This example shows that the method is still not terribly efficient: in fact the number of steps required is, if $n = ab$, $\frac{a+b}{2} - \lceil\sqrt{ab}\rceil$, in other words approximately equal to the difference between the arithmetic and the geometric means of a and b. When a and b differ by an order of magnitude, this difference becomes very large.[8]

Exercises

1. What pattern appears in the sequence $\{x_i^2 - 3973\}$ in the example? Could such sequences be used to speed up an implementation of the Fermat algorithm?
2. Use Fermat's method to find the factors of 706711.
3. Given that if $n = 31864033$, then $24779170^2 \equiv 1 \bmod n$, find a factor of n.

2.6 Solving Linear Congruences

We shall start with another example:

Observe that $7 \cdot 13 = 91 \equiv 1 \bmod 15$, and suppose that we are asked to find all x such that $13x + 5 \equiv 12 \bmod 15$.

Now we have observed that all arithmetical operations on $\mathbb{Z}$ "carry over" to modular operations. Thus we may subtract 5 from both sides of this congruence to obtain

$$13x \equiv 12 - 5 \equiv 7 \bmod 15$$

and we may multiply both sides of this congruence by 7, and get

$$7 \cdot 13x \equiv 7 \cdot 7 \equiv 49 \equiv 4 \bmod 15$$

and, since $7 \cdot 13 \equiv 1 \bmod 15$, we get the solution

$$x \equiv 4 \bmod 15$$

(or, if you are fussy, that $x \in \{\ldots, -26, -11, 4, 19, 34, \ldots\}$).

Note that if we had considered a similar congruence $10x + 5 \equiv 10 \bmod 15$, we would have been stumped, for, no matter how long we looked for it, we would not have been able to find a z such that $10z \equiv 1 \bmod 15$. (After all, this would mean that $10z = 1 + 15k$ for some

[8] In a recent paper (#2009/318) published on the IACR ePrint archive (eprint.iacr.org) Erra and Grenier prove that with a more sophisticated search technique the Fermat method will lead to a successful factorisation in polynomial time if $n = p \cdot q$, with p and q both prime, and $|p - q| < n^{1/3}$. "Polynomial time" means that the time taken increases with the length l of n in bits like a polynomial in l. As a general rule, algorithms for solving problems (like searching for a cryptographic key) are considered practicable or feasible if they run in polynomial time. This is a very simple approach, but may through ignoring the degree—and the coefficients—of the polynomials, lead to the rejection of some cryptological primitives which would, in practice, have been quite secure. But that's another story, and we should get back to Algebra. Considering all problems for which a problem can be computed in polynomial time to be easy is just the cryptologists' way of ensuring an adequate safety margin. Cryptologists are very conservative and like *big* safety margins.

integer k, but in that case we have the contradiction that the left-hand side of the equation is divisible by 5, and the right-hand side cannot be.)

Definition Let a and n be two nonzero integers. An integer a' is called a *multiplicative inverse* of a modulo n if $a \cdot a' \equiv 1 \bmod n$.

Thus 7 and 13 are each other's multiplicative inverses modulo 15, whereas 10 does not have a multiplicative inverse modulo 15.

Theorem *Let a, n be integers with $n \neq 0$. Then a has a multiplicative inverse modulo n if and only if a and n are relatively prime.*

Proof This follows immediately from Bezout's identity: a and n are relatively prime if and only if there exist integers s and t such that $as + nt = 1$ which is the case if and only if $as \equiv 1 \bmod n$.

Theorem *Let a, b, c, n be integers with $n \neq 0$. If a and n are relatively prime then $ab \equiv ac \bmod n$ if and only if $b \equiv c \bmod n$. More generally, if d is the greatest common divisor of a and n, then $ab \equiv ac \bmod n$ if and only if $b \equiv c \bmod n/d$.*

Proof Suppose that $\gcd(a, n) = 1$, and let a' be the multiplicative inverse of a modulo n. Then $ab \equiv ac \bmod n$ implies $a'ab \equiv a'ac \bmod n$ and since $aa' \equiv 1 \bmod n$, this means $b \equiv c \bmod n$.

To prove the general case, note that if $ab \equiv ac \bmod n$, then $ab = ac + kn$ for some integer k. Dividing both sides of this equation by $d = \gcd(a, n)$, we get

$$\frac{a}{d}b = \frac{a}{d}c + k\frac{n}{d}$$

so that

$$\frac{a}{d}b \equiv \frac{a}{d}c \mod \frac{n}{d}$$

and since $\frac{a}{d}$ and $\frac{n}{d}$ are relatively prime, the result follows from the first part of the proof.

The reader may have noticed that we have, rather deviously, changed from referring to *a* multiplicative inverse to calling it *the* multiplicative inverse. This is because a multiplicative inverse, if it exists, is essentially unique. For suppose that x and x' are both multiplicative inverses of a modulo n. Thus $ax \equiv 1 \bmod n$ and $ax' \equiv 1 \bmod n$. But then

$$x \equiv x \cdot 1 \equiv x(ax') \equiv (xa)x' \equiv 1 \cdot x' \equiv x' \mod n.$$

Note that, whatever n may be, the congruence class 0, i.e. the element $0 \in \mathbb{Z}_n$, will never have an inverse. Looking at the examples at the end of Sect. 2.4, we see that in $\mathbb{Z}_6$, only the elements 1 and 5 have inverses (and both are their own inverses, coincidentally, which follows from the facts that only 1 and 5 are relatively prime to 6, and that $5 \equiv -1 \bmod 6$). On the other hand in $\mathbb{Z}_5$ and $\mathbb{Z}_2$ every nonzero element has an inverse, which follows from the fact that 5 and 2 are primes: If p is a prime, then every integer a such that $p \nmid a$ is relatively prime to p. $\mathbb{Z}_5$ and $\mathbb{Z}_2$, and generally $\mathbb{Z}_p$, with p a prime, are our first examples of what are known as *finite fields*, about which more in later chapters, where we shall denote them as $GF(p)$, the "GF" standing for "Galois field".[9]

Exercises

1. Calculate in $\mathbb{Z}_{175}$: 37×19, 35×25, $13 \times (14 + 167)$, 4^{-1}.
2. Calculate in $\mathbb{Z}_{29}$: $7 \times 13, 7^{-1} \times 13^{-1}, 25^7$, $12 \times (13 + 14)$.
3. Solve for x:

$$333x + 129 \equiv 234 \mod 911.$$

4. Show that if $p > 2$ is a prime, then the congruence $x^2 \equiv 1 \mod p$ has exactly two solutions.
5. More generally, show that if p is a prime, then the congruence $x^2 \equiv a^2 \mod p$ has exactly two solutions. For how many values of a does the equation $x^2 \equiv a \mod p$ have a solution?
6. The congruence $x^2 \equiv 1 \mod 4033$ has the four solutions (mod 4033) $x = 1, 4032, 110, 3923$. Conclude that 4033 cannot be prime. Find the factorization of 4033, using the given data.
7. Prove: If the congruence $kx \equiv a \mod n$ has a solution, then it has s solutions, where $s = \gcd(k, n)$.
8. Show that as k runs through the values 1, 2, ..., 28, $2^k \mod 29$ runs through all the nonzero elements of $\mathbb{Z}_{29}$. (One wonders whether this sort of thing can be done for any prime. The answer is "yes", as we shall show later in Sect. 6.2. But not necessarily with 2 as base, though.)
9. Solve for x and y:

$$2x + 3y \equiv 8 \mod 17,$$
$$7x - y \equiv 7 \mod 17.$$

10. A test for divisibility by 9 is (if one uses decimal notation) to add up all the digits of the number: if this gives a multiple of 9, the number is divisible 9. Explain why this works and then devise a similar test for divisibility by 11. Can you devise a test for divisibility by 7? (*Hint*: If $n = 10a + b$ is divisible by 7, then so is $n - 21b$.) If you can, you'll easily find one for divisibility by 37.

[9] This notation is not standardised: many books use the notation $\mathbb{F}_p$, whereas it is also common to denote fields by the letter K and variations on that theme, following the German, where they are known as *Körper*. Much of the early work on these mathematical structures was done in Germany. The rather bland term "field" was introduced into English by American algebraists who had studied in Germany. If an applied mathematician tells you that he is interested in field theory, you will have to enquire further to find out what kind of fields he means: the algebraic ones or, say, electro-magnetic ones.

2.7 The Chinese Remainder Theorem

We next consider solving systems of congruences, or more precisely, solving for a single unknown which must satisfy (linear) congruence relations with respect to different moduli. The theorem and the method concerned, which are known as the *Chinese Remainder Theorem*, sometimes referred to as CRT for short,[10] allows for calculations to be performed modulo smaller integers, with the final result, i.e. the result modulo a large product of moduli, only being calculated at the very end. This is helpful, since arithmetical operations with very large integers (as required, for example, in RSA based systems) may be very time-consuming. (There is a price to pay for this increased efficiency in that use of the CRT leaves the operations more vulnerable to certain attacks, such as those known as "glitch attacks".)

Theorem (CRT) *Let $n_1, n_2, \ldots n_k$ be positive integers which are pairwise relatively prime,*[11] *and let $a_1, a_2, \ldots, a_k$ be arbitrary integers. Then there exists an integer z such that*

$$z \equiv a_i \bmod n_i \quad \forall\, i \in \{1, 2, \ldots, k\}.$$

Moreover, any other integer z' is also a solution to all these congruences if and only if $z \equiv z' \bmod n$, where $n = \prod_{i=1}^{k} n_i$.

Proof Let $n = \prod_{i=1}^{k} n_i$, and define, for $i = 1, 2, \ldots, k$

$$n'_i = n/n_i.$$

From the fact that the n_i are pairwise relatively prime, it is easy to see that $\gcd(n_i, n'_i) = 1$ for all $i = 1, 2, \ldots, k$. Therefore there exists, for each i, an integer m_i such that $m_i n'_i \equiv 1 \bmod n_i$.

Now put, for each $i = 1, 2, \ldots, k$,

$$w_i = m_i n'_i,$$

then we clearly have that

- $w_i \equiv 1 \bmod n_i$;
- $w_i \equiv 0 \bmod n_j$ for all $j \in \{1, 2, \ldots, k\} \setminus \{i\}$.

Hence, if we put

$$z = \sum_{i=1}^{k} w_i a_i$$

[10] The story goes that a Chinese emperor, wanting to find out how many troops he had, made them march past in rows of 3, then 5, and so on, noting in each case how many soldiers there were in the last incomplete row. He then applied the remainder theorem and computed the size of his army. Hence the name of the theorem. If you believe this story, you will be interested to know that the emperor concerned was taught the theorem by intergalactic aliens whose space ship had crash landed on the island of Atlantis.

[11] I.e., $\gcd(n_i, n_j) = 1$ whenever $i \neq j$.

then

$$z \equiv a_j \bmod n_j \quad \forall j \in \{1, 2, \ldots, k\}.$$

Moreover, if $z' \equiv z \bmod n$, then, since $n_i | n$ for all i, we also have that $z' \equiv z \equiv a_i \bmod n_i$ for all i, so that z' is also a solution of all the congruences.

Finally, if z' is any solution of all the congruences, then $z' \equiv z \bmod n_i$, for all i. In other words, $n_i | (z' - z)$ for all i. Since the n_i are relatively prime, this implies that $n | (z' - z)$, i.e. $z' \equiv z \bmod n$.

Examples

1. Consider the congruences

$$x \equiv 1 \bmod 7,$$
$$x \equiv 2 \bmod 23,$$
$$x \equiv 19 \bmod 29.$$

With the notation used in the theorem, we have $n = 7 \cdot 23 \cdot 29 = 4669$, $n'_1 = 667$, $n'_2 = 203$, $n'_3 = 161$, and we can compute

$$m_1 = n_1'^{-1} \bmod n_1 = 667^{-1} \bmod 7 = 4,$$
$$m_2 = n_2'^{-1} \bmod n_2 = 203^{-1} \bmod 23 = 17,$$
$$m_3 = n_3'^{-1} \bmod n_3 = 161^{-1} \bmod 29 = 20.$$

This yields

$$w_1 = n'_1 \cdot m_1 = 667 \cdot 4 = 2668,$$
$$w_2 = n'_2 \cdot m_2 = 203 \cdot 17 = 3451,$$
$$w_3 = n'_3 \cdot m_3 = 161 \cdot 20 = 3220,$$

so that

$$z \equiv 1 \cdot 2668 + 2 \cdot 3451 + 19 \cdot 3220 = 70750$$
$$\equiv 715 \bmod 4669.$$

2. In attempting to solve the congruence

$$x^2 \equiv 1 \bmod 323$$

we note that $323 = 17 \cdot 19$, so we are actually considering two simultaneous congruences:

$$x^2 \equiv 1 \bmod 17,$$
$$x^2 \equiv 1 \bmod 19.$$

Now, if p is a prime, then Exercise 5 of the previous section shows that the only solutions of a congruence $x^2 \equiv a^2 \bmod p$ are $x \equiv \pm a \bmod p$ for some a. So there are a total of four possibilities, which we solve by means of the Chinese Remainder Theorem:

x mod 17	1	1	−1	−1
x mod 19	1	−1	1	−1
x mod 323	1	18	305	322

Exercises

1. Solve for x: $x \equiv 1 \mod 7$, $x \equiv 2 \bmod 11$ and $x \equiv 3 \bmod 13$.
2. Find all solutions of the congruence $x^2 \equiv 13 \bmod 391$. (*Hint*: $391 = 17 \times 23$.)

2.8 Some Number-Theoretic Functions

2.8.1 Multiplicative Functions

In this section we shall briefly discuss some number-theoretic functions, of which the most important one, by far, is Euler's totient function, better known as Euler's ϕ-function. In order to make the discussion a little more general, we shall discuss this in the context of "multiplicative functions". The proofs involved are a little messy, but not *too* difficult.

But if you want to go straight through to the most important case (for our purposes), the value of $\phi(n)$ where $n = p \cdot q$, the product of two primes, feel free to go to Sect. 2.8.4, where we give an easy proof of our main result for that particular case.

Definitions A function defined on the non-negative (or the positive) integers whose codomain is the field of complex numbers or a subset thereof, is called a *number-theoretic* function. A number-theoretic function f is called *multiplicative* if for any positive integers a and b which are relatively prime

$$f(ab) = f(a) \cdot f(b).$$

The following are immediate consequences of the definition:

- If f is multiplicative and not identically 0, then for any positive integer a such that $f(a) \neq 0$, we have
$$f(a) = f(1 \cdot a) = f(1) \cdot f(a),$$
so that $f(1) = 1$.
- If f and g are multiplicative functions, then so is the function $f \cdot g$, defined by $(f \cdot g)(a) = f(a) \cdot g(a)$.

- If $n = p_1^{e_1} p_2^{e_2} \dots p_r^{e_r}$ is the factorization of n into a product of prime powers, and if f is multiplicative, then

$$f(n) = f(p_1^{e_1})f(p_2^{e_2}) \dots f(p_r^{e_r}).$$

  Thus any multiplicative function is completely determined by its values at the primes and their powers.

Any multiplicative function f gives rise to a related one, as in the following theorem.

Theorem *If f is a multiplicative function and F is defined by*

$$F(n) = \sum_{0<d, d|n} f(d)$$

then F is also multiplicative.

Proof If a, b are relatively prime positive integers, then any divisor d of ab is a product of a divisor of a and one of b, i.e. $d = a' \cdot b'$, where $a'|a$ and $b'|b$. (This factorization may be trivial, of course, with $a' = 1$ or $b' = 1$.) Hence

$$\begin{aligned} F(ab) &= \sum_{d|ab} f(d) = \sum_{d_1|a, d_2|b} f(d_1 d_2) \\ &= \sum_{d_1|a, d_2|b} f(d_1)f(d_2) \\ &= \left(\sum_{d_1|a} f(d_1) \right) \left(\sum_{d_2|b} f(d_2) \right) = F(a)F(b), \end{aligned}$$

as required.

Remarkably, the converse of this theorem is also true: If F is multiplicative, then so is f. This is a consequence of the Möbius inversion theorem, which we state and prove in the next subsection.

For now, let f be a multiplicative function, and define F as in the theorem. If p is a (positive) prime, then the only divisors of p^k are $1, p, p^2, \dots p^k$. We know that $f(1) = 1$, and therefore

$$F(p^k) = 1 + f(p) + f(p^2) + \cdots + f(p^k) = \sum_{i=0}^{k} f(p_i^k).$$

Hence

$$F(p_1^{e_1} p_2^{e_2} \dots p_r^{e_r}) = \prod_{i=1}^{r} \sum_{j=0}^{e_i} f(p_i^j).$$

Example Define f by $f(a) = a$. Trivially, f is multiplicative. If we define s by

$$s(a) = \sum_{d|a} f(d)$$

then $s(a)$ is simply the sum of all the divisors of a. By what we have done so far, we find that if $a = p_1^{e_1} \ldots p_r^{e_r}$, then

$$\begin{aligned} s(a) &= [1+p_1+\ldots+p_1^{e_1}] \times \cdots \times [1+p_r+\ldots+p_r^{e_r}] \\ &= \frac{p_1^{e_1+1}-1}{p_1-1} \times \cdots \times \frac{p_r^{e_r+1}-1}{p_r-1}. \end{aligned}$$

Thus, for example, the sum of the divisors of $1400 = 2^3 \times 5^2 \times 7$ is $\frac{2^4-1}{2-1} \times \frac{5^3-1}{5-1} \times \frac{7^2-1}{7-1} = 15 \times 31 \times 8 = 3720$.

Exercises

1. If f is any number-theoretic function, explain why

$$\sum_{d|a} f(d) = \sum_{d|a} f(a/d).$$

2. Define $d(a)$ by $d(a) = \sum_{d|a} 1$. Show that a is prime if and only if $d(a) = 2$ if and only if $s(a) = a + 1$ (where s is the function defined in the previous example).
3. A positive integer n is called *perfect* if $s(n) = 2n$, *deficient* if $s(n) < 2n$ and *abundant* if $s(n) > 2n$.

   (a) Show that any power of a prime is deficient.
   (b) Find the first four abundant natural numbers.
   (c) Perform a computer search to find the smallest odd natural abundant number.

   This terminology dates from antiquity, and is important in numerology,[12] but apart from providing some interesting number-theoretical problems (for example: no odd perfect number is known, but there is no proof that such a number cannot exist), there are few, if any, practical applications.

[12]From Oystein Ore's *Number Theory and its History:—*

> Alcuin (735–804), the adviser and teacher of Charlemagne, observes that the entire human race descends from the 8 souls in Noah's ark. Since 8 is a deficient number, he concludes that this second creation was imperfect in comparison with the first, which was based on the principle of the perfect number 6.

We are getting rather far from cryptology.

2.8.2 *The Möbius Function*

Definition The Möbius function μ is defined as follows: For any positive integer n

$$\mu(n) = \begin{cases} 1 & \text{if } n = 1, \\ 0 & \text{if } n \text{ is divisible by a square,} \\ (-1)^k & \text{if } n \text{ is the product of } k \text{ primes.} \end{cases}$$

Note that, from the definition, μ is a multiplicative function. Also recall that if f is any multiplicative function then $f \cdot \mu$ will also be multiplicative (so that $\mu(1)f(1) = 1$) and note that for a prime p

$$(\mu \cdot f)(p) = \mu(p)f(p) = -f(p),$$
$$(\mu \cdot f)(p^2) = 0.$$

In fact,

$$(\mu \cdot f)(p^k) = 0 \text{ whenever } k \geq 2.$$

Hence, if $n = p_1^{e_1} \ldots p_r^{e_r}$, then

$$\sum_{d|n} (\mu \cdot f)(d) = (1 - f(p_1))(1 - f(p_2)) \ldots (1 - f(p_r)).$$

In particular, if we take $f(a) = 1$ for all positive integers a, then we get

Property 1 For any integer $n > 1$

$$\sum_{d|n} \mu(d) = 0.$$

Similarly, by taking $f(n) = \frac{1}{n}$, we obtain

Property 2 For any integer $n > 1$

$$\sum_{d|n} \frac{\mu(d)}{d} = \left(1 - \frac{1}{p_1}\right) \ldots \left(1 - \frac{1}{p_r}\right).$$

The following lemma, which refines Property 1, will be needed in the proof of the main result in this section.

Lemma *Let a and b be integers greater than 1 such that $b|a$. Then*

$$\sum_{d:b|d|a} \mu\left(\frac{a}{d}\right) = \begin{cases} 1 & \text{if } b = a, \\ 0 & \text{otherwise.} \end{cases}$$

Proof Let $a = a' \cdot b$. If $b|d$ and $d|a$, put $d = d' \cdot b$. Then

$$\sum_{d\,:\,b|d|a} \mu(\frac{a}{d}) = \sum_{d'|a'} \mu(\frac{a'}{d'}) = \sum_{d'|a'} \mu(d')$$

according to the first exercise in the preceding subsection, and the last sum equals 0, by property 1, unless $a' = 1$, in which case the sum is just $\mu(1) = 1$.

Theorem (Möbius Inversion Theorem) *Let $f(n)$ be any number-theoretic function, and let F be defined by $F(n) = \sum_{d|n} f(d)$. Then*

$$f(n) = \sum_{d|n} \mu(d)F(n/d) = \sum_{d|n} \mu(n/d)F(d).$$

Proof The fact that the two sums are equal should not need explaining. Now observe that

$$\begin{aligned} \sum_{d|n} \mu(n/d)F(d) &= \sum_{d|n} \mu(n/d) \sum_{e|d} f(e) \\ &= \sum_{e|n} f(e) \sum_{d:e|d|n} \mu(n/d). \end{aligned}$$

By the lemma, the inner sum equals 0, except in the case where $e = n$, in which case it is 1. Thus the only nonzero term in the outer sum is the term for which $e = n$, and we have that

$$\sum_{d|n} \mu(n/d)F(d) = f(n),$$

as required.

2.8.3 *Euler's ϕ-Function*

Euler's *totient function*, usually referred to simply as "Euler's ϕ", is defined as follows:

Definition Let n be any integer greater than 1. Then

$$\phi(n) = \#\{a : 1 \leq a \leq n \text{ and } \gcd(a, n) = 1\}.$$

In other words, $\phi(n)$ is the number of positive integers not exceeding n which are relatively prime to n. Note that $\phi(1) = 1$. (The integers ± 1 are the only ones which are relatively prime to themselves.)

If p is a positive prime, then all the numbers in the set $\{1, 2, 3, \ldots, p-1\}$ are relatively prime to p, so $\phi(p) = p - 1$. Also $\phi(p^k)$ is the number of elements remaining in the set $\{1, 2, 3, \ldots, p^k - 1\}$ after all the multiples of p have been deleted; since there are p^{k-1} such multiples, we get that $\phi(p^k) = p^k - p^{k-1} = p^k(1 - 1/p)$.

ϕ is in fact a multiplicative function – as we'll prove in a moment. From that fact we get that if $n = p_1^{e_1} p_2^{e_2} \ldots p_r^{e_r}$, then

$$\begin{aligned}\phi(n) &= \phi(p_1^{e_1})\phi(p_2^{e_2})\ldots\phi(p_r^{e_r}) \\ &= p_1^{e_1}(1 - 1/p_1)p_2^{e_2}(1 - 1/p_2)\ldots p_r^{e_r}(1 - 1/p_r) \\ &= n \times \left(1 - \frac{1}{p_1}\right)\left(1 - \frac{1}{p_2}\right)\ldots\left(1 - \frac{1}{p_r}\right).\end{aligned}$$

Here is the promised theorem:

Theorem *If m and n are relatively prime, then $\phi(mn) = \phi(m) \cdot \phi(n)$.*

Proof Denote the set of residue classes modulo k which have multiplicative inverses by $\mathbb{Z}_k^*$. There are $\phi(k)$ such classes.

Thus we need to prove that if m and n are relatively prime, then

$$\#\mathbb{Z}_{mn}^* = (\#\mathbb{Z}_m^*)(\#\mathbb{Z}_n^*).$$

We do this by simply defining a function

$$p: \; \mathbb{Z}_{mn}^* \longrightarrow \mathbb{Z}_m^* \times \mathbb{Z}_n^*$$

by $p(x \bmod mn) = (x \bmod m, x \bmod n)$. Since $\gcd(x, mn) = 1$ implies that $\gcd(x, m) = \gcd(x, n) = 1$, the function p is certainly into $\mathbb{Z}_m^* \times \mathbb{Z}_n^*$. Moreover, the Chinese Remainder Theorem guarantees that $p(x \bmod mn) = p(y \bmod mn)$ if and only if $x \bmod mn = y \bmod mn$, so that p is one-to-one. Finally, for any element of $\mathbb{Z}_m^* \times \mathbb{Z}_n^*$, there exists, by the same theorem, an element of $\mathbb{Z}_{mn}^*$ which gets mapped onto it by p. Thus p is one-to-one and onto, and the two sets have the same cardinality.

Exercises

1. Find $\sum_{k=1}^{n} \phi(p^k)$, where p is a prime.
2. Show that

   (a) $\phi(4n) = 2\phi(2n)$;
   (b) $\phi(4n + 2) = \phi(2n + 1)$;
   (c) $\phi(n^2) = n\phi(n)$.

3. Prove that there does not exist a positive integer n such that $\phi(n) = 2p$ where p is a prime and $2p + 1$ is composite. Conclude that $\phi(n)$ can never equal 14.

2.8.4 The Case $n = p \cdot q$

If you followed our suggestion and skipped the previous few subsections, or if, in the less likely case that you didn't but have already forgotten: if n is a positive integer then $\phi(n)$ is the number of integers in the interval $[1, n]$ which are relatively prime to n.

It is obvious that if p is a prime then $\phi(p) = p - 1$.

In many cryptological applications, such as RSA, numbers of the form $n = p \cdot q$ are used, where both p and q are primes. Now in the interval $[1, n]$ there are q multiples of p, namely $p, 2p, 3p, \ldots, (q-1)p, qp$ and p multiples of q, namely $q, 2q, 3q, \ldots, (p-1)q, pq$. So the number of integers in $[1, pq]$ relatively prime to n is

$$\phi(n) = n - p - q + 1,$$

where the "+1" occurs because we twice counted pq itself. Hence

$$\phi(pq) = pq - p - q + 1 = (p-1)(q-1).$$

Exercise Use the facts that $n = 12509443$ is known to be the product of two primes and $\phi(n) = 12501720$ to find the two prime factors.

3 Groups, Rings and Ideals

An algebraic structure generally consists of a set, and one or more binary operations on that set, as well as a number of properties that the binary operation(s) has (have) to satisfy.

In the following pages we shall discuss the most important algebraic structures, viz groups, rings and fields. For the purposes of our applications, viz applications to Cryptology, it will be sufficient if in all such structures we restrict ourselves to the commutative instances, i.e. we look only at groups and rings in which the binary operations are commutative. Thus where we discuss groups we shall only consider the so-called "Abelian groups", and where we discuss rings, we limit ourselves to those which are in the literature referred to as (unsurprisingly) "commutative rings". More precisely they are "commutative rings with identity", since even the existence of an identity element is not a requirement for rings in general.

3.1 Groups

Definitions A *group* is a pair $\{G, *\}$ consisting of a set G and a binary operation $*$ on G, such that the following are satisfied:

1. $\forall a, b, c \in G,\ a * (b * c) = (a * b) * c$ (i.e. the operation $*$ is *associative*).
2. $\exists e \in G$ such that $\forall a \in G,\ e * a = a * e = a$. e is called the *identity element* of the group.
3. $\forall a \in G\ \exists x \in G$ such that $x * a = a * x = e$. x is called the *inverse* of a and denoted by a^{-1} (in our present choice of notation).

   If, in addition to the above, the binary operation satisfies

4. $\forall a, b \in G,\ a * b = b * a$, i.e. the operation $*$ is *commutative*,

then the structure is called an *Abelian* group.

Before giving any examples, we note a few immediate consequences of this definition:

- The identity element is unique. For suppose e and e' are both identity elements. Then

$$e = e * e' = e',$$

A.R. Meijer, *Algebra for Cryptologists*, Springer Undergraduate Texts in Mathematics and Technology,
DOI 10.1007/978-3-319-30396-3_3

where the first equality holds because e' is an identity element, and the second equality holds because e is an identity element.

- The inverse of an element a is unique. For suppose x and x' are both inverses of a. Then

$$x = x * e = x * (a * x') = (x * a) * x' = e * x' = x'.$$

- $(a * b)^{-1} = b^{-1} * a^{-1}$ for any $a, b \in G$, as is easily verified.

The word "Abelian" is derived from the name of the Norwegian mathematician Niels Henrik Abel. In general *groups*, condition 4 of our definition need not apply, but (almost) all the groups with which we shall be concerned will be Abelian. It is customary in the literature to write "+" for the operation in Abelian groups. For such groups, we shall use the notation

$$na = \underbrace{a + a + \ldots a}_{n \text{ times}}$$

for repeated addition of a group element a with itself. We denote in an Abelian group the inverse of a by $-a$, and introduce the further notation $a - b$ as shorthand for $a + (-b)$. Then we can also write

$$n(-a) = \underbrace{-a - a - \cdots - a}_{n \text{ times}} \stackrel{\text{def}}{=} (-n)a$$

where n is again a positive integer. We define

$$0a = 0,$$

being careful to distinguish the two zeros in this equation: the one on the left is the natural number 0, the one on the right is the identity element of the group we are working with. In this way we can multiply, in a meaningful way, an element a of an Abelian group by any integer of our choice. In the first two of the following examples, we see what happens to $\mathbb{Z}$, looked at in the group-theoretic way.

Examples

1. The pair $\{\mathbb{Z}, +\}$ is an Abelian group in which the identity element is 0, and the inverse of $a \in \mathbb{Z}$ is $-a$.
2. The pair $\{n\mathbb{Z}, +\}$, $n\mathbb{Z} = \{na | a \in \mathbb{Z}\}$, where n is any integer, is an Abelian group in which the identity element is $0 = n \cdot 0$, and the inverse of $na \in n\mathbb{Z}$ is $n(-a)$.
3. The set of integers is *not* a group under the operation of multiplication. While $a \cdot 1 = a$ for all integers a, only the elements 1 and -1 have inverses.
4. The set of all rational numbers is *not* a group under multiplication, since 0 does not have an inverse.
5. The set of all nonzero rational numbers is an Abelian group under multiplication: The identity element is 1, and the inverse of $\frac{a}{b}$ (where $a, b \neq 0$) is $\frac{b}{a}$.

6. The pair $\{\mathbb{Q}, +\}$, where $\mathbb{Q}$ denotes the set of all rational numbers, is an Abelian group.
7. The set $\mathbb{Z}_n$ of equivalence classes modulo $n \neq 0$ is an Abelian group under the operation of addition of equivalence classes, as defined in Sect. 2.4.
8. Let n be any integer, and consider the set of equivalence classes

$$\{\overline{a} \mid gcd(a, n) = 1\}.$$

This set, with the definition of multiplication as defined in Sect. 2.4, forms an Abelian group. We shall denote this group by $\mathbb{Z}_n^*$. The number of elements of this group is $\phi(n)$. Thus, for example, $\mathbb{Z}_{15}^*$ consists of the set $\{1, 2, 4, 7, 8, 11, 13, 14\}$ of $8 = \phi(15)$ elements and the group operation of multiplication (denoted by "$\cdot$") is as given by the following table:

$\cdot$	1	2	4	7	8	11	13	14
1	1	2	4	7	8	11	13	14
2	2	4	8	14	1	7	11	13
4	4	8	1	13	2	14	7	11
7	7	14	13	4	11	2	1	8
8	8	1	2	11	4	13	14	7
11	11	7	14	2	13	1	8	4
13	13	11	7	1	14	8	4	2
14	14	13	11	8	7	4	2	1

9. In particular, if p is a prime, then $\mathbb{Z}_p \backslash \{\overline{0}\}$ forms an Abelian group under multiplication. (The crucial point being that every integer not divisible by p has a multiplicative inverse modulo p, as observed in Sect. 2.6.)
10. A class of (Abelian) groups which is important for cryptographic purposes consists of the points on elliptic curves defined over finite fields. An "addition" operator can be defined on the set of these points which results in an abelian group structure. We shall discuss such groups very briefly in Sect. 6.6.
11. Let $\{G, *\}$ and $\{H, \cdot\}$ be groups. We can turn the cartesian product $G \times H$ of the two underlying sets into a group by defining a group operation (let's denote it by $\odot$) componentwise:

$$(g_1, h_1) \odot (g_2, h_2) = (g_1 * g_2, h_1 \cdot h_2).$$

The result is called the *direct product* (but, in the case of Abelian groups, it is frequently called the *direct sum*) of the groups G and H. It is easy to see how this construction can be extended to any number (finite or infinite) of groups.
12. The set of all $n \times n$ matrices with rational entries and nonzero determinant, forms a *non-abelian*[1] group under matrix multiplication.

[1]Provided $n > 1$, in case you want to split that particular hair.

13. Let Ω be any finite set, and let $\mathcal{P}$ be the set of all permutations of Ω, i.e. the set of all one-to-one onto mappings from Ω to Ω. If we define a binary operation on $\mathcal{P}$ by composition (i.e. the product $f * g$ of $f, g \in \mathcal{P}$ is the permutation that maps $x \in \Omega$ onto $f(g(x)) \in \Omega$), then $\mathcal{P}$ gets the structure of a (non-Abelian) group. The identity element in this group is the identity map that maps every element of Ω onto itself, and the inverse of any element f is the function that maps s onto t if and only if $f(t) = s$. The order of this group is the number of possible permutations of the elements of Ω, namely $(\#\Omega)!$. If $\#\Omega = n$, the corresponding group $\mathcal{P}$ is called the *symmetric group* of degree n.

    If, for example, Ω is the alphabet {A,B, ..., Z} then $\mathcal{P}(\Omega)$ is a group with $26! \approx 4.0 \times 10^{26}$ elements, each of which could be used as a cipher. Such a cipher is called a *substitution cipher*. More than one person has fallen into the trap of believing that, because the group $\mathcal{P}(\Omega)$ is so large, i.e. because there are so many keys (approximately $4.0 \times 10^{26} \approx 2^{88}$) to choose from, such ciphers must be very strong. In fact, substitution ciphers are extremely weak because of the ease of doing frequency analysis: the most frequently occurring symbol must represent the letter E, etc.
14. Consider in particular the set $\{0,1\}^{128}$ of all 128-bit strings of 0s and 1s. A (128-bit) block cipher is a subset of the set of permutations of the set $\{0,1\}^{128}$, indexed by a key set $\mathcal{K}$ that associates with every $k \in \mathcal{K}$ one of the permutations ψ_k, say. If it is a "good" block cipher, it should be impossible to find k when given $\pi = \psi_k$ for some $k \in \mathcal{K}$, except by running through all values of k systematically, and verifying whether or not the permutations ψ_k and π are indeed equal. This technique is known as *exhaustive search*; by taking the set $\mathcal{K}$ large enough, exhaustive search can be made infeasible in practice.[2]

    More generally, it should not be possible to decide whether any given permutation belongs to the set $\{\psi_k | k \in \mathcal{K}\}$ or not. If this is not the case, the cipher is said to have succumbed to a *distinguishing attack*. It may be argued that some distinguishing attacks are merely of theoretical interest, but this is a dangerous attitude to take: a successful distinguishing attack indicates that there is a weakness in the design of the block cipher, and someone else may find a way of exploiting that weakness.

    Another question that may be asked is whether the set $\{\psi_k | k \in \mathcal{K}\}$ is itself a group (in which case it would be a *subgroup* of the group of all permutations, as we shall shortly define that concept). In the case of a well-designed block cipher, it should not be![3]

The reader will appreciate that these examples cover quite a few mathematical "objects", seemingly quite different. The beauty of Algebra lies in its abstraction: as long as one uses only properties which can somehow be deduced from the definition of "group", the resulting

[2]Common key lengths for block ciphers are 128, 192 and 256 bits, this means that in an exhaustive search one would have to sift through $2^{128}, 2^{192}$ or 2^{256} possible keys. This is not only exhausting, it's downright impossible, no matter how fast your computer is running.

[3]The reason, very roughly, is that if the set $\{\psi_k | k \in \mathcal{K}\}$ is a group, then applying two of the (secret) permutations, one after the other, would not make breaking the cipher any more difficult than breaking just a single application, because there would be a single permutation in the set having the same effect anyway.

theorem will be true for any one of them. The definition abstracts crucial properties which are shared by all these objects. In order to derive some of these theorems, we need to define a few more terms.

Definitions

1. As noted, a group consists of both a set, and a binary operation on that set. If the set is finite, we refer to the number of elements in the set as the *order of the group*. Thus the group of Example 8 above has order 8. If the set is infinite, we say that the group is of infinite order.
2. If G is a group, with binary operation $*$, and if $a \in G$ and n any integer, then we define $a^n = \underbrace{a * a * a * \cdots * a}_{n \text{ times}}$, if $n > 0$, $a^0 = e$, and $a^n = (a^{-1})^{-n}$ if $n < 0$.

We leave the proofs of the following as exercises:

Theorem *Let G be an Abelian group, with binary operation $*$. For all $a, b, c \in G$ and for all $m, n \in \mathbb{Z}$*

1. $a * b = a * c \Longrightarrow b = c$;
2. *The equation $a * x = b$ has a unique solution;*
3. $(a * b)^{-1} = a^{-1} * b^{-1}$;
4. $(a^{-1})^{-1} = a$;
5. $a^{-n} \stackrel{\text{def}}{=} (a^n)^{-1} = (a^{-1})^n$;
6. $(a^m)^n = a^{mn}$;
7. $(a * b)^n = a^n * b^n$.

Exercise Prove the above theorem.

3.2 Subgroups

Definition Let $\{G, *\}$ be a group, and let H be a nonempty subset of the set G such that

- $\forall a, b \in H,\ a * b \in H$;
- $\forall a \in H,\ a^{-1} \in H$.

Then H is called a *subgroup* of G.

The important thing is that H is then itself a group, using exactly the same binary operation as is used in the larger ("supergroup") G. This follows quite easily once one realises that the identity element e of G must also belong to H (and will, of course, behave like the identity element there). For, since H is nonempty, we can choose $a \in H$, but then, by the second condition $a^{-1} \in H$, so that by the first condition $e = a * a^{-1} \in H$. The other requirements for a group structure are either in the definition of subgroup or are inherited from G.

We give three important examples of subgroups:

1. Let $\{G, *\}$ be a group, let $m > 0$ be an integer, and define

$$H = \{g^m \mid g \in G\}.$$

Then H is a subgroup of G. For instance, if we take the additively written group of all integers, then the subgroup we get is the group of all integers divisible by m.

On the other hand, if we take the multiplicatively written group of Example 8 of the previous section, and take m=2, we get the set $\{a^2 \bmod 15 | a \in \{1, 2, 4, 7, 8, 11, 13, 14\}\} = \{1, 4\}$, with multiplication table

	1	4
1	1	4
4	4	1

But if we were to consider $\{a^3 \bmod 15 | a \in \{1, 2, 4, 7, 8, 11, 13, 14\}\}$, we would find ourselves with the whole group again.

$\{a^3 \bmod 15 | a \in \{1, 2, 4, 7, 8, 11, 13, 14\}\}$ is therefore not a *proper* subgroup.

2. Let G be any Abelian group, and let m again be any positive integer. The set

$$H = \{a \,|a^m = e\}$$

is easily seen to be a subgroup of G. This subgroup is denoted by $G[m]$ (and called the m-torsion subgroup of G).

3. Let X be a finite set, with n elements, and let S_n be the group of permutations of X. Now it can be proved that every permutation can be written as a product of *transpositions*, where a transposition is a permutation which interchanges two of the elements of the set S and leaves all the others in place. For example, let $X = \{a, b, c, d\}$ and consider, say, the permutation

$$\tau = \begin{pmatrix} a\ b\ c\ d \\ c\ a\ d\ b \end{pmatrix}$$

where we use an obvious notation: $\tau(a) = c$, $\tau(b) = a$, etc. We could also write $\tau = (a\ c\ d\ b)$, indicating the cyclic nature of $\tau : a \mapsto c \mapsto d \mapsto b \mapsto a$, which in turn can be written as the product $(a\ b)(a\ d)(a\ c)$ where $(x\ y)$ indicates that x and y are swopped (and everything else left unchanged) with the functions being applied from right to left.

Now the representation of a permutation as a product of transpositions is not unique (for example, we could have written τ as $(d\ c)(d\ a)(c\ a)(c\ a)(d\ b))$, but it can be shown that the parity of the number of transpositions is fixed: any permutation is either even or odd. Thus, in our example, no matter how you express τ as a product of a number of transpositions, it must always be an odd number.

If we consider just the even permutations, it is not hard to see that they form a subgroup of S_n. This subgroup is called the *alternating group* and denoted by A_n. In the exercises you will be asked to prove that $\#(A_n) = \frac{1}{2}\#(S_n)$.

Exercises

1. Find all the subgroups of the additive group $\mathbb{Z}_{15}$.
2. Show that the two conditions on a nonempty subset of a group which need to be checked can be reduced to one: A nonempty subset H of a group G is a subgroup if and only if

$$\forall a, b \in H, \ ab^{-1} \in H.$$

3. Let G be a group and let $a \in G$ be a fixed element. Show that

$$H_a = \{x \in G | x * a = a * x\}$$

   is a subgroup of G. (If G is Abelian $H_a = G$, of course.)
4. Let G be an Abelian group, and let p and q be two distinct prime numbers. Prove that, in the notation of Example 2 above, $G[p] \cap G[q] = \{e\}$.
5. Prove that exactly half the elements in S_n are even permutations.
6. Let H be a subgroup of $\{\mathbb{Z}, +\}$. Show that there exists an integer d such that $H = d\mathbb{Z}$.
7. Show that $m\mathbb{Z}$ is a subgroup of $n\mathbb{Z}$ if and only if $n|m$.
8. Let H_1 and H_2 be subgroups of an Abelian group G and define $H_1 * H_2 = \{h_1 * h_2 | h_1 \in H_1, \ h_2 \in H_2\}$. Show that $H_1 * H_2$ and $H_1 \cap H_2$ are subgroups of G. What about $H_1 \cup H_2$?

3.3 The Lattice of Subgroups

The last exercise of the previous section raises an interesting question: If H_1 and H_2 are subgroups of a group G, then $H_1 \cap H_2$ is also a subgroup. It might be the trivial subgroup $\{e\}$, of course. But the argument that works for the intersection of the two sets does not work for the union $H_1 \cup H_2$. So another approach is needed.

A way to look at it is as follows: $H_1 \cap H_2$ is the largest subgroup of G that is contained in both H_1 and H_2. What one might call the question *dual* to this statement is: what is the *smallest* subgroup of G which *contains* both H_1 and H_2? We shall call this the subgroup generated by H_1 and H_2 and denote it for the moment by $H_1 \vee H_2$.

In the case of Abelian groups, $H_1 \vee H_2$ is easily found: any subgroup of G (with $*$ as the group operation) which contains both H_1 and H_2 must contain all elements of the form $h_1 * h_2$ with $h_1 \in H_1$ and $h_2 \in H_2$. But as you showed in Exercise 8 of the preceding section, the set of all such elements already forms a subgroup of G so it must be that smallest subgroup we were looking for.

We therefore have that in the Abelian case, $H_1 \vee H_2 = H_1 * H_2$, or, since the group operation is customarily written as addition,

$$H_1 \vee H_2 = H_1 + H_2.$$

Although not really relevant to anything that follows in this book, the reader may find it interesting to confirm that the following hold for any subgroups H_i of an Abelian group G. To make our formulae look pretty, we shall write "$\wedge$" instead of "$\cap$".

1. $H_i \wedge H_i = H_i \,, H_i \vee H_i = H_i$.
2. $H_1 \wedge H_2 = H_2 \wedge H_1, H_1 \vee H_2 = H_2 \vee H_1$.
3. $H_1 \wedge (H_2 \vee H_3) = (H_1 \vee H_2) \wedge (H_1 \vee H_3), H_1 \vee (H_2 \wedge H_3) = (H_1 \wedge H_2) \vee (H_1 \wedge H_3)$.
4. $H_1 \wedge (H_! \vee H_2) = H_1 \vee (H_1 \wedge H_2) = H_1$.

If these expressions look familiar, they are: $\wedge$ and $\vee$ could have been replaced by $\cap$ and $\cup$ respectively if we had been talking about subsets of a given set, or by AND (&&) and OR (||) respectively if we had been talking about Boolean operations in the C language.

Any set with binary operations $\wedge$ and $\vee$ satisfying these four requirements is called a *lattice*[4]

Exercise Let n be a number and consider the set of S divisors of n. Define, for $a, b \in S$,

$$a \wedge b = \text{greatest common divisor of } a \text{ and } b,$$

$$a \vee b = \text{least common multiple of } a \text{ and } b.$$

Show that S with these operations forms a lattice.

3.4 Cosets

Let H be a subgroup of a group G, and let $g \in G$. The *coset* gH is the set defined by

$$gH = \{g * h | h \in H\}$$

which consists of all products of g by elements of H. Because $e \in H$, we have that for every $g \in G, g = g * e \in gH$. In other words, the cosets completely cover the set G: $G = \bigcup_{g \in G} gH$.

Moreover, the relation $\equiv$ defined by $g_1 \equiv g_2$ if and only if $g_1 H = g_2 H$ is (almost trivially) an equivalence relation. Accordingly we have, from the nature of equivalence relations, that for any two elements $g_1, g_2 \in G$, either $g_1 H = g_2 H$ or else $g_1 H \cap g_2 H = \phi$. Note that if the

[4] We need to warn here that the word *lattice* has two entirely different meanings in Mathematics. The type of lattice that we have defined here is related to the concept of (partial) order. Another unrelated definition of a lattice is as a discrete subgroup of the vector space $\mathbb{R}^n$, which spans $\mathbb{R}^n$. This second concept is the more important one in Number Theory, and therefore in Cryptology; in fact an important technique in this kind of Lattice Theory, the so-called L^3 algorithm, named after Lenstra, Lenstra and Lovász, was fundamental in breaking an early public key system which was based on the so-called knapsack problem. More recently, hard problems in lattice theory (of this second kind) have also been used in order to construct new public key schemes. See, e.g., the paper *The two faces of lattices in cryptology* by Nguyen and Stern in the Proceedings of the Cryptography and Lattices Conference 2001, edited by J.H. Silverman, LNCS 2146, Springer Verlag. Since the publication of that paper some major developments have taken place in lattice based cryptology, such as Craig Gentry's discovery of a fully homomorphic encryption scheme, 2009. See Sect. 11.2.

first of these holds, then, since $g_1 \in g_1H = g_2H$, there exists an $h \in H$ such that $g_1 = g_2 * h$, so that $g_2^{-1} * g_1 \in H$. Conversely, if $g_2^{-1} * g_1 \in H$, say $g_2^{-1} * g_1 = h$, then for any $h_1 \in H$, $g_1 * h_1 = (g_2 * h) * h_1 = g_2 * (h * h_1) \in g_2H$, so that $g_1H \subseteq g_2H$. Similarly $g_2H \subseteq g_1H$, and therefore $g_1H = g_2H$. We therefore have that

$$\forall\, g_1, g_2 \in G \ \left(g_1H = g_2H \Longleftrightarrow g_2^{-1} * g_1 \in H\right).$$

Next, observe that every coset contains the same number of elements: for, given two cosets g_1H and g_2H we can set up a one-two-one correspondence by associating the element $g_1 * h$ of the one with $g_2 * h$ of the other. (This is a one-to-one correspondence because $g_2 * h_1 = g_2 * h_2$ implies that $h_1 = h_2$—and therefore $g_1 * h_1 = g_1 * h_2$—and every $g_2 * h \in g_2H$ is associated with some $g_1 * h \in g_1H$.)

How many elements are there in each coset? Easy: look at the coset eH: this is precisely the subgroup H itself, so $\#(gH) = \#(H)$ for every $g \in G$.

Now suppose that the order of G is finite, and let H be a subgroup of G. We have seen that the cosets cover all of G and that each of them has $\#(H)$ elements. But this implies the following

Lagrange's Theorem *If G is a finite group and H is a subgroup of G then $\#(H)|\#(G)$.*

Perhaps even more remarkable is the fact that if the group G is Abelian, then the set of cosets can itself be turned into a group, under a binary operation "induced" by the binary operation of the group itself:

Definition Let $\{G, *\}$ be an Abelian group (not necessarily finite) and let H be a subgroup of G. Denote by G/H the set of all cosets gH, $g \in G$. On G/H define a binary operation, which we shall (rather naughtily, and, we hope, not confusingly) also denote by "$*$", by

$$(g_1H) * (g_2H) = (g_1 * g_2)H.$$

Then $\{G/H, *\}$ is a group, called a *quotient group* or *factor group*.

We remark here that, since our main interest is in Abelian groups, we shall not prove a similar result for non-abelian groups. For such groups, the stated result is not true for *all* subgroups H of G, but only for the so-called "normal" subgroups.[5] A subgroup H of a group G is normal if and only of $aH = Ha\ \forall a \in G$, i.e. if and only if for any $a \in G$

$$\{ah | h \in H\} = \{ha | h \in H\}.$$

In the case of Abelian groups, this is obviously true, whatever H may look like, since $ah = ha$ for all h. In the non-Abelian case, one might have that even if H is normal then if $ah_1 = h_2a$, $h_1 = h_2$ does not follow in general.

[5]This terminology is typical of the usage among mathematicians. If some property is convenient, or allows one to do things one would like to do, it is given some pleasant sounding name like "normal" or "regular". In actual fact, in non-abelian groups, "normal" subgroups are in the minority among subgroups, and should therefore be considered somewhat abnormal.

But let us prove the assertion for Abelian groups:

Proof One should, first of all, show that the operation "$*$" is well-defined on G/H. After all, a coset gH may be represented in lots of different ways: whenever $g_2^{-1} * g_1 \in H, g_1H = g_2H$.

So suppose that $a_1H = a_2H$ and $b_1H = b_2H$. We need to show that then $(a_1 * b_1)H = (a_2 * b_2)H$, for the definition above to make sense. But this follows from $a_2^{-1} * a_1 \in H$ and $b_2^{-1} * b_1 \in H$, which implies that $(a_1 * a_2^{-1}) * (b_1 * b_2^{-1}) \in H$, i.e. $(a_1 * b_1) * (a_2^{-1} * b_2^{-1}) = (a_1 * b_1) * (a_2 * b_2)^{-1} \in H$, so $(a_1 * b_1)H = (a_2 * b_2)H$.

Verifying the group axioms is straightforward:

- $aH * (bH * cH) = (a * (b * c))H = ((a * b) * c)H = (aH * bH) * cH$;
- $(eH) * (aH) = (ea)H = aH = (ae)H = (aH) * (eH)$, so the identity element of G/H is the coset eH (which is the set H itself);
- $aH * a^{-1}H = (a * a^{-1})H = eH = (a^{-1}a)H = a^{-1}H * aH$, so the inverse of aH is simply $a^{-1}H$;
- $(aH)(bH) = (a * b)H = (b * a)H = (bH) * (aH)$.

Note that we have (perhaps without realising it) used this construction before when we discussed congruence. Let $G = \mathbb{Z}$ and consider the subgroup H consisting of all multiples of the integer $n \neq 0$. Thus $H = n\mathbb{Z}$, and for any $a, b \in \mathbb{Z}$, we have $a + H = b + H$ if and only if $a - b \in H$ if and only if $n|a - b$. Thus G/H is precisely the set of all residue classes modulo n.

It is easy to see that if G is finite, say $\#(G) = n$, and H is a subgroup of G of order m (and we have already proved that then $m|n$), then the order of $G/H = n/m$.

n/m is called the *index* of the subgroup H in G. This term is in fact in more general use: if G is a group (not necessarily finite) and H is a subgroup such that the order of the group G/H is finite, say $|G/H| = k$, then k is called the index of H in G. For example: the alternating group A_n has index 2 in the symmetric group S_n, as you proved in Exercise 5 of Sect. 3.2. The subgroup $n\mathbb{Z}$ has index n in the (infinite) group $\mathbb{Z}$.

Examples

1. Consider the group $G = \mathbb{Z}_6$ with addition as the operation. The set $H = \{0, 2, 4\}$ is a subgroup of G, and G/H consist of the two sets H and $1 + H$ with table

$+$	H	$1+H$
H	H	$1+H$
$1+H$	$1+H$	H

2. Consider again the group $G = \mathbb{Z}_{15}^*$ (see Example 8 of Sect. 3.1). We have already seen that $H = \{1, 4\}$ is a subgroup of G. There are $8/2 = 4$ congruence classes, viz

$$1H = \{1, 4\}$$
$$2H = \{2, 8\}$$

$$7H = \{7, 13\}$$
$$11H = \{11, 14\}$$

and the table for G/H is

	$1H$	$2H$	$7H$	$11H$
$1H$	$1H$	$2H$	$7H$	$11H$
$2H$	$2H$	$1H$	$11H$	$7H$
$7H$	$7H$	$11H$	$1H$	$2H$
$11H$	$11H$	$7H$	$2H$	$1H$

Exercises

1. Let H and H' be subgroups of an Abelian group G such that $H' \subseteq H$. Prove that H/H' is a subgroup of G/H'.
2. Let G be an Abelian group, H a subgroup of G and $a \in G$ an element of order n. If m is the order of $aH \in G/H$, what can you say about m?

3.5 Cyclic Groups

Let G be a group, and let $a \in G$. We have already noted that the set $\{a^m | m \in \mathbb{Z}\}$ is a subgroup of G. It is, in fact, the smallest possible subgroup of G which contains the element a, because any group which contains a must also contain all the powers (both positive and negative) of a. This subgroup is called the subgroup *generated* by a. We shall denote this subgroup by $< a >$, so, in multiplicative notation

$$<a> = \{e, a, a^2, a^3, \dots\} = \{x | x = a^i \text{ for some } i \in \mathbb{Z}\}$$

and in additive notation

$$<a> = \{0, a, 2a, 3a, \dots\} = \{x | x = ia \text{ for some } i \in \mathbb{Z}\}.$$

Now suppose that G is a finite group. Then the elements $e, a, a^2, a^3, \dots$ cannot all be different, so for some i and j (with $j > i$, say) we must have that

$$a^i = a^j.$$

But this would mean (multiplying both sides of this equation by a^{-i}) that $a^{j-i} = e$. Let n be the smallest positive integer such that $a^n = e$. Then n is called the *order* of a.[6]

[6]It is unfortunate that the word "order" is used in what are really two different senses. On the one hand the order of a group is the number of elements in the group. On the other, the order of an element is as we have

Note, by the way, that our definition of the order of an element implies that if for some group element a one has that $a^m = e$, then m is a multiple of the order of the element. For let the order of a be n and suppose that $n \not| m$. Then $m = qn + r$, for some q and some $0 \leq r < n$. But then, as is very easily shown, one would have that $a^r = e$, contradicting the minimality of n.

If a has order n, then the subgroup generated by a consists of the elements $e, a, a^2, \ldots a^{n-1}$, and multiplication of these elements is done simply by adding the exponents modulo the order n:

$$a^i * a^j = a^{(i+j) \bmod n}.$$

Example In the group $\mathbb{Z}_{15}^*$ of Example 8 of Sect. 3.1 the subgroup generated by 2 consists of $2, 2^2 = 4, 2^3 = 8, 2^4 = 1$, so the order of 2 in this group is 4. Also $2^3 \cdot 2^2 = 2^5 = 2 \cdot 2^4 = 2 \cdot 1 = 2 = 2^{(2+3) \bmod 4}$.

We now note an important consequence of Lagrange's theorem (Sect. 3.4): If G is a finite group, and $a \in G$ has order n, then the subgroup generated by a has order n and therefore $n|\#(G)$. In other words, the order of any element of the group is a divisor of the order of the group.

This leads to the following theorem:

Theorem *If G is a finite group, and $a \in G$, then $a^{\#G} = e$.*

Proof If a has order n, then n is a divisor of $\#G$, say $\#G = nq$. Then

$$a^{\#G} = a^{nq} = (a^n)^q = e^q = e.$$

Cyclic groups are extremely important in Abelian group theory. It can be shown that every finite Abelian group is in some sense the direct sum of a number of cyclic groups, so that the cyclic groups may be considered the fundamental building blocks of finite Abelian groups. For example if we consider the group $\{G, +\} = \{\mathbb{Z}_{24}, +\}$ and consider the subgroups $G[3] = \{a|3a = 0\}$ and $G[8] = \{a|8a = 0\}$ (recall Example 2 of Sect. 3.2), then

$$G = G[3] + G[8],$$
$$G[3] \cap G[8] = \{0\}.$$

This can be shown to lead to the conclusion that G is "isomorphic" (i.e. is "of the same form") as the direct sum $G[3] \oplus G[8]$.[7]

---

just defined it. The two are related: the order of an element a is equal to the order of the subgroup generated by a.

[7]Here the symbol "$\oplus$" does not stand for the exclusive or (XOR) operation, but for the operation which we first came across in Example 11 of Sect. 3.1: We recall that if $\{G, *\}$ and $\{H, \cdot\}$ are groups, then the set $G \times H$ can be turned into a group, called the direct product, which is denoted by $G \otimes H$. In the Abelian case this is called the direct sum of G and H and denoted by $G \oplus H$. In both cases the group operation on the set $G \times H$ is defined component-wise.

We note two specific properties of cyclic groups:

1. Every subgroup of a cyclic group is itself cyclic. For let G be generated by a and let H be a subgroup of G. If H is a trivial subgroup, i.e. if $H = \{e\}$ or $H = G$, then clearly H is cyclic. Assume therefore that H is non-trivial.
   Then all the elements of H are powers of a, and let k be the smallest positive power of a such that $a^k \in H$. Thus
   $$k = \min\{i | i > 0 \text{ and } a^i \in H\}.$$
   Then H is generated by a^k. For if $a^n \in H$, then, by the division algorithm, there exist q and r with $0 \leq r < k$ such that $n = qk + r$. But then $a^r = a^{-qk} * a^n$ belongs to H because both a^{qk} and a^n do. But this contradicts the choice of k, unless $r = 0$. Hence a^k generates H.
2. If a generates a cyclic group G of order n, and $b = a^s$, then the cyclic group generated by b has $\frac{n}{\text{gcd}(s,n)}$ elements. To see this, put $d = \text{gcd}\ (n, s)$, then $b^{n/d} = a^{s \cdot n/d} = (a^n)^{s/d} = e^{s/d} = e$, so that the order of b is a divisor of n/d. On the other hand, if $b^k = e$, then $a^{sk} = e$, so that $n|sk$, and therefore $\frac{n}{d}|k\frac{s}{d}$. But since n/d and s/d are relatively prime, this implies that $\frac{n}{d}|k$. In particular, n/d is a divisor of the order of b.
3. A particular instance of the second phenomenon occurs when $\text{gcd}(s, n) = 1$. In that case a^s generates the whole cyclic group of order n. Thus we have the following important

Observation A cyclic group of order n has $\phi(n)$ distinct generators.

Exercises

1. Let $a, b \in G$, where G is an Abelian group, with the group operation written multiplicatively. If a has order m, and b has order n, where $\text{gcd}(m, n) = 1$, show that ab has order mn. (Equivalently: show that $< a > \cap < b > = \{e\}$ and that $< ab >$ is the smallest subgroup of G that contains both a and b.)
2. Show, by means of a counterexample, that the following "converse" of the first of the statements above is false: "If every proper subgroup of a group G is cyclic, then G is cyclic."
3. Show that every group of prime order is cyclic.
4. Show that for any positive integer $\sum_{d|n} \phi(d) = n$. (*Hint*: think in terms of generators of cyclic groups.)
5. Use the result of the previous exercise and the Möbius Inversion Theorem of Sect. 2.8.2 to give another proof of the fact that ϕ is a multiplicative function.

3.6 Fermat's Little Theorem

Consider the multiplicative group $\mathbb{Z}_p^*$, where p is a prime: $\mathbb{Z}_p^* = \{1, 2, 3, \ldots, p-1\}$. This clearly has order $p - 1$, so by Lagrange's theorem $a^{p-1} = 1$ for all $a \in \mathbb{Z}_p^*$. Since multiplication in this group is essentially just multiplication modulo p, we could also write this as

$$a^{p-1} \equiv 1 \bmod p \ \ \forall a \in \{1, 2, \ldots, p-1\}.$$

If we take any integer a which is not divisible by p, then $a = kp + r$ for some integer k and some r in the range $\{1, 2, \ldots, p-1\}$, and

$$a^{p-1} \equiv (kp + r)^{p-1} \equiv r^{p-1} \equiv 1 \bmod p,$$

where the congruence in the middle follows from the binomial expansion of $(kp + r)^{p-1}$: all the terms in that expansion, except the last one, contain powers of p.

As an immediate corollary to this, we can say that, if p is a prime, then for any integer a

$$a^p \equiv a \bmod p.$$

This is Fermat's so-called "Little Theorem".[8]

The group $\mathbb{Z}_p^*$ is in fact a cyclic group, but the proof is postponed to Sect. 6.2.

Exercises

1. Find $13^{70} \bmod 23$.
2. Show that $n^{31} - n$ is divisible by 341 for any n.

3.7 Primality Testing

The question of whether an integer is prime or composite is of relatively frequent occurrence, in, for example, public key cryptography. Fermat's Little Theorem gives us a property which all prime numbers have. We would therefore like to use it as a test for primality: given an integer n, test whether n is prime or not. However, there is a difficulty:

Fermat's Little Theorem may be put in the following form

$$\gcd(a, p) = 1 \text{ and } a^{p-1} \not\equiv 1 \bmod p \Longrightarrow p \text{ is composite.}$$

But note that the implication only goes in one direction: it is *not* an "if and only if" statement. For example $2^{628} \equiv 305 \not\equiv 1 \bmod 629$, so we may safely conclude that 629 is composite. But, on the other hand $2^{560} \equiv 1 \bmod 561$, but we cannot conclude anything about whether 561 is prime or composite.

However, we do know that if p is prime, then the congruence $x^2 \equiv 1 \bmod p$ has exactly two solutions, since if $p|(x-1)(x+1)$ then either $p|x-1$ or $p|x+1$, i.e. either $x \equiv 1 \bmod p$ or

[8]It is called his "little" theorem, to distinguish it from his famous (or infamous) "Last Theorem".

$x \equiv -1 \bmod p$. Thus, to continue the example, if 561 were prime, then we should have that

$$2^{280} \equiv \pm 1 \bmod 561.$$

Since 2^{280} is in fact congruent to 1 mod 561, we still haven't proved anything. But, by the same argument as before, if 561 were prime, then we should have that $2^{140} \equiv \pm 1 \bmod 561$. But, in fact, $2^{140} \equiv 67 \bmod 561$, so we may conclude that 561 is not a prime.[9]

Let us call the first step of this procedure "Fermat's test": An integer n for which $a^{n-1} \equiv 1 \bmod n$, will be said to have "passed" Fermat's primality test for that particular a. Thus a prime number will pass Fermat's test for all possible values of a, but if n passes the test for some value[10] of a that does not necessarily mean that n is prime.

An odd composite number n which "passes" Fermat's test for *primality* for a given a is called a *base a pseudoprime*.

3.7.1 Miller's Test

The method we used to show that 561 is composite shows the basis for Miller's test for *proving* that a given number n is *composite*:

Given n, choose (randomly) an integer $b \in \{2, 3, \ldots, n-1\}$. Let $n - 1 = 2^s q$ where q is odd.

$$\begin{array}{ll} \text{If} & b^q \not\equiv 1 \bmod n \\ \text{and} & b^q \not\equiv -1 \bmod n \\ \text{and} & b^{2q} \not\equiv -1 \bmod n \\ \text{and} & b^{4q} \not\equiv -1 \bmod n \\ \text{and} & \ldots \\ \text{and} & b^{2^{s-1}q} \not\equiv -1 \bmod n \end{array}$$

then n is not prime.

Examples

1. $n = 41, n - 1 = 2^3 \cdot 5$. Take $b = 21$. Then

$$21^5 \equiv 9 \bmod 41,$$

$$21^{2 \cdot 5} \equiv 40 \equiv -1 \bmod 41,$$

so we may conclude that n may or may not be composite. (41 is prime, in case you didn't know.)

[9] If you had done or had remembered Exercise 8 of Sect. 2.6, you would have spotted immediately that $11|561$. But then you would have missed the point for which we chose this example, which comes next.

[10] Or even for *all* values of a! See our definition of Carmichael numbers in the next subsection.

2. $n = 20381, n - 1 = 2^2 \cdot 5095$. Take $b = 131$, then

$$131^{5095} \equiv 17365 \bmod 20381,$$

$$131^{2 \cdot 5095} \equiv 6330 \bmod 20381,$$

and we conclude that 20381 is not prime.

3. $n = 2047, n - 1 = 2 \cdot 1023$. Take $b = 2$. Then

$$2^{1023} \equiv 1 \bmod 2047$$

so Miller's test shows that 2047 may or may not be composite. (It is: $2047 = 23 \cdot 89$.)

An odd composite number n which passes Miller's test (when used as a test for *primality*) for a given b is called a *base b strong pseudoprime*. Explicitly:

- If $b^{n-1} \equiv 1 \bmod n$, then n is a base b pseudoprime.
- If $n - 1 = 2^s \cdot q$ and, furthermore, $b^q \equiv \pm 1$ or $b^{2q} \equiv -1$ or $b^{4q} \equiv -1$ or ... or $b^{2^{s-1}q} \equiv -1$ then n is a base b strong pseudoprime.

We don't seem to have made all that much progress: we are still dealing with a property that all primes have, but unfortunately one that is also a property of some composites. In fact, for the sake of completeness, we should mention the existence of the so-called *Carmichael numbers*: n is a Carmichael number if it passes the Fermat test for every possible base b, i.e. n is a base b pseudoprime for every possible base b ($1 < b < n$) which is relatively prime to n. Fortunately they are quite rare (561 is the smallest one, the next one is 1105) and of no practical relevance to us.

And help, of some kind, is at hand.

3.7.2 The Miller–Rabin Primality Test

Here is the help:

An integer n is prime if and only if it is a base b strong pseudoprime for every $b \in \{2, 3, \cdots, n - 1\}$.

The "only if" part of this statement is obvious, we shall not prove the "if" part. This observation, together with the following theorem, leads to the Miller–Rabin test for primality, which is essentially the Miller test repeated with as many different bs as required to give a sufficiently low probability of getting a "prime" answer when n is actually composite.

Theorem (Rabin) *If n is not a prime, then n is a base b strong pseudoprime for at most $\lfloor \frac{n}{4} \rfloor$ bases $b \in \{2, 3, \ldots, n - 1\}$.*

This implies that if n "passes" Miller's test (used as a test for primality) for a *randomly* selected base b, then the probability that n is nevertheless composite is less than $\frac{1}{4}$. Repeating the test m times (with different randomly selected b), with n "passing" each time, gives a probability of less than $\frac{1}{4^m}$ that n is actually composite.

In practice, this estimate for getting a wrong answer is very pessimistic, and in the range of numbers used in cryptography, five tests can usually be considered adequate. But perhaps one should always remember that the Miller-Rabin test is a probabilistic test. Exact tests ("if-and-only-if" theorems) do exist, but are impracticable.

Exercises

1. Show that 341 is a base 2 pseudoprime, but not a base 2 strong pseudoprime.
2. Use Miller's test to prove that 437 and 18467 are composite numbers.
3. Prove that if n is prime or a base 2 pseudoprime, then $2^n - 1$ is a base 2 pseudoprime. Conclude that there are infinitely many base 2 pseudoprimes.

3.8 Rings and Ideals

After this entertaining and cryptologically highly relevant detour into primality testing, let us return to algebraic structures.

The reader will have observed that the structures we have been discussing so far, viz groups, allow for only one binary operation, and may object that much of the algebra he or she has been concerned with in his or her life actually involved more than one operation. For example, when dealing with the integers, one may add (and subtract) them as well as multiply (and sometimes divide) them. This brings us to, firstly, the subject of rings[11] and, as a special subclass of rings, we have the important structures known as fields. And fields, as we'll show in Chaps. 7 and 9, are fundamental in the design of both stream ciphers and modern block ciphers. While Number Theory underlies a lot of asymmetric cryptography (as we'll show in the next chapter, returning to it in the final chapter), Algebra may, with only slight exaggeration, be considered to be the foundation on which current symmetric cryptography is built.

Definition A (commutative) *ring* (with identity) is a set R on which two binary operations, denoted by $+$ (and called *addition*) and $\cdot$ (and called *multiplication*) are defined, in such a way that

- $\{R, +\}$ is an abelian group.
- The operation $\cdot$ satisfies the following conditions[12]:
  - $a(bc) = (ab)c \;\; \forall a, b, c \in R$;
  - $ab = ba \;\; \forall a, b \in R$;

[11]We have already referred to ring-theoretic properties when dealing with the integers in the previous chapter.

[12]The reader must try not to be offended by what is virtually universal practice among algebraists: We denote the operation by '$\cdot$', and then immediately proceed by writing 'ab' instead of '$a \cdot b$'.

– There exists an element, denoted by 1, such that

$$1 \cdot a = a \cdot 1 = a \ \forall a \in R.$$

- These two operations are connected ("intertwined") by the *distributive property*:

$$a(b + c) = ab + ac \ \forall a, b, c \in R.$$

These rings are called "commutative" because the multiplication operation is commutative ($ab = ba$ for all a and b). Addition in rings is always commutative, even in non-commutative rings! In our discussions we shall never need to deal with rings in which the multiplication is not commutative. (Actually, strictly speaking, rings don't even have to have a multiplicative identity, but let us not pursue that matter. As far as we are concerned, the above definition of the concept "ring" will serve us admirably.)

Examples

1. The set of all integers, with the usual addition and multiplication defined on it.
2. The set $\mathbb{Z}_n$ of residue classes modulo the integer n.
3. The sets of all rational numbers, real numbers, or complex numbers, with the usual definitions of addition and multiplication, all form rings.
4. The set of all $n \times n$ diagonal matrices with real entries.
5. As an example of a non-commutative ring, consider the set of all $n \times n$ matrices with real entries. All the requirements for a ring are satisfied, except for the commutativity of multiplication. We repeat that we shall not need to consider such rings.

Note that the structure $\{R, \cdot\}$ is *not* a group: while there exists (in the type of ring that we are concerned with) an identity element, 1, for the multiplicative operation $\cdot$, we do not insist on the existence of inverses. In fact, we cannot, since it is impossible to define a meaningful structure in which 0 (the identity element for addition) has a multiplicative inverse. In $\mathbb{Z}$ only the elements ± 1 have inverses; in the ring $\mathbb{Z}_n$ we observed that the residue classes $\bar{s}$ such that s and n were relatively prime were the only invertible elements.

It is easy to see that if R is a ring then the set R^* of the invertible elements of R forms a group under the operation of multiplication. This group may be trivial in the sense that it consists only of the elements $\pm 1 \in R$; on the other hand, it may consist of all the nonzero elements, i.e. $R^* = R \backslash \{0\}$. In the latter case, the ring is called a *field*. We shall have more to say about fields shortly.

In the following theorem we list some elementary properties; before doing so we note, perhaps unnecessarily, because we've seen this kind of thing before, the following matter concerning notation. Let R be a ring, $a \in R$ and let n be an integer. Then, if $n > 0$ we denote by na the sum $a + a + \cdots + a$, n times. If $n < 0$ then na stands for $-a - a \cdots - a$, $-n$ times. By convention $0a = 0$, where the 0 on the left-hand side is the natural number 0, and the one on the right is the 0-element of R (i.e. the identity element of the group $\{R, +\}$). But see the second statement of the next theorem.

Theorem *Let R be a ring with additive and multiplicative identities 0 and 1 respectively. Then, for all $a, b \in R$ and any integer n*

1. *The additive and multiplicative identities are unique;*
2. $0 \cdot a = 0$ *(where this time round '0' is the 0-element of the ring on both sides of the equation);*
3. $(-a)b = -(ab) = a(-b)$*;*
4. $(-a)(-b) = ab$*;*
5. $(na)b = n(ab) = a(nb)$*.*

Proof

1. The uniqueness of the additive identity follows from the group structure of $\{R, +\}$. The uniqueness of the multiplicative identity is proved in the same way as in the case of groups.
2. $a = 1 \cdot a = (1 + 0) \cdot a = 1 \cdot a + 0 \cdot a = a + 0 \cdot a$. Now add $-a$ to both sides.
3. $0 = 0 \cdot b = (a - a)b = ab + (-a)b$, so adding $-(ab)$ to both sides gives $-(ab) = (-a)b$.
4. $0 = (-a)(b + (-b)) = -(ab) + (-a)(-b)$ so adding ab to both sides gives $ab = (-a)(-b)$.
5. If $n > 0$ we have $(na)b = (a + a + \cdots + a)b = ab + ab + \cdots + ab = n(ab)$. The rest is left as a rather uninteresting exercise.

In our discussion of the integers in the previous chapter we introduced the concept of ideals in the ring $\mathbb{Z}$; this concept carries over without change to rings in general:

Definition Let R be a ring. A nonempty subset $I \subseteq R$ is called an *ideal* in R if

- $\forall a, b \in I \; a - b \in I$;
- $\forall r \in R \; \forall a \in I \; ra \in I$.

The first of these conditions shows that I is a subgroup of the additive group $\{R, +\}$, so that it is meaningful to discuss the additive group R/I, or, in other words, the addition operation carries over from R to R/I. The second condition ensures that the same is true for the multiplicative operation, as we shall show shortly.

An almost trivial, but nevertheless important, kind of ideal is formed by the so-called principal ideals. Analogous to what we had in the ring $\mathbb{Z}$, a principal ideal in a ring R consists of all the multiples of an element $a \in R$: thus the principal ideal *generated* by a, usually written as Ra, is the set $\{ra | r \in R\}$. It is easy to see that Ra is the smallest ideal that contains a.

More generally, one may consider *finitely generated* ideals: An ideal I in the ring R is finitely generated if there exists a subset $\{a_0, a_1, \ldots, a_{n-1}\} \subseteq R$, such that every element $x \in I$ can be written as a "linear combination" of the a_i, i.e. if for every $x \in I$ there exist $r_k \in R$, $k = 0, \ldots, n-1$, such that

$$x = \sum_{k=0}^{n-1} r_k a_k.$$

Again, one can easily prove that the smallest ideal containing the set $\{a_0, a_1, \ldots, a_{n-1}\}$ is precisely the set of all sums of this form.

We shall presently consider rings which have the property that every ideal is a principal ideal; not entirely surprisingly such rings are called *principal ideal rings*.[13] $\mathbb{Z}$ is therefore a principal ideal ring.

In analogy with what happens in the case of addition, we would like to define the multiplication on the set R/I by

$$(r + I)(s + I) = rs + I$$

but in order to do so, we need to be certain that this definition is independent of the representation of the cosets: it is, after all, perfectly possible that $a + I = b + I$ without a and b being equal. So suppose that $r_1 + I = r_2 + I$ and that $s_1 + I = s_2 + I$. We need to show that $r_1 s_1 + I = r_2 s_2 + I$. Now we have that $r_1 - s_1 \in I$ and $r_2 - s_2 \in I$ and we note that

$$r_1 s_1 - r_2 s_2 = r_1(s_1 - s_2) + (r_1 - r_2)s_2.$$

By the second condition defining an ideal, both the terms on the right-hand side of this equation belong to I, and therefore (according to the first condition) so does their sum. Hence $r_1 s_1 + I = r_2 s_2 + I$.

Verifying that the set R/I with these definitions of addition and multiplication is a ring is now a purely routine matter. R/I is called a *quotient ring*.

Following our custom in arithmetic, we shall sometimes write $a \equiv b \bmod I$ to indicate that $a - b \in I$, or, equivalently, that $a + I = b + I$.

Suppose that R is a finite ring, i.e. the set R has a finite number of elements, and let $a \in R$. Since $\{R, +\}$ is a finite abelian group, there must exist an n such that $na = a + a + \cdots + a = 0$. The smallest positive integer n such that $na = 0$ for all elements $a \in R$ is called the *characteristic* of the ring. More conveniently, it may equivalently be defined as the smallest positive integer such that $n \cdot 1 = 0$. (In Exercise 4 below we leave it to the reader to convince himself or herself that this gives the same result.)

In the next chapter we'll discuss how some of the concepts that we have discussed so far have immediate application in Cryptography.

Exercises

1. Show that the set $\mathbb{Z} \times \mathbb{Z}$ with the ring operations defined componentwise is a ring. Find a subring of this which is not an ideal.
2. Let R be a ring and let $a \in R$. Define the *annihilator* $\text{Ann}(a)$ of a by $\text{Ann}(a) = \{x | x \in R \text{ and } ax = 0\}$. Show that $\text{Ann}(a)$ is an ideal in R. What can you say about the element $a + \text{Ann}(a)$ in the ring $R/\text{Ann}(a)$?

[13] Considerably less obvious is the fact that rings in which every ideal is finitely generated are called *Noetherian rings*, after the German mathematician Emmy Noether. So a principal ideal ring is Noetherian, but not every Noetherian ring is a principal ideal ring. Just in case you wanted to know.

3. If R is a ring, a nonzero element $a \in R$ is called a zero-divisor if there exists a nonzero $b \in R$ such that $ab = 0$.

   (a) Show that an element which has a multiplicative inverse cannot be a zero-divisor.
   (b) Show that in $\mathbb{Z}_n$ the coset $\overline{s} \neq 0$ is a zero-divisor whenever $\gcd(s, n) \neq 1$.
   (c) Conclude that $\mathbb{Z}_p$ has no zero-divisors if p is a prime.
   (d) Show that the set consisting of 0 and all the zero-divisors in R need not be an ideal in R.

4. Let R be a ring and n a positive integer. Show that $na = 0$ for all $a \in R$ if and only if $n \cdot 1 = 0$.
5. Let G be an Abelian group. An *endomorphism* of G is a function $f : G \longrightarrow G$ which maps sums to sums, i.e. such that $f(g_1 + g_2) = f(g_1) + f(g_2) \ \forall \ g_1, g_2 \in G$. Define the sum of two endomorphisms in the usual way and the product as their composition. Show that the resulting algebraic structure is a ring (non-commutative in general).
6. Continuing: Suppose G is a cyclic group of order n. Describe the ring of endomorphisms as it was defined in the previous exercise.
7. A Boolean ring is a ring R in which $a^2 = a$ for all $a \in R$. Prove that every Boolean ring is commutative.
8. Let S be a nonempty set, and let $P = \mathcal{P}(S)$ be the set of all subsets of S. Define on P the operations $+$ and $\cdot$ by

$$A + B = A\Delta B \ \ (= (A\backslash B) \cup (B\backslash A)),$$

$$A \cdot B = A \cap B.$$

   Show that P is a ring under these operations.
   If $a \in S$ show that the set $\mathcal{J}$ of all subsets of S not containing a is an ideal in the ring P.
9. Let P' be the set of all binary strings of length n, and define on P' the operations $+$ and $\cdot$ as follows: If $A = (a_0, a_1, \ldots, a_{n-1})$ and $B = (b_0, b_1, \ldots, b_{n-1})$, $(a_i, b_i \in \{0, 1\})$, then (where AND is the logical operator and in what we hope is otherwise an obvious notation) for all $i = 0, \ldots n - 1$:

$$(A + B)_i = a_i \oplus b_i,$$

$$(AB)_i = a_i \text{ AND } b_i.$$

   Show that under these operations P' is a ring. What is the connection between this ring and the ring of the previous exercise?

4 Applications to Public Key Cryptography

In this chapter we describe, at an elementary level, some of the applications of the Group Theory and Number Theory we have developed so far to Cryptology. We emphasise that these "textbook versions" of the applications do not do justice to the complexities that arise in practice, and warn the reader that implementing the mechanisms that we discuss in the form given here would lead to severe vulnerabilities of the schemes.[1] The reader is encouraged to start by reading the paper on *Why textbook ElGamal and RSA encryption are insecure*.[2]

Our discussion falls into two components: in the first we exploit the difficulty of factoring integers into their prime factors. In the second we use large cyclic groups in order to establish secrets known only to the participating parties.

4.1 Public Key Encryption: The RSA Mechanism

The notion of public-key encryption goes back to the seminal paper by Diffie and Hellman[3] which appeared in 1976 and in which it was suggested for the first time that it might be possible to design a cipher algorithm in which the keys for encryption and decryption are different: in such a way that neither key could be obtained or computed even if the other one was known. One of these (the encryption key) could then be made public and only a holder of the matching decryption key would be able to obtain the plaintext from ciphertext computed using that public key. Of course, interchanging the roles, a secret private key can be used for signing messages: if, on decryption using the public key, a message is obtained which contains some text or data of a kind previously agreed upon, the receiver may conclude that

[1]And if those sentences read like just another of those painful "Disclaimers" one finds everywhere, I apologise.

[2]Boneh, D., Joux, A. and Nguyen, P.: *Why textbook ElGamal and RSA encryption are insecure*, Proc. Asiacrypt 2000, LNCS 1976, Springer-Verlag.

[3]Diffie, W. and Hellman, M.E.: New Directions in Cryptography, *IEEE Trans. on Information Theory* **22** (1976), 644–654.

A.R. Meijer, *Algebra for Cryptologists*, Springer Undergraduate Texts in Mathematics and Technology,
DOI 10.1007/978-3-319-30396-3_4

the message was sent by someone in possession of the private key.[4] If the cipher is used for both secrecy and for authenticity, there will be a need for two sets of keys for each user.

At the time when Diffie and Hellman's paper appeared, no practical scheme of this nature was known. Such a scheme appeared about a year later, viz the RSA scheme, named after its inventors Rivest, Shamir and Adleman.[5] This scheme depends on some elementary number theory, which, in turn, depends on group theory as we discussed it. Its security depends on the fact that while multiplication of integers (even large ones) is a simple process (and in fact one for which a simple algorithm exists), the inverse operation, namely that of determining the factors of an integer, is considerably more difficult.

4.1.1 RSA

Let p and q be distinct prime numbers, let $n = pq$ be their product and consider the group $\mathbb{Z}_n^*$ of all integers relatively prime to n, under multiplication modulo n. This group, obviously, has order

$$\phi(n) = \phi(pq) = (p-1)(q-1)$$

and, by Lagrange's little theorem,

$$a^{\phi(n)} = a^{(p-1)(q-1)} \equiv 1 \ mod \, n$$

for any $a \in \{1, 2, \ldots, n-1\}$.

Now let e be an integer relatively prime to $\phi(n)$. Then e has a multiplicative inverse modulo $\phi(n)$ which we shall call d. Thus:

$$de \equiv 1 \mod \phi(n)$$

or

$$de = 1 + k \cdot \phi(n) \text{ for some } k \in \mathbb{Z},$$

from which it follows, in turn, that

$$(a^e)^d = a^{1+k\cdot\phi(n)} = a \cdot (a^{\phi(n)})^k \equiv a \cdot 1^k \equiv a \mod n$$

[4]I used the words "of course" in this sentence, but it is perhaps not all that obvious: the British cryptologists at GCHQ, who anticipated what later became the RSA scheme, but whose work was only declassified in 1997, seem to have missed this point.

[5]Rivest, R.L., Shamir, A. and Adleman, L.M.: A method for obtaining digital signatures and public-key cryptosystems; Communications of the ACM, **21** (1978), pp. 120–126. Prior to that, the system had already appeared in Martin Gardner's "Mathematical Games" column in *Scientific American*, August 1977.

for all $a \in \mathbb{Z}_n^*$. Of course, one also has that for all $a \in \mathbb{Z}_n^*$

$$(a^d)^e \equiv a \mod n.$$

This is really the heart of the RSA algorithm. Assume that some method exists for identifying plaintexts with integers in the range $\{1, \ldots, n-1\}$, and assume that n and e are known. Then anyone can, given a message a, compute $a^e \mod n$, but only someone who knows d can obtain a from a^e.

Thus the function

$$E : \mathbb{Z}_n \longrightarrow \mathbb{Z}_n : a \mapsto a^e \mod n$$

is a *trapdoor one-way function* if n is large enough to make knowledge of d (given n and e) impossible to obtain: only someone who knows the 'trapdoor' (d) knows that its inverse is

$$D : \mathbb{Z}_n \longrightarrow \mathbb{Z}_n : a \mapsto a^d \mod n.$$

Anyone else would, given $a \in \mathbb{Z}_n^*$ find it very hard to find $D(a)$.[6]

All this means that a user (and let's call her Alice) of the RSA algorithm can make n and e public. If any other user (and let's call him Bob) wants to send her a message a, he merely computes $a^e \mod n$ and sends that to her.

But it is interesting, and important, to note that RSA can also be used in order to prove the identity of the sender of a message, i.e. to *sign* a message. This is done by reversing the roles of the two keys. If Alice wants to send a signed (but unencrypted) message a to Bob, she sends $a^d \mod n$, and Bob computes a which *if it makes sense, i.e.* contains enough redundancy of an expected kind, Bob will accept as having originated with Alice, since only she (presumably) knows d. This assumes that Bob can with a high degree of certainty be sure that e and n are indeed associated with Alice. This is, in practice, achieved by Alice providing a certificate, issued by a trusted third party, asserting that the public key used is indeed hers.

The above is a very crude description of RSA. For example, if all $a \in \mathbb{Z}_n^*$ are acceptable messages, so that there is no redundancy, the signature scheme will fail. (This could be the case if, for example, the messages consist of random numbers to be used as keys in symmetric encryption.) In any case, as a general principle, it is bad practice to use cryptographic keys for more than one purpose: Alice should not use the same d for both decryption and for signing. However, these practical considerations, while very important, lie somewhat outside our present discussion of applicable (rather than applied) algebra and number theory. For the same reason, we shall not worry unduly about the lengths (in bits or in decimals) of the integers p and q which are needed to make the system secure.[7]

---

[6]For *most* a, anyway. Of course, there will be values of a for which there is no problem. D(1) = 1, for example!

[7]Bear in mind that we can only speak of *computational security*, since an attacker with "infinite" computing power can, of course, factor the modulus.

Another problem may arise if there are only a limited number of messages that Bob might want to send to Alice, in other words if the message space is too small. An attacker can (because the entire encryption mechanism is public) make a list of all these messages and their encryptions, and sort these pairs by ciphertext. When he intercepts a message from Bob to Alice, he merely compares it with the ciphertexts in his list and finds the plaintext. Because, as described here in its "textbook" version, the process is completely deterministic, it cannot be *semantically secure*: semantic security of a cryptosystem being defined by its property that no probabilistic algorithm, running in polynomial time, can determine any information about the plaintext (other than its length) from its encryption. This has been shown to be equivalent to the property that, given two messages m_1 and m_2, and the encryption c of one of them, an adversary is unable to determine which one. Clearly textbook RSA fails this test!

In addition, an adversary can modify an intercepted message: it follows from the definition that if c is the RSA encryption of m, then $2c \bmod n$ is the encryption of $2m$.

Because of these weaknesses, the textbook version is in practice replaced by more complex versions which involve some random padding.

We reiterate that it is in general not a good idea to use a cryptographic key for more than one purpose. Thus Alice will have different values of n, d, and e depending on whether the keys are used for encryption/decryption or signing/verification.

Finally, there is the problem of ensuring that the public key Alice uses for communicating with Bob really belongs to Bob, and not to some impostor. This really needs the existence of a so-called *Public Key Infrastructure* (PKI) which we shall very briefly discuss at the end of this section.

4.1.2 Breaking RSA

The most immediate way of "breaking" RSA, by which we, for this discussion, mean obtaining one of the two RSA keys, given the modulus n as well as the other key, is to find the factorisation of the public modulus.

It is clear from the foregoing exposition that if an attacker can factor the modulus n then he or she knows (obviously) what $\phi(n)$ is, and can therefore find the secret d as the inverse of the (public) e. Thus, being able to factor n means breaking RSA. Whether the converse is true, i.e. whether breaking RSA (in the sense of being able to decrypt data without being given the secret key d) implies being able to factor n is an open question. It is widely, possibly universally, *believed* to be true, but a mathematical proof to support this belief has never been found.

What is true, is that if n is known and is known to be the product of two primes, then if $\phi(n)$ is known, the factorization of n can be found. This is easily seen from the fact that

$$\phi(n)=(p-1)(q-1)=pq-p-q+1=n-(p+q)+1$$

so that

$$p + q = n + 1 - \phi(n) \quad \text{and} \quad pq = n,$$

which leads to a quadratic equation in p (or q) which is solved by the high school method.

Again, assume that n is given, and is known to be the product of two primes. We shall show that, given e and n, finding d is equivalent to finding $\phi(n)$. Obviously, if $\phi(n)$ is known, then d is easy to find, but proving the converse is (just a little) more difficult.

We know, of course, that $de = 1 + r \cdot \phi(n)$ for some r. Now write $de - 1$ as $de - 1 = 2^k \cdot \psi$, where k is some integer and ψ is odd. We now try to find a non-trivial square root of 1 modulo n as follows:

1. Pick randomly $a \in \{2, 3, \ldots, n-1\}$.
2. Find the smallest value of j such that $a^{2^j \psi} \equiv 1 \mod n$.
3. If $a^{2^{j-1} \psi} \not\equiv -1 \mod n$, return $a^{2^{j-1} \psi}$, which is the required non-trivial square root of 1. Else go to step 1.

As we mentioned when we discussed the rather primitive Fermat method of factoring in Sect. 2.5, once one has a non-trivial square root of 1, one has a factorization of the modulus. For if $b^2 \equiv 1 \mod n$, then $(b-1)(b+1) \equiv 0 \mod n$, so that the greatest common divisor of $b - 1$ and n (which we find efficiently using the Euclidean algorithm) is a factor of n.

Note that this implies that if one has a group of users, every member of the group *must* have his or her own modulus n. It might be tempting to use the same modulus for all the members, because this might make both implementation and key management easier, but since every user knows both his e and his d, he can factor n and therefore compute the secret keys of all the other members of the group.

A more serious problem, and one which should in theory have been avoidable with ease, has arisen in practice through a poor choice of factors p and q. If n and n' are the moduli used by two entities and if n and n' share a common factor p, then this is easily found by simply using the Euclidean algorithm to find $p = \gcd(n, n')$. Assuming that all users generate their prime factors properly, and that the moduli are large enough, say 1024 to 2048 bits long, this would be very unlikely to occur. But in fact, Arjen K. Lenstra et al. found that of 11.4 million (1024-bit) RSA keys in use on the Internet, no fewer than 26965 were vulnerable to this attack.[8] Astonishingly, there was one prime which was used in 12720 different keys!

In order to avoid other factoring attacks, it is desirable to choose p and q reasonably far apart and such that neither $p - 1$ nor $q - 1$ has any small factors. Additionally, there is an attack by Wiener[9] which, using the continued fraction expansion of n/e, yields d. A sufficient condition for this attack to succeed is that $d < \frac{1}{3}n^{\frac{1}{4}}$. This result was improved in 1999 by Boneh and Durfee, who showed that such an attack would succeed if $d < n^{0.292}$. More complicated attacks can successfully be mounted if the value of e is too small: $e = 3$ would

[8]Lenstra, A.K. et al.: *Ron was wrong; Whit is right*, IACR ePrint archive, https://eprint.iacr.org/2012/064.pdf.

[9]Wiener, M.: *Cryptanalysis of short RSA secret exponents*, IEEE Trans. Info Th. **36** (1990), 553–558.

be a good choice for efficiency in encryption, but $e = 65537$, of equally low weight, is a lot safer.[10] The interested reader may refer to the exposition by Boneh.[11]

We should note that, while we are here merely discussing the mathematics behind RSA, in practice RSA is never used to encrypt or to sign long messages. Even with a moderate length modulus (n of length 1024 bits or some 310 decimal digits, say, which is currently about the shortest modulus giving any sense of security,[12]) encrypting with RSA would take about 1000 times as long as encrypting with a decent block cipher (which would be a secret key algorithm, of course). Thus RSA is, in practice, used for encrypting short data strings, such as the keys to be used in subsequent communication using a secret key (or *symmetric*) encryption function. Similarly, RSA would never be used to sign anything other than a very short message; if the message to be signed is too long—and it almost invariably is—a *message digest* or a "fingerprint" of the message is signed instead. Such message digests are obtained using a *cryptographic hash function.*[13]

Moreover, as noted earlier, RSA is in practice not used without some randomised padding, RSA-OAEP (RSA Optimal Asymmetric Encryption Padding) was introduced in 1995 by Bellare and Rogaway[14] and has been adopted as a standard.

4.1.3 Using the Chinese Remainder Theorem

The main reason for the lack of speed in RSA encryption is the fact that performing arithmetical operations with very large integers is called for. If one instead performs the calculations modulo p and modulo q separately, and then combines the two results using the Chinese Remainder Theorem (see Sect. 2.7), most of the calculations are done with integers of length approximately only half that of n, resulting in less time being needed. Of course, this can only be done if the factorisation of n does not need to be protected (because it is done

[10] The (Hamming) weight of an integer is just the number of 1s in its binary representation. The relevance of this is that in exponentiation of an integer, a multiplication and a squaring is required whenever a 1 appears in the exponent. In the case of a 0, only squaring is required. This is the same as exponentiation in a finite field, and we return to this matter in Sect. 6.4.4, where a complete explanation is more appropriate than here.

[11] Boneh, D.: *Twenty years of attacks on the RSA cryptosystem*, Notices of the Am. Math. Soc. **46** (1999), 203–213.

[12] Current (2015) recommendations from such bodies as eCrypt and the German *Bundesamt für Sicherheit in der Informationstechnik* are that the length of n should be at least 3072 bits. This kind of requirement needs updating regularly as (e.g.) factoring techniques improve as does the computing power of the adversary.

[13] Some kind of definition is required. A cryptographic hash function is a function defined on the set of binary strings (of all finite, but unbounded, lengths) and has as output strings of a fixed length. As a minimum cryptographic requirement one has *one way-ness*; while it should be easy to compute $Y = h(X)$ when given X, it should be computationally infeasible to find X when given Y. Moreover, for most purposes it is also necessary that it should be computationally infeasible to find two strings X_0 and X_1 such that $h(X_1) = h(X_2)$. This property is called *collision resistance.* From the definition it is clear that collisions *must* occur; it must just be computationally impossible to find them.

[14] Bellare, M. and Rogaway, P.: Optimal Asymmetric Encryption—How to encrypt with RSA. Extended abstract in Proc. Eurocrypt '94, LNCS 950, Springer-Verlag, 1995.

where the operations are under the control of someone who is allowed access to the private key) or if the computations are done in a thoroughly protected, and physically secure "black box" (assuming that this can be provided!).

Let us explain the mechanism involved with a toy **example**:

Suppose $n = 391 = 17 \cdot 23$, $c = 100$ and $d = 7$. To compute $c^d \mod n$, we compute

$$c^d \mod 17 = 100^7 \mod 17 = 8,$$
$$c^d \mod 23 = 100^7 \mod 23 = 12.$$

Since $17^{-1} \mod 23 = 19$ and $23^{-1} \mod 17 = 3$, the CRT gives us that

$$\begin{aligned} c^d \mod n &\equiv 8 \cdot 23 \cdot 3 + 12 \cdot 17 \cdot 19 = 4428 \\ &\equiv 127 \mod 391. \end{aligned}$$

Using the CRT exposes the system to a kind of attack which is (also) interesting from an arithmetical point of view. Suppose the attacker is able to disrupt the computation at the stage where the second part (the "modulo q part") of the computation is about to start. In fact, suppose that he is able to flip a bit of the input into that part. Then he can compute the factorisation of n, thereby breaking the entire system!

To continue the toy example, changing the input into the second congruence by one bit, we get the congruences

$$x \equiv 8 \mod 17,$$
$$x \equiv 4 \mod 23.$$

Solving these two, using the CRT again, we get $x \equiv 280 \mod 391$.

The attacker knows that his answer and the correct answer (127) are congruent to each other modulo the first factor of n. By computing the gcd of $280 - 127 = 153$ and $n = 391$, he gets the factor 17 of n.

4.1.4 Public Key Infrastructures

In this book we emphasise the mathematical aspects of cryptography and by and large we ignore the practical aspects. In the case of RSA, however, we should mention, even if only briefly, the problem of authenticating the keys. There is little point in Alice publishing her modulus n and public key e on the Internet, since Bob has no means of knowing that those actually belong to Alice, and have not been placed there by the evil Eve, who did so in the hope of intercepting and decrypting highly sensitive information meant for Alice. The way of circumventing this problem is to get a mutually trusted party Trent to issue a certificate, basically to the effect that

> This is Alice's modulus n and encryption[15] exponent e. I have verified that she is who she claims to be and that she indeed possesses the private key which matches this public one. And here is my signature on this data.
> (signed) Trent

The astute reader will recognise that we have merely shifted the problem: even if we trust Trent, how do we know that it is indeed he who signed Alice's certificate? Trent in turn has to present a certificate signed by someone who is even more trustworthy than he. But this cannot continue indefinitely!

In practice, in closed systems, the ascending chain terminates with a certificate issuer who is, or whose certificates are, indeed trusted by everyone in the network. The network of certificate issuers (who are known as Certificate Authorities or CAs, and do more than just issue certificates, but we'll let that pass) is known as a Public Key Infrastructure or PKI.

In open systems, the top level certificates are installed and signed by the developer of the browser, who signs his own certificate. (We all trust that organisation, don't we?).

4.2 The Discrete Logarithm Problem (DLP)

RSA was the first of our applications of Algebra and Number Theory to asymmetric (public key) cryptography. There are many others, the oldest one of which is the family of Diffie–Hellman protocols which enables Alice and Bob to *establish* a secret which only they share. Most, if not all, Diffie–Hellman type protocols depend for their security on some version of the Discrete Logarithm Problem, which we shall discuss in this section. We shall return to the Diffie–Hellman protocols in Sect. 4.3. For the moment we shall deal with the problem as it appears in groups which are multiplicatively written. The case of the Elliptic Curve Discrete Logarithm Problem will be mentioned briefly in the final section of this chapter.

4.2.1 Statement of the Problem

Let G be a (multiplicative) cyclic group, generated by an element g. (G may be a subgroup of a—possibly also cyclic—group, but we shall not concern ourselves with this larger group.) Let the order $|G|$ of G be n, and let $h \in G$. The discrete logarithm problem is the problem of finding $x \in \{0, 1, \ldots n-1\}$ such that

$$g^x = h.$$

x is called the discrete logarithm of h; this is sometimes written as $x = \log_g h$.

[15] or "signature verification exponent e'", of course. Alice, and everyone else, needs at least two certificates.

It is obvious that the following three properties, equivalent to properties of the "classical" logarithm, hold—the main difference being that all calculations are now done modulo n:

- $\log_g(h_1 h_2) = (\log_g(h_1) + \log_g(h_2)) \mod n$;
- $\log_g(h^s) = (s \log_g h) \mod n$; and
- $\log_{g_2} h = (\log_{g_1} h)(\log_{g_1} g_2)^{-1} \mod n$,

if g_1 and g_2 are both generators of G. This last one may be interpreted as saying that the difficulty of the discrete logarithm problem is independent of the choice of generator.

Example In the additive group $\mathbb{Z}/n\mathbb{Z}$ of order n, taking logarithms to the base 1 is trivial: Since $m = 1+1+\cdots+1$, so $\log_1(m) = m$. It follows that $\log_k(m) = log_1(m) \times (\log_1(k))^{-1} = m \times k^{-1} \mod n$. Apart from having to compute an inverse modulo n, that's about the same amount of work for logarithms to the base 1 or to the base k. But how are you going to tackle the discrete logarithm if the cyclic group is, say, a subgroup of order n of the symmetric group S_{2n}? (See Examples 13 and 14 in Sect. 3.1 for the definition of *symmetric group*.) From the algebraist's point of view, all cyclic groups of order n are isomorphic, and you may as well regard them as all being the same. From the computational point of view there may well be big differences, and we are, again, dealing with security in the computational sense.[16]

The most common applications use for G the multiplicative group of nonzero elements of a finite field or a subgroup of such a group. In the next chapter we shall define a field as an algebraic structure in which some of the elements (all the nonzero ones) form a multiplicative group and we shall later prove that this group is in fact cyclic, as are therefore all its subgroups. The order of the group selected depends on the level of security required.

The three methods which we shall outline in the following subsections are *generic* in the sense that they may, in principle, be used for solving the discrete logarithm problem in any group. There exist methods, however, which are more efficient for specific groups.[17] In the final section of this chapter we shall, as we have mentioned earlier, briefly describe the (additive) groups of points on elliptic curves over finite fields; these were seen as having the advantage over the two others mentioned that (until recently) only the generic methods appeared to work or, in other words, it appeared that there are no "short cuts". Recent research has cast some doubt on this optimistic view.

4.2.2 Exhaustive Search

To solve $g^x = h$, one simply computes $g^2, g^3, \ldots$, until one reaches h. In the case of large n, this is, obviously, infeasible. (But how large "large" has to be in order to offer security depends on the computing power of the adversary, and will change over time, as computers get faster and faster.)

---

[16] "If you ever hear anyone talk of *the* cyclic group of order n, beware!"—Carl Pomerance.

[17] Running ahead of our development: some attacks work better for prime fields ($GF(p)$) while others are better if the field is of characteristic 2 ($GF(2^n)$).

4.2.3 Shanks' Method

This technique is frequently referred to under the rather twee name of "Baby Step-Giant Step Method", and amounts to a time-memory trade-off version of the exhaustive search method.

Let $m = \lceil \sqrt{n} \rceil$. Compile two tables: In the first, list all pairs (j, g^j) for j running from 1 to m, and sort these according to the second entry. For the second, list all pairs (k, hg^{mk}), for $k = 0, 1, 2, \ldots$, until a pair is found such that the second entry appears as a second entry in the first table, i.e. until a k is found such that

$$g^j = hg^{mk}$$

for some j. Then $h = g^{j-mk}$, i.e. $\log_g h = (j - mk) \bmod n$.

4.2.4 Pollard's Rho Method

This method requires approximately the same running time as Shanks', but considerably less storage.

Partition the set G into three subsets S_1, S_2, S_3 of approximately equal size. Define a sequence z_i iteratively by $z_0 = 1$, and

$$z_{i+1} = f(z_i) = \begin{cases} hz_i \text{ if } z_i \in S_1, \\ z_i^2 \text{ if } z_i \in S_2, \\ gz_i \text{ if } z_i \in S_3. \end{cases}$$

Now it may be verified that for each i, $z_i = g^{a_i} h^{b_i}$, where the sequences $\{a_0, a_1, \ldots\}$ and $\{b_0, b_1, \ldots\}$ are defined iteratively by $a_0 = b_0 = 0$ and

$$a_{i+1} = \begin{cases} a_i & \text{if } z_i \in S_1, \\ 2a_i \bmod n & \text{if } z_i \in S_2, \\ a_i + 1 \bmod n & \text{if } z_i \in S_3, \end{cases}$$

$$b_{i+1} = \begin{cases} b_i + 1 \bmod n & \text{if } z_i \in S_1, \\ 2b_i \bmod n & \text{if } z_i \in S_2, \\ b_i & \text{if } z_i \in S_3. \end{cases}$$

One continues until for some i, $z_{2i} = z_i$, in which case $a_{2i} + b_{2i} \log_g h = a_i + b_i \log_g h \bmod n$, from which $\log_g h$ may be calculated (except in the unfortunate case where $\gcd(b_{2i} - b_i, n) \neq 1$).

Exercises

1. Explain why in Shanks' method there will always be an element of the group which will appear in both lists.

2. Solve, using Shanks' method:

$$2^x \equiv 518 \mod 547.$$

3. Why is Pollard's rho method called that?

4.3 Diffie–Hellman Key Establishment

Because of the perceived difficulty of the discrete logarithm problem, exponentiation in a group may be regarded as a *one-way* function: the function $f : \mathbb{Z} \longrightarrow G$, defined by $x \mapsto g^x$ (where g is some element of the, suitably chosen, group G) is easy to evaluate, but hard to invert.[18]

We next describe the original and simplest version of the Diffie–Hellman key establishment protocol.[19] The purpose of this protocol is to establish a key which Alice and Bob, the two participants, can use for encrypting, using a symmetric cipher, their subsequent communications. The wonderful aspect of this method is that Alice and Bob do not need any previously shared secret key: they need never have met at all.[20] The protocol runs as follows:

Given: A (multiplicatively written) group G of order n and a generator g of G.

1. Alice randomly selects an element $x_A \in \{1, 2, \ldots, n-1\}$.
2. She transmits g^{x_A} to Bob.
3. Bob randomly selects an element $x_B \in \{1, 2, \ldots, n-1\}$.
4. He transmits g^{x_B} to Alice.

Result: Alice and Bob can now both compute the otherwise secret value $g^{x_A x_B}$. After all, Alice knows g^{x_B} as well as x_A and can therefore compute $(g^{x_B})^{x_A}$ and similarly Bob can compute $(g^{x_A})^{x_B} = (g^{x_B})^{x_A}$. This they can now use as a key for their symmetric cipher, or they can apply some key derivation function to this secret value to obtain such a key.

An adversary eavesdropping on the conversation knows only g^{x_a} and g^{x_B} and, unless she can somehow derive x_A or x_B, does not have the ability to compute the agreed upon secret $g^{x_a x_B}$.

---

[18]In the case of elliptic curves, discussed in Sect. 6.6, the group under consideration is a cyclic subgroup of an additively written Abelian group. So the notation changes, but the principle remains the same.

[19]Though we have used the word "protocol" before, it may by now need a bit of an explanation. Van Tilborg's *Encyclopedia of Cryptography and Security* gives, in an article by Berry Schoenmakers, the following description:

> A cryptographic protocol is a distributed algorithm describing precisely the interaction of two or more entities to achieve certain security objectives. The entities interact with each other by exchanging messages over private and/or public communication channels.

[20]Sadly, things go wrong in their relationship, in Sect. 10.9.3 they are going through a divorce.

At first sight this appears to be a very satisfactory state of affairs. Without actually having to meet with each other, Alice and Bob have arrived at a secret which only they know. Moreover, they have both had a say in deriving it (so that Alice, for example, does not have to fear that Bob deliberately chose a very weak secret), and in particular both can be assured that the secret is "fresh"; the serious risks involved in re-using old keys are therefore avoided.

Nevertheless, the above form of the Diffie–Hellman key establishment protocol is totally insecure; not, we hasten to add, because of a weakness in the mathematical part—as far as anyone knows, the secret value can only be found by an attacker who can solve the discrete logarithm problem[21] in G—but because of a weakness in the protocol itself. Let Eve be an attacker who has the capability of intercepting messages on the network and substituting other messages for them. Eve now perverts the protocol run as follows, where $E(X)$ denotes Eve, masquerading as, or pretending to be, X:

$$\begin{array}{lll} 1 & A \longrightarrow E(B): & g^{x_A}, \\ 1' & E(A) \longrightarrow B: & g^{x_E}, \\ 2 & B \longrightarrow E(A): & g^{x_B}, \\ 2' & E(A) \longrightarrow A: & g^{y_E}. \end{array}$$

Eve now shares the secret $g^{x_A y_E}$ with Alice, who thinks she shares it with Bob, and the secret $g^{x_E x_B}$ with Bob, who thinks he shares it with Alice. When Alice and Bob start communicating, Eve can read every message, re-encrypt it and pass it on. This is the classic example of a *man-in-the-middle attack*.[22] The problem in the protocol is, of course, that neither Alice nor Bob has any guarantee that the entity they are communicating with is actually the entity they think it is. A more sophisticated version, known as the Station-to-Station (or STS) Protocol, requires digital signatures, which in turn require certificates issued by a trusted third party guaranteeing that Alice's public key does indeed belong to her (in the sense that Alice is the unique entity which is in possession of the corresponding private/signing key).

The STS protocol runs as follows, where Cert_X denotes such a certificate issued to entity X, and $\mathrm{sig}_X(D)$ denotes the signature (using X's private key, corresponding to the public key which appears on Cert_X) on the data D. $||$ denotes concatenation of data blocks.

Alice and Bob have previously agreed on a symmetric encryption algorithm E. We denote the ciphertext using E with K as key to encrypt plaintext data D by $E_K(D)$.

$$\begin{array}{lll} 1 & A \longrightarrow B: & g^{x_A}, \\ 2 & B \longrightarrow A: & g^{x_B}||\mathrm{Cert}_B||E_K(\mathrm{sig}_B(g^{x_B}||g^{x_A})), \\ 3 & A \longrightarrow B: & \mathrm{Cert}_A||E_K(\mathrm{sig}_A(g^{x_A}||g^{x_B})), \\ & \text{where } K = & g^{x_A x_B}. \end{array}$$

[21]But an expert would point out that the problem of finding $g^{x_A x_B}$ when g^{x_A} and g^{x_B} are known is *not* the same as that of finding x_A or x_B. See also some comments in Sect. 4.4.1.

[22]In the case we described, it looks more like a "woman-in-the-middle" attack. Let's try to be serious about this!

The fact that the STS Protocol is a lot more complicated than the basic version of Diffie–Hellman exemplifies a fact which a cryptologist ignores at his or her peril: Designing cryptographic protocols is much harder than one might think. History is full of protocols which were broken, sometimes only after having been in actual use for years. The lessons here are: Do not invent your own protocols and distrust any protocol which a supplier recommends as being a simpler version of a secure one.[23] Rather insist that only protocols which have been analysed in the open literature and not found wanting be used. International standards such as those issued by ISO/IEC are a good source.[24]

4.4 Other Applications of the DLP

4.4.1 ElGamal Encryption

RSA is not the only public key encryption scheme. Various other proposals for such schemes have been made, based on hard mathematical problems. One early, but unsuccessful (though interesting!) scheme was Merkle and Hellman's encryption scheme based on the knapsack (or subset-sum) problem. A more successful scheme, "successful" in the sense that with a suitable choice of parameters it has not been broken, is ElGamal's scheme.

In this scheme $< g >$ generates a cyclic group of sufficiently large order q. (Alternatively, q may be a multiple of the order of g.) The groups frequently used are the multiplicative groups of nonzero elements of a finite field, as in Sect. 6.2. Alice picks an integer x, which will constitute her secret key, and publishes $y_A = g^x$. Under the Diffie–Hellman assumption that finding discrete logarithms is computationally infeasible, x will remain secret. The complete public key will thus be the triple $< g, q, y_A >$, although if a set of users all use the same group and generator, only y_A needs to be published, of course.

If Bob wants to send Alice a message m ($m \in < g >$), he picks, randomly, an integer k in the interval $[0, q-1]$, and computes $c_1 = g^k$ and $c_2 = m \times y_A^k$. The message he then transmits is the pair $< c_1, c_2 >$.

Alice can find m, because she can compute $y_A^k = (g^x)^k = (g^k)^x = c_1^x$, and she knows the values of both c_1 and x. She then finds $m = c_2 \times ((y_A)^K)^{-1}$.

The security of this scheme actually depends on something slightly different from the simple assumption of intractability of the discrete logarithm problem. The same applies to our form of the Diffie–Hellman key establishment protocol. On going through the description

---

[23] Or worse, as one developed "in-house" by one of their experts. Even if the "expert" has an international reputation.

[24] This was originally written before it became known, through Wikileaks, that the National Security Agency had succeeded in getting a weak pseudo-random generator approved by ISO/IEC and included in one of its standards. Which goes to show that in applied cryptology suffering from paranoia is a distinct advantage. We refer the reader to the article by Thomas C. Hales: *The NSA back door to NIST*, which appeared in the Notices of the American Mathematical Society (volume **61** (2014), pp. 190–192) for some of the history and the nature of this embarrassing episode. Also read the resulting correspondence in subsequent issues of the Notices.

of the schemes as given above, one sees that the assumption is really that someone who knows g, g^α and g^β will not be able to compute $g^{\alpha\beta}$. This is known as the *computational Diffie–Hellman (CDH) assumption.* Obviously anyone who can find discrete logarithms can solve the computational Diffie–Hellman problem, but it is not known whether the converse holds.[25]

There is a further property of the ElGamal encryption scheme which may make it unattractive for use: the transmitted message is twice the length of the plaintext message m. In addition, encryption is computationally quite demanding, hence slow, and ElGamal encryption is therefore in practice only used in hybrid schemes: the key of a symmetric encryption scheme is transmitted encrypted under ElGamal, but the rest of the conversation uses the symmetric scheme. ElGamal is not unique in this respect: the same applies to RSA and other asymmetric/public key schemes.

Another potential weakness in the ElGamal scheme lies in the fact that any ciphertext message can to some extent be modified: Given the encryption of a message m, an adversary can trivially compute the encryption of the message $2m$ or even of $nm \bmod q$.

4.4.2 The ElGamal Digital Signature Scheme

As we noted in our discussion on RSA in Sect. 4.1, encryption and signing in that scheme are just each other's inverse: the secret key can be used for signing or for decryption and the public key can be used for signature verification or decryption. This is not true for the ElGamal schemes: although both the ElGamal encryption scheme and the ElGamal signature scheme are based on the DLP, they are not inverse to each other.

To generate a private key, Alice finds a prime p of adequate length (adequate for her security requirements, that is), a random generator g for the (multiplicative) cyclic group $\mathbb{Z}_p^*$ and a random integer a, $0 < a < p - 2$. Her public key also includes the specification of a hash function H whose input can be any string and whose output is within the range $(0, p-1)$. The complete public key is then the list $< p, g, y, H >$, where $y = g^a$. Her private key is $< p, g, a, H >$. So far, the situation is not dissimilar to what we had for the encryption.

Now to sign a message m, Alice picks a random integer $k \in [1, p-2]$ which is relatively prime to $p-1$, and computes $r = g^k \bmod p$, and then $s = k^{-1}(H(m) - ar) \bmod (p-1)$. (If s should turn out to be zero, try another value for k.)

Now the signature consists of the pair $< r, s >$.

Verification consists of computing $v = y^r r^s \bmod p$. If $v = g^{H(m\|r)}$, the signature is accepted as valid, but otherwise it must be rejected. But even before that, it should be verified that $r \in [1, p-1]$ and if that is not the case, rejection should also be immediate.[26]

---

[25]In a more theoretical approach, we should also have mentioned the *Decisional Diffie–Hellman* (DDH) assumption. This is the assumption that, for a given group, it is not possible to determine with a probability of success significantly better than 50 % which of two arbitrarily chosen triples (g^a, g^b, g^{ab}) and (g^a, g^b, g^c) (with $c \neq ab$) is a Diffie–Hellman triple. It is currently not known whether DDH and CDH are equivalent assumptions in general.

[26]Failing to verify this first can be quite catastrophic.

The ElGamal signature scheme is not widely used. In order to provide an acceptable level of security p should be at least 1024 bits long, which makes for a signature of length 2048 at least. This is much longer than an equivalent signature created under the Digital Signature Algorithm (DSA), which we discuss next.

Exercises

1. Verify that the condition stated should hold if the signature is genuine.
2. Explain why it is vital that a new random value k must be picked for each new message.

4.4.3 The Digital Signature Algorithm

The Digital Signature Algorithm (DSA) was proposed by the U.S. National Institute of Standards and Technology (NIST) in 1991 to be included in their Digital Signature Standard specification. At the time there was some concern that the RSA signature scheme had not been selected, a decision which was ascribed to the fact that RSA can be used for both signatures and encryption, while the U.S. government at the time was very emphatic about restricting the use of strong encryption, especially in the case of exports. The DSA, which is based on the ElGamal algorithm, shares with that algorithm the property that it is purely for signatures.

As in the case of the ElGamal scheme (and most if not all signature schemes) a hash function H is required to "condense" the message to a bit length suitable as input into the DSA. Sometimes the SHA-1 hash function is specified, but we won't pursue the precise specification of the DSA, nor discuss the bit lengths of the used constants required for security. We shall, as before in this section, merely outline the algebraic construction of the algorithm.

In order to generate the keys for DSA, randomly generate large primes p and q such that $q|p-1$ and find an integer g of order q modulo p. Randomly choose $x \in [2, q-1]$ and put $y = g^x \bmod p$. The public key now consists of p, q, g and y and the private key is x. So far, the similarity to ElGamal is obvious.

For suppose the adversary has a genuine message m with a valid signature $< r, s >$. He wishes to forge a signature on a message M of his own. He puts

$$u = H(M) \cdot (H(m))^{-1} \bmod (p-1),$$
$$S = su \bmod (p-1)$$

and chooses (using the CRT) R such that $R \equiv ru \bmod (p-1)$ and $R \equiv r \bmod p$. Then, as is easily verified, $y^R R^S = g^{H(m) \cdot u} = g^{H(M)}$, so $< M, R, S >$ will be accepted as genuine.

It may also be useful to use $H(m||r)$ (where "||" denotes concatenation) in place of simply $H(m)$, as this makes it very hard for the adversary to find a suitable R.

Now to sign a message m, firstly find its hash value $H(m)$. Choose, randomly, $k \in [1, q-1]$, and compute

1. $r = (g^k \bmod p) \bmod q$;
2. k^{-1}, the inverse of k modulo q;
3. $s = k^{-1}(H(m) + xr) \bmod q$.

The signature is the pair $< r, s >$, so the signed message will be $< m, r, s >$.

To verify the signature on a message $< M, R, S >$, the recipient also starts by finding $H(M)$ and then computes:

1. S^{-1}, the inverse of S modulo q;
2. $U_1 = S^{-1}H(M) \bmod q$;
3. $U_2 = S^{-1}R \bmod q$;
4. $V = (g^{U_1}y^{U_2} \bmod p) \bmod q$.

The receiver accepts the message as valid if and only if $V = R$ and rejects it otherwise.

Again, we leave it to the reader to check that this algorithm will indeed work as advertised.

The main difference between the original ElGamal signature and the DSA lies in the reduction modulo q which yields a signature of a more manageable length. Note, by the way, that when signing the message m its hash value is only required in the final (third) step. This implies that much of the work required can be done beforehand "off-line", thus speeding up the actual signing process.

Exercise Prove that the verification process works.

4.4.4 *A Zero Knowledge Proof*

The discrete logarithm problem also provides an easy example of how one party, whom we'll call Alice, can prove to another (Bob) that she knows a secret, without divulging it. In this case the public information is $A = g^x$, where g is a generator of a group G of very large order, and the secret information is x, which Alice wants to prove she knows, without letting Bob in on it.

The protocol for doing so runs as follows:

1. Alice chooses at random an integer r, computes $C = g^r$ and sends it to Bob.
2. Bob flips a coin. If it shows "heads" he asks Alice to return r. If it shows "tails", he asks Alice for $x + r$.
3. Alice returns the required value.
4. Bob verifies: If he asked for r, he computes g^r himself and verifies that $g^r = C$. If he asked for $x + r$ he verifies that $g^{x+r} = CA$.

The essential point here is that Alice, if she is cheating, can fake an answer to either one of the questions Bob poses, but not to both. Hence Alice, if she is cheating, is at a 50 % risk of being caught out. By repeating the test as many times as he wants to, Bob can reduce the probability of being cheated by Alice to as small a value of 2^{-n} as he is comfortable with.

A protocol like the above is called an *interactive proof* and Alice is referred to as the *prover*, while Bob in this case is the *verifier*. An interactive proof should be both *complete* in the sense that an honest prover should be able to convince an honest verifier, and *sound* in the sense that a dishonest prover should *not* be able to do so.

But where does the "zero knowledge" come in? The answer to that lies in the *transcript* of the data exchanges, the (possibly hypothetical) record of all the messages exchanged in the process of executing the protocol. In the case of the above protocol, this consists of Bob's request in step 2, and Alice's response in step 3. But a moment's reflection convinces one that Bob could have written such a record of those steps all by himself, should he, for some strange reason want to convince Jerry the judge that a conversation had taken place between him and Alice. Moreover, Jerry would not be able to decide whether the transcript Bob shows him is a genuine one or one forged by Bob.

The consequence of this argument is that, no matter how many times the protocol is run, Bob at the termination knows nothing that he didn't know before, or, in other words, he is still as uncertain about the value of x as he was before. In Chap. 7 we shall equate the providing of information with the removal of uncertainty; in this case, therefore, no information has been provided, although Bob now has a high degree of certainty that Alice does indeed know the value of x.

Exercise Show how Alice can cheat if she knows beforehand what question Bob is going to pose.

4.4.5 What Groups to Use?

We list a few groups for which the Diffie–Hellman assumption is believed to be valid.[27] Our list is from a paper by Dan Boneh[28]:

1. Let $p = 2q + 1$, where both p and q are prime. Then p is called a *safe prime*. Use the subgroup of $\mathbb{Z}_p^*$ consisting of all elements which are squares. These are known as the quadratic residues; more about them in the final chapter. The quadratic residues form a cyclic group of order q.
2. More generally, let $p \doteq aq + 1$, where $q > \sqrt[10]{p}$. Use the cyclic subgroup of order q of $\mathbb{Z}_p^*$.
3. Take $N = pq$ where both p and q are safe primes. Use a cyclic subgroup of $\mathbb{Z}_n^*$ of order $\frac{(p-1)(q-1)}{2}$.
4. Use a group of prime order of points on an elliptic curve, defined over a prime field $GF(p)$. More about fields in Chaps. 5 and 6. More (but not much more) about elliptic curves in the final section of Chap. 6.

---

[27]Strictly speaking, the Decisional D.–H. assumption, as in footnote 23 above. But it is not hard to see that if DDH holds, then so does CDH, and certainly DLP will be intractable.

[28]Boneh, D.: *The Decision Diffie–Hellman Problem*, Proceedings of the Third Algorithmic Number Theory Symposium, LNCS 1423, Springer-Verlag, pp. 48–63, 1998.

The Digital Signature Standard, which mandates DSA, is specific about the use of the second of these options: p is a randomly selected prime (at least 512 bits long in the standard, which is rather inadequate by modern standards), and q a large prime divisor of $p - 1$. In practice it would be easier to select q, and find an integer a such that $p = aq + 1$ is prime.

Then randomly choose, as base for the discrete logarithm, an integer g such that $g^{\frac{p-1}{q}} \neq 1 \bmod p$, which guarantees that g has order q.

5 Fields

In this chapter and the next we consider fields, which are rings in which the nonzero elements also form a group, under multiplication. All the classical fields, such as $\mathbb{R}$ and $\mathbb{C}$, are infinite. We shall instead concentrate on the finite ones and in this chapter show how one may construct them.

As we shall show in subsequent chapters, finite fields form a fundamental part of the construction of stream ciphers and modern block ciphers. In addition, their multiplicative groups and their subgroups are, as noted in the previous chapter, useful when it comes to constructions of the Diffie–Hellman and ElGamal types.

In this chapter we shall develop the underlying Algebra which enables us to construct finite fields; in the next we deal with some of their rather elegant properties.

5.1 Rings "Without" Ideals

We have observed that the elements of a ring cannot form a group under multiplication, because 0 (the identity element of the ring for addition) cannot have an inverse.[1] On the other hand it is possible that the set of all *nonzero* elements of a ring forms a group under multiplication: we recall that such a ring is called a *field*.

Definition A field is a set $\mathbb{F}$ on which two binary operations + and $\cdot$ are defined such that

- $\{\mathbb{F}, +\}$ is an abelian group. We denote the identity element of this group by 0, and the inverse of an element a by $-a$.
- $\{\mathbb{F}\backslash\{0\}, \cdot\}$ is an abelian group. We denote the identity element of this group by 1, and the inverse of an element a by a^{-1}.

[1] If 0 had an inverse, ζ say, then it would follow that for any two elements a, b of the ring $0 \cdot a = 0 = 0 \cdot b$ and therefore $a = \zeta \cdot 0 \cdot a = \zeta \cdot 0 \cdot b = b$, so that the ring can have only one element. Since any ring must contain 0 and 1, we have 0 = 1. Such a completely trivial ring we exclude from our discussion!

A.R. Meijer, *Algebra for Cryptologists*, Springer Undergraduate Texts in Mathematics and Technology,
DOI 10.1007/978-3-319-30396-3_5

- The distributive law holds:

$$\forall a, b, c \in \mathbb{F} \ a(b + c) = ab + ac.$$

Examples The familiar sets of rational numbers, real numbers, and complex numbers with the usual addition and multiplication are all fields. So is the slightly less familiar example of the set $\{a + b\sqrt{2} | a, b \in \mathbb{Q}\}$ as may easily be verified (essentially by checking that sums and products, as well as inverses, of expressions of the form $a + b\sqrt{2}$ are again of this form). A further, important, class of fields are of the type first encountered in Sect. 2.4, where we had $\mathbb{Z}_5$ and $\mathbb{Z}_2$ as examples: studying the "multiplication tables" for these structures we note that the nonzero elements form groups—cyclic groups, at that. In fact, if p is any prime, then every nonzero element of $\mathbb{Z}_p$ has an inverse:

Definition Let p be a prime number and let $< p >$ be the principal ideal in $\mathbb{Z}$ generated by p. Then $\mathbb{Z}_p = \mathbb{Z}/ < p >$ is a field, called the prime field of order p, and denoted by $GF(p)$.

These are not the only finite fields; we shall discuss later how larger finite fields may be constructed. However, recalling a definition at the end of Sect. 3.8 we can at this moment prove the following

Theorem *Every finite field has prime characteristic.*

Proof Suppose $\mathbb{F}$ is a finite field of characteristic n. n is therefore the smallest positive integer such that $n \cdot 1 = 0$. If n were composite, say $n = rs$ (with $n > r, s > 0$), then $n \cdot 1 = (r \cdot 1)(s \cdot 1)$. By definition of the characteristic, $r \cdot 1 \neq 0$, so it has a multiplicative inverse and hence $s \cdot 1 = 0$, in contradiction to the fact that $s < n$.

Now let $\mathbb{F}$ be a field of characteristic p and consider the set $\{k \cdot 1 | k \in \{0, 1, 2, \ldots, p - 1\}\}$. (Here $k \cdot 1$ means $\underbrace{1 + 1 + \cdots + 1}_{k \text{ times}}$, of course, as it did before.) This is really nothing but the ring $\mathbb{Z}_p$, because addition and multiplication in these two sets behave in exactly the same way. Thus we observe:

Theorem *Any field of characteristic p "contains" the field GF(p).*

Definition Let $\mathbb{F}$ be a field, and let E be a subset of $\mathbb{F}$ such that, under the operations of addition and multiplication E is itself a field. Then E is called a *subfield* of $\mathbb{F}$.

For the sake of completeness we should mention that there exist infinite fields of finite (prime) characteristic; also that there exist infinite fields in which every element (except 0) has infinite order in the additive group; in fact all the "classical fields" of rationals, reals, etc. are of the latter kind.

The following theorem states a charming property of fields of prime characteristic:
"Freshman Theorem"[2]: Let $\mathbb{F}$ be a field of characteristic p, and let $a, b \in F$. Then

$$(a + b)^p = a^p + b^p.$$

[2] A convenient way of referring to this result. It is also sometimes referred to as "the freshman's dream".

Proof By the binomial theorem

$$(a+b)^p = a^p + \sum_{k=1}^{p-1} \binom{p}{k} a^k b^{p-k} + b^p.$$

But all the coefficients of the middle terms are divisible by p and therefore all these terms vanish, leaving only the two outer terms.

Of course, we know that in $GF(p)$ the equation $a^p = a$ (and therefore $(a+b)^p = a+b = a^p + b^p$) holds, by Fermat's Little Theorem. In fields of characteristic p, the Freshman Theorem shows that the set $\{a|a^p = a\}$ forms a subfield of any such field: it is trivial to verify that this set is closed under addition and multiplication, that the inverse of any nonzero element of the set again belongs to the set, and so on. Moreover, as we shall show shortly, the equation $x^p = x$ has at most p solutions, so $\{a|a^p = a\}$ is precisely the subfield $GF(p)$ described above. Our statement that *Every field of characteristic p contains the prime field GF(p)* can also be phrased as *Every field of characteristic p is an* extension *of the prime field GF(p)*. How to construct such extensions (and in particular, how to construct extensions of $GF(2)$) will form the subject of later discussions. But before we can get there, we need to look at another family of rings, which, in many ways, have properties similar to those we covered in our discussion of the ring of integers. And even before that, we shall, as an example, describe the world's smallest non-prime field.

Example Consider the following set: $\{0, 1, \alpha, \beta\}$ on which we define addition and multiplication according to the following tables:

$+$	0	1	α	β
0	0	1	α	β
1	1	0	β	α
α	α	β	0	1
β	β	α	1	0

$\cdot$	0	1	α	β
0	0	0	0	0
1	0	1	α	β
α	0	α	β	1
β	0	β	1	α

It is easy to see that the nonzero elements $1, \alpha, \beta$ form a cyclic group under multiplication, and (with some effort) that all the requirements of a field are met. Thus we have a field with 2^2 elements, and obviously this field contains the prime field $GF(2) = \{0, 1\}$.

Exercise Show that in $GF(p)$ the elements 1 and -1 are the only ones which are their own inverse. Use this to prove one half of *Wilson's Theorem*, which states that if $n \in \mathbb{Z}$ is prime then

$$(n-1)! \equiv -1 \mod n.$$

(The other half of Wilson's theorem states that the converse is also true: if $(n-1)! \equiv -1 \mod n$, then n is prime. For obvious reasons, this is not a practical test for primality.)

5.2 Vector Spaces

We briefly recall a few concepts from Linear Algebra. Let $\mathbb{F}$ be a field. A vector space V (or ${}_{\mathbb{F}}V$ if the relevant field needs to be made explicit) is a set V on which an operation $+$ is defined such that $\{V, +\}$ is an abelian group, and such that, moreover, for every $\alpha \in \mathbb{F}$ there is a function

$$v \mapsto \alpha v$$

from V to V, called "scalar multiplication by α", such that for all $\alpha, \beta \in \mathbb{F}$ and all $v_1, v_2 \in V$

$$\alpha(v_1 + v_2) = \alpha v_1 + \alpha v_2,$$
$$(\alpha + \beta)v_1 = \alpha v_1 + \beta v_1,$$
$$(\alpha\beta)v_1 = \alpha(\beta v_1).$$

The elements of a vector space may be called *vectors* and the elements of the field $\mathbb{F}$ over which the vector space is defined would be called *scalars*.

Among the properties of scalar multiplication that follow immediately from the definition are

- $1 \cdot x = x$ for every $x \in V$;
- $0 \cdot x = 0$ for every $x \in V$, where the 0 on the left denotes the zero (the additive identity) of the field, and the one on the right the zero element of V.

Let now ${}_{\mathbb{F}}V$ be a vector space. Any sum of the form $\sum_{i=0}^{n-1} \alpha_i v_i$, where the α_i are scalars and the v_i are vectors, is called a *linear combination* of the v_i.

Let $A = \{v_0, v_1, \ldots, v_k\} \subset V$ be a set of vectors. The set A is called *linearly independent* if the only solution to the equation $\sum_{i=0}^{k} x_i v_i = 0$ (where again the right-hand side represents the zero vector, i.e. the additive identity element of the group V, and where the $x_i \in \mathbb{F}$) is the trivial solution $x_0 = x_1 = \cdots = x_k = 0$ (0 this time denoting the zero element of $\mathbb{F}$). A set of vectors that is not linearly independent is called *linearly dependent*.

A subset $B = \{v_i\}_{i \in I}$ of V is called a *basis* for V if it satisfies one of the following equivalent conditions:

- Every vector in V can be written in a unique way as a linear combination of elements of B;
- The set B is a maximal linearly independent set (i.e. any set of vectors properly containing B is linearly dependent);

- The set B is minimal with respect to the property that every vector in V can be written as a linear combination of elements in B.[3]

Every vector space has a basis.[4] If a basis is a finite set, then the vector space is called finite dimensional, and it can be shown that every basis for that space will contain the same number of elements, which is then called the *dimension* of the vector space.

For our purposes the importance of these ideas lies in the following

Observation If E is a subfield of the field $\mathbb{F}$, then $\mathbb{F}$ has the structure of a vector space with E as the field of scalars.

This follows immediately from the fact that scalar multiplication obeys exactly the same rules as multiplication of elements of $\mathbb{F}$ by elements of E.

Example In the example at the end of the previous section we had the field $GF(2) = \{0, 1\}$ as a subset of the field $GF(2^2) = \{0, 1, \alpha, \beta\}$. Any element in $GF(2^2)$ can be written as a linear combination of 1 and α (with the coefficients coming from $GF(2)$):

$$0 = 0 \cdot 1 + 0 \cdot \alpha,$$
$$1 = 1 \cdot 1 + 0 \cdot \alpha,$$
$$\alpha = 0 \cdot 1 + 1 \cdot \alpha,$$
$$\beta = 1 \cdot 1 + 1 \cdot \alpha.$$

Moreover, the only solution to the equation $x \cdot 1 + y \cdot \alpha = 0$ with $x, y \in GF(2)$ is (obviously) the trivial solution $x = y = 0$, so the set $\{1, \alpha\}$ is linearly independent, so it forms a basis for ${}_{GF(2)}GF(2^2)$, which is therefore 2-dimensional.

Combining the observation above with the fact that any finite field is of prime characteristic p, therefore contains the prime field $GF(p)$, and is therefore a vector space over $GF(p)$ of dimension n, say, we conclude that it must contain p^n elements. After all, there will be p^n different linear combinations of the n basis vectors. We have therefore proved that *every finite field has p^n elements for some prime p and some integer $n \geq 1$*. We shall show that, in fact, any two fields with p^n elements are, for all practical purposes, identical, differing, if at all, only in the naming of the elements. Thus it makes algebraic sense to speak of *the* finite field $GF(p^n)$.

Harking back to our use of finite fields in the Diffie–Hellman constructions of the previous chapter, the question arises whether $GF(2^n)$ or $GF(p^n)$, for some $p \neq 2$ will offer better security: It appears that a field of characteristic $p \neq 2$ with approximately 2^n elements offers a slightly greater level of security than $GF(2^n)$. But to some extent this advantage of prime

[3] "Minimal" in the sense that no proper subset of B has this property.

[4] We accept this as an axiom. Proving that every finite dimensional vector space has a basis is easy, and is sufficient for our purposes anyway. But proving the statement as given needs an invocation of the Axiom of Choice.

($\neq 2$) fields is offset by the fact that operations in $GF(2^n)$ can be performed more efficiently than operations in $GF(p^m)$ when these fields are of approximately the same size.

Exercises

1. Find all other bases for ${}_{GF(2)}GF(2^2)$.
2. Find bases for the following vector spaces: ${}_{\mathbb{R}}\mathbb{R}$, ${}_{\mathbb{R}}\mathbb{C}$.

5.3 Rings of Polynomials

In order to be able to construct extensions $GF(p^n)$ of the prime fields $GF(p)$, we need to develop a bit more Algebra. Regrettably, this is, on first acquaintance, a rather boring process.

5.3.1 Definitions

Let $\mathbb{F}$ be any field. We assume that everyone has a good idea of what a polynomial is, but we define it explicitly as a *formal* finite sum of the form

$$f(x) = \sum_{i=0}^{n} a_i x^i,$$

where the *coefficients* $a_i \in \mathbb{F}$. In order to avoid difficulties here we shall assume that $a_n \neq 0$.[5] We emphasize the word "formal", because, contrary to what you might have expected, the x appearing in this expression does not have any meaning. Thus, in particular, $f(x)$ is not to be considered a function. In fact, we could have defined $f(x)$ merely as the finite string $(a_0, a_1, a_2, \ldots, a_n)$, but for the purposes of defining the ring operations, it is more convenient to bring in the symbols x^i (and easier to understand as well).

The *leading coefficient* of the polynomial $f(x)$ above is a_n (which, we reiterate, is nonzero), and n is called the *degree* of this polynomial. The coefficient a_0 is called the *constant term* of the polynomial. In order to make things work, we also need to include the zero polynomial, in which the constant term is 0, and there are no other (nonzero) terms. We assign the degree $-\infty$ to the zero polynomial. We shall denote the degree of a polynomial $f(x)$ by $\deg(f(x))$.

We consider the (infinite) set $\mathbb{F}[x]$ of all polynomials with coefficients from $\mathbb{F}$, using the above notation. We could also have considered the set $\mathbb{F}[y]$, where we use y instead of x as the "indeterminate". This is really the same set, since the x and y are there only for notational convenience. We now turn this set into a ring by defining the sum of two polynomials

[5] Unless *all* the a_i are zero! As we shall point out in a moment, we need a zero element in order to turn the set of polynomials into a ring—every ring has a zero element, after all—and the only way we can do so is by allowing for a polynomial in which all the coefficients are zero.

componentwise:

$$\text{If } f(x) = \sum_{i=0}^{n} a_i x^i \text{ and } g(x) = \sum_{i=0}^{n} b_i x^i \text{ then}$$

$$f(x) + g(x) = \sum_{i=0}^{n} (a_i + b_i) x^i.$$

If f and g are of different degree, we just add zero terms to the shorter one, so that we have equal numbers of terms in both.

It is a trivial exercise, and excruciatingly boring, to verify that $\{F[x], +\}$ now forms an abelian group.

Multiplication is defined as follows:

$$\text{If } f(x) = \sum_{i=0}^{n} a_i x^i \text{ and } g(x) = \sum_{i=0}^{m} b_i x^i \text{ then } f(x)g(x) = \sum_{j=0}^{m+n} c_j x^j,$$

where

$$c_j = \sum_{i=0}^{j} a_i b_{j-i}.$$

It is easy, but tedious and messy, to verify that all the ring axioms are satisfied. Essentially, these definitions of addition and multiplication correspond to what all of us have been used to since our happy days in high school. One must get used, however, to the fact that the fields from which the coefficients come are not necessarily old friends like the reals or the rationals.

It is also easy (and neither tedious nor messy) to check that if one multiplies two polynomials $f(x)$ and $g(x)$ of degrees n_1 and n_2 respectively, then their product $f(x)g(x)$ has degree $n_1 + n_2$.

Example Consider the polynomials $1 + x^2, 1 + x^2 + x^4 \in GF(2)[x]$. Their sum is x^4 and their product is $1 + x^2 + x^4 + x^2 + x^4 + x^6 = 1 + x^6$. Note that, following again what we did during the aforementioned happy days, we write x^n if the coefficient of x^n is 1. This leads to $x^2 + x^2$ being equal to 0, since $x^2 + x^2$ stands for $1.x^2 + 1.x^2 = (1 + 1)x^2 = 0.x^2 = 0$.

In $GF(p)[x]$ we have $(f(x) + g(x))^p = (f(x))^p + (g(x))^p$, as in the Freshman Theorem, and for the same reason: all the intermediate terms in the binomial expansion vanish because $p \cdot 1 = 0$. The same is true in any field of characteristic p, i.e. in any extension of the field $GF(p)$. Hence $(\sum_{i=0}^{n} a_i x^i)^p = \sum_{i=0}^{n} a_i^p x^{pi}$. In particular, if $f(x) \in GF(p)[x]$, say $f(x) = \sum_{i=0}^{n} a_i x^i$, with $a_i \in GF(p)\ \forall i$, then $(f(x))^p = \sum_{i=0}^{n} a_i x^{pi}$.

Example In $GF(2)[x]$, $(1 + x + x^2)^8 = (1 + x^2 + x^4)^4 = (1 + x^4 + x^8)^2 = 1 + x^8 + x^{16}$.

5.3.2 The Euclidean Algorithm Again

In many ways a ring of polynomials over a field behaves like the ring of integers. The fundamental property which we exploit is again the possibility of "dividing with remainder":

Let $\mathbb{F}$ be any field, and let $f(x), g(x) \in \mathbb{F}[x]$. Then there exist unique polynomials $q(x), r(x) \in \mathbb{F}[x]$ such that

$$f(x) = q(x)g(x) + r(x) \text{ and } \deg(r(x)) < \deg(g(x)).$$

Here $q(x)$ is (surprise, surprise!!) called the *quotient* and $r(x)$ is called the *remainder*.

When it comes to ideals, we have exactly the same situation as we had for the integers:

Theorem *Let $\mathbb{F}$ be a field. Then every ideal in the ring $\mathbb{F}[x]$ is principal. (In other words, if I is any ideal in this ring, then there exists a polynomial $a(x)$ such that $I = \{c(x)a(x)|c(x) \in F[x]\}$, the set of all multiples of $a(x)$.)*

The proof is of the same form as the analogous proof for the ring of integers and is left as an exercise. It depends on the fact that any ideal $I \neq \{0\}$ contains a nonzero element of lowest degree, and this polynomial generates the ideal. (And the ideal $\{0\}$ is principal, of course.)

Note, by the way, that the polynomial generating the ideal is not unique. If all the elements in I are multiples of $a(x)$, say, then they are also multiples of $c \cdot a(x)$, where c can be any nonzero element of $\mathbb{F}$. To get round this inconvenience, it is frequently convenient to choose the unique *monic* polynomial that generates the ideal, where a monic polynomial is defined to be one whose leading coefficient is 1.

We can now repeat our discussion on greatest common divisors (and least common multiples): Let I, J be ideals and consider the smallest ideal D in $\mathbb{F}[x]$ which contains both I and J. Suppose that I and J are generated by $f(x)$ and $g(x)$ respectively. Since, again, D must contain all sums of the form $a(x)f(x) + b(x)g(x)$, and since the set of all such expressions already forms an ideal, we have $D = \{a(x)f(x) + b(x)g(x)|a(x), b(x) \in \mathbb{F}[x]\}$. But we have seen that D is itself a principal ideal, so $D =< d(x) >$ for some polynomial $d(x)$. But then $d(x)$ must be a divisor of both $f(x)$ and $g(x)$, and

$$d(x) = a(x)f(x) + b(x)g(x)$$

for some polynomials $a(x), b(x) \in \mathbb{F}[x]$,[6] so any common divisor of $f(x)$ and $g(x)$ must be a divisor of $d(x)$. Hence $d(x)$ may be considered the *greatest common divisor* of $f(x)$ and $g(x)$.

How does one find the greatest common divisor of two polynomials? The analogy with the ring of integers carries on, we just adapt the Euclidean algorithm, as in the following

Example Let $f(x) = x^5 + x^4 + x^2 + 1$ and $g(x) = x^4 + x^2 + x + 1$ be polynomials in $GF(2)[x]$. To find their greatest common divisor we perform long division to get $q_1(x)$ and $r_1(x)$:

$$x^5 + x^4 + x^2 + 1 = (x + 1)(x^4 + x^2 + x + 1) + (x^3 + x^2)$$

[6]This is really Bezout's identity all over again. And this time the result is actually due to Bezout himself.

and then

$$x^4 + x^2 + x + 1 = (x+1)(x^3 + x^2) + (x+1)$$
$$x^3 + x^2 = x^2(x+1) + 0$$

so the gcd is $x + 1$. To get the terms required in Bezout's identity, we perform the same steps as for integers:

i	-2	-1	0	1	2
$q_i(x)$			$x+1$	$x+1$	x^2
$P_i(x)$	0	1	$x+1$	x^2	x^4+x+1
$Q_i(x)$	1	0	1	$x+1$	x^3+x^2+1

This shows that

$$x^2 g(x) + (x+1) f(x) = x + 1.$$

Exercises

1. Find the greatest common divisor of $x^3 - 6x^2 + x + 4$ and $x^5 - 6x + 1 \in \mathbb{Q}[x]$.
2. Find the greatest common divisor of $x^4 + x^3 + x^2 + 1, x^6 + x^4 + x^2 + 1 \in GF(2)[x]$.
3. Find the least common multiple of the two polynomials in the previous exercise.

5.3.3 The "Remainder Theorem"

Somewhere we need to point out an obvious consequence of the fact that division with remainder is always possible, when faced with two polynomials in $\mathbb{F}[x]$, where $\mathbb{F}$ is some field. Let us do so now.

Suppose that $f(x) \in \mathbb{F}[x]$, and that $\alpha \in F$ is any field element. Then $x - \alpha$ is a polynomial of degree 1, so we can divide:

$$f(x) = q(x)(x - \alpha) + r(x)$$

where $r(x)$ is a polynomial of degree less than 1, i.e. a polynomial of degree either 0 or else $-\infty$, i.e. an element of $\mathbb{F}$ in either case. So:

$$f(x) = q(x)(x - \alpha) + \rho$$

for some $\rho \in \mathbb{F}$. If we "evaluate" both sides of this equation, that is to say, if we replace all the xs by αs and perform the necessary arithmetic, we see that $x - \alpha$ is a factor of $f(x)$ if and only if $f(\alpha) = 0$.[7] Then α is called a *root* of $f(x)$.

An immediate consequence of the Remainder Theorem is that a polynomial $f(x) \in \mathbb{F}[x]$ of degree k can have at most k roots, since every root α_i contributes a factor $(x - \alpha_i)$ to the factorisation of $f(x)$, and the degree l of $(x - \alpha_1)(x - \alpha_2) \dots (x - \alpha_l)$ cannot exceed k. Note that a root α may be *repeated*: this occurs when $(x - \alpha)^e$ is a factor of $f(x)$ with $e > 1$.

The diligent and/or obnoxious reader may object to the above "proof" in that we seemed to use the polynomial $f(x)$ as a function, even though we earlier specifically said that polynomials are purely formal expressions. We can get round this by introducing, for every $\alpha \in \mathbb{F}$, a function (let's call it ψ_α) from $\mathbb{F}[x]$ to $\mathbb{F}$, which maps $\sum_{i=0}^{n} a_i x^i$ onto $\sum_{i=0} a_i \alpha^i$. It is very easy to see that this is a well-behaved function in that it maps sums onto sums and products onto products:

$$\psi_\alpha(f(x) + g(x)) = \psi_\alpha(f(x)) + \psi_\alpha(g(x)),$$
$$\psi_\alpha(f(x) \cdot g(x)) = \psi_\alpha(f(x)) \cdot \psi_\alpha(g(x)).$$

Such a function which maps sums to sums and products to products is called a (ring-) *homomorphism*. Applying the homomorphism ψ_α is what we called "evaluation".

The concept of homomorphism is an important one. The following exercises serve as an introduction, but we'll need a separate section (Sect. 5.4) of this chapter to deal with the matter a little more thoroughly.

Exercises

1. Show that the mapping ϕ from $\mathbb{C}$ to the ring of 2×2 real matrices, defined by

$$\phi(a + bi) = \begin{pmatrix} a & b \\ -b & a \end{pmatrix}$$

   is a homomorphism. Is it one-to-one? Is it onto?
2. Let $\phi : R \longrightarrow S$ be a homomorphism from the ring R to the ring S. Let 0_R and 0_S be the zero elements of R and S respectively, and 1_R and 1_S the identity elements.
   (a) Prove that $\phi(0_R) = 0_S$.
   (b) Is it necessarily true that $\phi(1_R) = 1_S$?
   (c) Prove that the set $\{a | a \in R \text{ and } \phi(a) = 0_S\}$ is an ideal in R. (This ideal is known as the *kernel* of the homomorphism.)
   (d) With ψ_α as defined above, how would you describe its kernel?
   (c) Show that a ring-homomorphism defined on R is one-to-one if and only if its kernel $\mathrm{Ker}(\phi) = \{0_R\}$.

[7] In the author's schooldays this was one of the few theorems in high school algebra, and it went under the grand name of "The Remainder Theorem". Of course the field referred to was always $\mathbb{Q}$ or at worst $\mathbb{R}$.

(e) Conclude that a ring-homomorphism defined on a field is either one-to-one or the constant function which maps every element onto 0.

5.3.4 Irreducible Polynomials

Definition Let $\mathbb{F}$ be a field. A polynomial $f(x) \in \mathbb{F}[x]$ is called *irreducible* if it cannot be written as the product of two polynomials (in $\mathbb{F}[x]$) both of which are of lower degree than itself. We hardly need to add that a polynomial which is not irreducible is said to be *reducible*.

Examples

1. The polynomial $x^5 + x^4 + 1 \in GF(2)[x]$ is reducible, since $x^5 + x^4 + 1 = (x^2 + x + 1)(x^3 + x + 1)$.
2. The polynomial $f(x) = x^2 + x + 1 \in GF(2)[x]$ is irreducible. For if it were reducible, then it would have to be the product of two polynomials, belonging to $GF(2)[x]$, each of degree 1 (i.e. it would have two *linear* factors), so we would have $x^2 + x + 1 = (x + \alpha)(x + \beta)$ for some $\alpha, \beta \in GF(2) = \{0, 1\}$. But then, if we substituted α for x, we would have $f(\alpha) = 0 \cdot (\alpha + \beta) = 0$. But neither 0 nor 1 is a root of the polynomial.
3. But if we consider the same polynomial as a polynomial over the field we considered at the end of Sect. 5.1, we see that there we have the factorization $f(x) = (x + \alpha)(x + \beta)$. Thus an irreducible polynomial $f(x) \in \mathbb{F}[x]$ may well be reducible in a field $\mathbb{F}' \supset \mathbb{F}$. In fact, we shall show that such a field $\mathbb{F}'$ *always* exists.
4. In the same way one can show that $x^3 + x + 1 \in GF(2)[x]$ is irreducible. For if it were reducible, it would have to have at least one linear factor. Again, trying all (both) the elements of $GF(2)$ one sees that this isn't the case. In fact, this polynomial is irreducible also over the larger field considered as an example at the end of Sect. 5.1.

Irreducible polynomials play the same role in $\mathbb{F}[x]$ that prime numbers play in the ring of integers, as we shall show in the next section. We shall see that in order to construct finite fields other than the prime fields one needs irreducible polynomials and we therefore include the following algorithm for testing polynomials over $\mathbb{Z}_p$ for irreducibility. (The following description of the algorithm is taken verbatim from the *Handbook of Applied Cryptography*.)[8] The algorithm depends on the fact, which we shall note in the next chapter (Sect. 6.1), that every irreducible polynomial in $GF(p)[x]$, of degree m, is a factor of $x^{p^m} - x$.

[8] Menezes, A.J., van Oorschot, P.C. and Vanstone, S.A.: *Handbook of Applied Cryptography*, CRC Press, Boca Raton, 1996.

Algorithm for testing irreducibility of a polynomial

INPUT: A prime p and a polynomial $f(x)$ of degree m in $\mathbb{Z}_p[x]$.
OUTPUT: An answer to the question: "Is $f(x)$ irreducible over $\mathbb{Z}_p$?"

1. Set $u(x) \longleftarrow x$.
2. For i from 1 to $\lfloor \frac{m}{2} \rfloor$ do the following:
   - Compute $u(x) \longleftarrow u(x)^p \bmod f(x)$. (Note that $u(x)$ is a polynomial in $\mathbb{Z}_p[x]$ of degree less than m.)
   - Compute $d(x) = \gcd(f(x), u(x) - x)$.
   - If $d(x) \neq 1$ then return "Reducible".
3. Return "Irreducible".

Note that if $f(x)$ is reducible, then it must have an irreducible factor $g(x)$ of degree less than $\frac{m}{2}$, so for some $j \leq \frac{m}{2}$, $g(x)$ will be a divisor of $x^{p^j} - x$, and consequently $f(x)$ and $x^{p^j} - x$ will have a greatest common divisor not equal to 1. Since $\gcd(f(x), x^{p^j} - x) = \gcd\,(f(x), (x^{p^j} - x) \bmod f(x))$, the arithmetic can be kept within reasonable bounds by working modulo $f(x)$ throughout.

If you need an irreducible polynomial in order to construct a finite field, as we shall do later in this chapter, you may wonder how difficult it is to obtain one. The *Handbook of Applied Cryptography*, offering some reassurance, states:

> [O]ne method for finding an irreducible polynomial of degree m in $\mathbb{Z}_p[x]$ is to generate a random monic polynomial of degree m in $\mathbb{Z}_p[x]$, test it for irreducibility, and continue until an irreducible one is found. ... [T]he expected number of polynomials to be tried before an irreducible one is found is approximately m.

Exercises

1. Prove that $x^2 + 1 \in GF(11)[x]$ is irreducible.
2. Prove that $x^2 + 1 \in GF(13)[x]$ is reducible.
3. The derivative of a polynomial $f(x) = \sum_{i=0}^{n} a_i x^i \in \mathbb{F}[x]$ is defined to be the polynomial $f'(x) = \sum_{i=1}^{n} i a_i x^{i-1}$. Suppose that $f(x)$ and $f'(x)$ are relatively prime. Show that $f(x)$ has no repeated roots.

5.4 Ring Homomorphisms

We recall the following definition: If R and S are rings, a function $\psi : R \longrightarrow S$ is called a (ring-) homomorphism if

- for all $x, y \in R$, $\psi(x + y) = \psi(x) + \psi(y)$;
- for all $x, y \in R$, $\psi(xy) = \psi(x)\psi(y)$.

We already have examples in the evaluation functions ψ_α of Sect. 5.3.3 above. Other examples are the functions that map ring elements onto their equivalence classes modulo some ideal: if R is a ring and I is an ideal in R, then the function

$$r \mapsto r + I$$

is a ring homomorphism. In fact, we shall see in a moment that every ring homomorphism is really a mapping of this type.

A ring homomorphism has been described as a mapping which is compatible with the ring operations. This explains why we included the word *ring* (admittedly, in parentheses) in the name: algebraists also frequently use homomorphisms in the study of other algebraic structures. For example, a group-homomorphism is a mapping, from one group to another, which is compatible with the group operation. (Or rather, with the group operations—plural—, since the operations in the two groups are distinct. The same applies in the case of rings, obviously.) The reader will be familiar with the concept of *linear mappings* between vector spaces over some field $\mathbb{F}$: these are mappings which map sums onto sums and scalar multiples onto scalar multiples: If $\psi : V_1 \longrightarrow V_2$ is such a mapping, then

$$\psi(\mathbf{x} + \mathbf{y}) = \psi(\mathbf{x}) + \psi\mathbf{y}) \quad \forall \ \mathbf{x},\mathbf{y} \in V_1,$$
$$\psi(c\mathbf{x}) = c\psi(\mathbf{x}) \ \ \forall \ c \in \mathbb{F}, \ \forall \mathbf{x} \in V_1.$$

In other words, a linear mapping could really be called a "vector space homomorphism": a mapping which is compatible with the basic vector space operations.

A homomorphism which is one-to-one is called a *monomorphism*, a homomorphism that is onto is an *epimorphism*, and one that is both is called an *isomorphism*. If an isomorphism between two rings R and S exists, we call R and S isomorphic, denoted by $R \cong S$.

It may be convenient to think of an isomorphism as merely a renaming of the elements, or as an alternative description of the same ring. Thus, by way of example, we could regard the rings $\mathbb{Z}_{12}$ and $\mathbb{Z}_4 \times \mathbb{Z}_3$ as one and the same since the function ψ defined by $\psi(a+ < 12 >) = (a+ < 4 >, a+ < 3 >)$ is an isomorphism. (We leave the detailed verification of this fact to the reader as an exercise.)

In fact, the Chinese Remainder Theorem may be interpreted as a statement of the fact that if $n = n_1 n_2 \ldots n_k$ where the n_i are relatively prime integers, then

$$\mathbb{Z}_n \cong \mathbb{Z}_{n_1} \times \mathbb{Z}_{n_2} \times \cdots \times \mathbb{Z}_{n_k}.$$

In essence we used this isomorphism when we proved that Euler's ϕ is multiplicative, in Sect. 2.8.3.

Now let $\psi : R \longrightarrow S$ be a ring homomorphism. We define the *kernel* $\mathrm{Ker}(\psi)$ to be the set $\{x \in R | \psi(x) = 0\}$. In a previous exercise the reader showed that $K = \mathrm{Ker}(\psi)$ is an ideal in the ring R. It makes sense to define a mapping $\overline{\psi} : R/K \longrightarrow S$ by

$$\overline{\psi}(r + K) = \psi(r),$$

where by it "making sense" we mean that it is a well-defined function from one set to the other. Moreover: it is a monomorphism.

For it is, firstly, not hard to see that a ring homomorphism ϕ is a monomorphism if and only if its kernel is the set $\{0\}$. (In a standard[9] abuse of notation we shall write $\text{Ker}(\phi) = 0$.) For suppose ϕ is a monomorphism. Since $\phi(0) = 0$, and only one element can be mapped onto 0, we must have that $\text{Ker}(\phi) = 0$. Conversely, if $\text{Ker}(\phi) = 0$, then $\phi(a) = \phi(b)$ implies $\phi(a - b) = 0$, which implies $a - b = 0$, i.e. $a = b$.

Now, to return to the mapping $\overline{\psi}$, we have: $\overline{\psi}(r + K) = 0$ if and only if $\psi(r) = 0$ if and only if $r \in \text{Ker}(\psi)$ if and only if $r + K = 0 + K$.

In particular, of course, if $\psi : R \longrightarrow S$ is an epimorphism, then $\overline{\psi}$ is an isomorphism, so that $S \cong R/\text{Ker}(\psi)$. This result is sometimes called the *Fundamental Homomorphism Theorem*.[10]

Exercises

1. Let R be a ring, and let I, J be ideals in R with $J \subset I$. Show that
   (a) I/J is an ideal in the ring R/J;
   (b) $(R/J)/(I/J) \cong R/I$.

2. An ideal M in a ring R is called *maximal* if whenever N is an ideal such that if $M \subseteq N \subseteq R$, then either $N = M$ or $N = R$. Prove that M is a maximal ideal if and only if R/M is a field.[11]

   Hints:
   If M is maximal, and $a \notin M$ then the ideal $\{x = m + ra | m \in M, r \in R\} = R$. And if R/M is a field and N is an ideal with $M \subset N \subseteq R$, what can you say about the ideal N/M of R/M?
3. Let $f(x) \in \mathbb{F}[x]$, where $\mathbb{F}$ is a field, be an irreducible polynomial. Show that the ideal generated by $f(x)$ is maximal in $\mathbb{F}[x]$.

We shall exploit the consequences of the last two exercises in the final section of this chapter, and many times thereafter.

[9] Meaning: everyone does it.

[10] Results like this are not restricted to rings, but are common throughout algebra. If your algebraic structure is defined as a set with some operations defined on it, and you consider only functions which behave themselves with respect to these operations, you will get the corresponding homomorphism theorem. An example of this kind is the well-loved theorem in linear algebra: If T is a linear mapping from vector space V_1 to vector space V_2, then the dimensions satisfy the equation $\dim(\text{Ker}(T)) + \dim(\text{Image}(T)) = \dim(V_1)$, because $V_1/\text{Ker}(T)$ is isomorphic to $\text{Im}(T)$. Recall, by the way, that the dimension of the kernel of T is called the *nullity* of T. A proper discussion of these things would lead us too far from our main route, which remains tying up algebra and cryptology, so we abandon it, with some regret.

[11] The reader with previous experience of algebra is requested to recall that all our rings are commutative.

5.5 The Chinese Remainder Theorem Again

We have said it before, but we'll nevertheless say it once more: The role of irreducible polynomials in a ring $\mathbb{F}[x]$ of polynomials over a field $\mathbb{F}$ is very similar to the role of prime numbers in the ring $\mathbb{Z}$ of integers. This goes as far as the Chinese Remainder Theorem: the CRT for integers has an exact analogue in the ring $\mathbb{F}[x]$.

Theorem *Let $\mathbb{F}$ be a field and let $e_i(x) \in \mathbb{F}[x]$ $(i = 1, \ldots, n)$ be distinct irreducible polynomials. If $a_1(x), \ldots, a_n(x)$ are any elements of $\mathbb{F}[x]$, then there exists a polynomial $f(x) \in \mathbb{F}[x]$ such that, for all $i = 1, \ldots n$*

$$f(x) \equiv a_i(x) \mod e_i(x).$$

Moreover, $f(x)$ is uniquely determined modulo $\prod_{i=1}^{n} e_i(x)$.

We shall omit the proof of this theorem, since it is the exact analogue of the number-theoretic version of the CRT.

As in that case, we can use the CRT to establish a ring isomorphism. In the notation of the theorem, we define an isomorphism

$$\overline{\psi} : \mathbb{F}[x]/I \longrightarrow \mathbb{F}[x]/E_1 \times \mathbb{F}[x]/E_2 \times \cdots \times \mathbb{F}[x]/E_n$$

where I is the principal ideal generated by $\prod_{i=1}^{n} e_i(x)$, and (for each i) E_i is the ideal generated by $e_i(x)$. Define ψ by

$$\psi(f(x)) = (f(x) + E_1, f(x) + E_2, \ldots, f(x) + E_n).$$

This is a homomorphism and the CRT tells us that it is an epimorphism. Its kernel is, as one easily sees, the ideal $I =< \prod_{i=1}^{n} e_i(x) >$, so that $\overline{\psi}$ is indeed an isomorphism, by the Fundamental Homomorphism Theorem.

5.6 Construction of Finite Fields

Yet again we repeat it: irreducible polynomials play the same role that prime numbers play in the ring of integers, or, more precisely, the principal ideals generated by irreducible polynomials play in $\mathbb{F}[x]$ the same role that ideals generated by primes play in the ring $\mathbb{Z}$. If we take a prime p in $\mathbb{Z}$, the ring $\mathbb{Z}$ modulo the ideal $< p >$ is a field; similarly if we have a field $\mathbb{F}$ and we take the ring $\mathbb{F}[x]$ modulo the ideal $< p(x) >$ which is generated by the irreducible polynomial $p(x)$ we get a field. This will be the main result of this section; this method is the way in which finite fields, other than the prime fields $GF(p)$, are constructed.

Theorem *Let $p(x) \in \mathbb{F}[x]$ be an irreducible polynomial. If a product $r(x)s(x) \in \mathbb{F}[x]$ is divisible by $p(x)$ then $r(x)$ or $s(x)$ must be divisible by $p(x)$.*

Proof The proof is an exact analogue of the corresponding proof for divisibility by prime numbers. If $r(x)$ is not divisible by $p(x)$ then the greatest common divisor of $r(x)$ and $p(x)$ must be 1, since $p(x)$ has no divisors other than 1 and itself.[12] Hence there exist $a(x), b(x) \in \mathbb{F}[x]$ such that

$$a(x)p(x) + b(x)r(x) = 1$$

and therefore

$$a(x)s(x)p(x) + b(x)r(x)s(x) = s(x).$$

Since both terms on the left are divisible by $p(x)$, the term on the right must also be divisible by $p(x)$.

This leads to the following analogue of the fundamental theorem of arithmetic.

Theorem *Let $\mathbb{F}$ be a field. Any monic polynomial $f(x) \in \mathbb{F}[x]$ has a factorization*

$$f(x) = \prod_{i=0}^{n} (p_i(x))^{e_i},$$

where the $p_i(x) \in \mathbb{F}[x]$ are monic and irreducible, and the $e_i > 0$ are integers. This factorization is unique, apart from the order in which we write the factors.

Again, we shall omit the (dull and laborious) proof of this theorem.

Theorem *Let $\mathbb{F}$ be a field and let $p(x) \in \mathbb{F}[x]$ be irreducible over $\mathbb{F}$. If $a(x) \in \mathbb{F}[x]$ is not divisible by $p(x)$, then there exists a polynomial $b(x) \in \mathbb{F}[x]$ such that $a(x)b(x) - 1$ is divisible by $p(x)$, i.e.*

$$a(x)b(x) \equiv 1 \bmod\ < p(x) > .$$

Thus if $a(x)+ < p(x) > \neq 0+ < p(x) >$ then $a(x)+ < p(x) >$ has a multiplicative inverse in $\mathbb{F}[x]/ < p(x) >$.

Proof Again, the proof is analogous to the one we had for primes in the ring of integers: If $a(x) \notin < p(x) >$, then $\gcd(a(x), p(x)) = 1$, so there exist polynomials $b(x), s(x) \in \mathbb{F}[x]$ such that

$$a(x)b(x) + s(x)p(x) = 1$$

[12] Strictly speaking, one should say: no divisors other than nonzero constants and nonzero constant multiples of itself.

so that

$$a(x)b(x) - 1 = -s(x)p(x) \in < p(x) >$$

or

$$(a(x)+ < p(x) >) \cdot (b(x)+ < p(x) >) = 1+ < p(x) > .$$

Theorem *Let $\mathbb{F}$ be a field, and let $p(x) \in \mathbb{F}[x]$ be a polynomial of degree n which is irreducible over $\mathbb{F}$. Let $< p(x) >$ be the principal ideal generated by $p(x)$. Then $\mathbb{F}[x]/ < p(x) >$ is a field.*

Proof Actually, all the required work has been done:

We know that $\mathbb{F}[x]/ < p(x) >$ is a ring, and the only other requirement to make it a field is that every nonzero element must have a multiplicative inverse. And that's what we have just done.

We conclude with what we hope is an illuminating

Example Consider the field $GF(2)$, and the polynomial $x^4+x+1 \in GF(2)[x]$. It can be shown that this polynomial is irreducible, but we won't bother to do so (hoping that you'll take our word for it, or, even better, that you will try out the algorithm testing for irreducibility given in Sect. 5.3.4). Let I denote the principal ideal generated by $x^4 + x + 1$, and let us consider what the elements of $GF(2)[x]/I$ look like.

First of all, note that every element is of the form

$$a_3x^3 + a_2x^2 + a_1x + a_0 \ + I$$

where all the a_i belong to $GF(2)$ (i.e. they are either 0 or 1). This is because if we have any polynomial $a(x) \in GF(2)[x]$, we can divide $a(x)$ by $x^4 + x + 1$ until we get a remainder $r(x)$ of degree 3 or less, and $a(x) + I = r(x) + I$. Since there are $16 = 2^4$ possible choices for the a_i, we should be getting a field with 2^4 elements. Two of these are obviously going to be $0(= 0 + I)$ and $1(= 1 + I)$. What about the others?

One of these is clearly $x+I$. Let us call this α. We must also have the elements x^2+I, x^3+I and so on. But note that $x^2 + I = (x + I)(x + I) = \alpha \cdot \alpha = \alpha^2$ and $x^3 + I = \alpha^3$. What about $x^4 + I$? $x^4 + I = -(x + 1) + I = (-x + I) - (1 + I)$, because $x^4 + x + 1 \in I$, so we get $\alpha^4 = \alpha + 1$. (We have dropped the minus sign, because in $GF(2)$, $-1 = 1$, which makes arithmetic a lot easier!). Continuing in this way we get the following listing of *all* the elements of our field (the meaning of the last column will be made clear presently):

0				0	0000
1				1	0001
α			α		0010
α^2		α^2			0100
α^3	α^3				1000
α^4			α	$+1$	0011
α^5		α^2	$+\alpha$		0110
α^6	α^3	$+\alpha^2$			1100
α^7	α^3		$+\alpha$	$+1$	1011
α^8		α^2		$+1$	0101
α^9	α^3		$+\alpha$		1010
α^{10}		α^2	$+\alpha$	$+1$	0111
α^{11}	α^3	$+\alpha^2$	$+\alpha$		1110
α^{12}	α^3	$+\alpha^2$	$+\alpha$	$+1$	1111
α^{13}	α^3	$+\alpha^2$		$+1$	1101
α^{14}	α^3			$+1$	1001

If we were to add one more row to this, we would get $\alpha^{15} = 1$, so we would, in fact, start all over again. This is not surprising, since the multiplicative group of (the nonzero elements of) this field has order $16 - 1 = 15$, so every element in this group would be a root of the equation $x^{15} = 1$. Talking of which: what we have constructed is a field in which the equation $z^4 + z + 1 = 0$ (the left-hand side of which was irreducible over $GF(2)$) has a solution, viz $z = \alpha$, so $z - \alpha$ is now a factor of $z^4 + z + 1$.

In fact: the equation $z^4 + z + 1 = 0$ has four solutions. This follows from the following

Observation If ζ is a root of a polynomial $\sum_{i=0}^{n} a_i z^i \in GF(2)[z]$, then so is ζ^2. More generally, if ζ is a root of a polynomial in $GF(p)[z]$, then so is ζ^p. To prove this is easy, as it follows from Fermat's Little Theorem and the Freshman Theorem: If

$$\sum_{i=0}^{n} a_i \zeta^i = 0$$

then

$$\left(\sum_{i=0}^{n} a_i \zeta^i \right)^p = 0$$

so

$$\sum_{i=0}^{n} a_i^p \zeta^{pi} = 0.$$

But since the a_i belong to $GF(p)$, $a_i^p = a_i$, and we have

$$\sum_{i=0}^{n} a_i(\zeta^p)^i = 0.$$

Returning to our example: Since α is a root of $z^4 + z + 1$, so are α^2, α^4 and α^8. Hence we have in our field that

$$x^4 + x + 1 = (x + \alpha)(x + \alpha^2)(x + \alpha^4)(x + \alpha^8).$$

In order to familiarise himself/herself with the arithmetic in our field, the reader is invited to multiply out the right-hand side of this equation.

It is easy to see how addition of elements takes place: they are, after all just linear combinations (with coefficients from $GF(2)$) of $1, \alpha, \alpha^2$ and α^3, and addition takes place "componentwise". Thus, for example $\alpha^6 + \alpha^{14} = \alpha^2 + 1$. In fact, the field of our example, which we may as well call $GF(16)$ or $GF(2^4)$, is, as far as addition is concerned, a *vector space* of dimension 4 over the field $GF(2)$. In our example, we have written everything in terms of the *basis* $\{\alpha^3, \alpha^2, \alpha, 1\}$: these four elements are linearly independent and span the whole "space" $GF(2^4)$. The last column in the table above gives the coefficients of each element in terms of that basis.

Of course, checking for linear independence, we might, somewhat arbitrarily, have selected the basis $\{\alpha^7, \alpha^{10}, \alpha^{11}, \alpha^{12}\}$, even if there does not appear any good reason for that choice. An interesting (and possibly useful) choice of basis is $\{\beta_3, \beta_2, \beta_1, \beta_0\} = \{\alpha^9, \alpha^{12}, \alpha^6, \alpha^3\}$ (in that order), where each "basis vector" is the square of the next one. With respect to this basis, squaring an element $\gamma = \sum_{i=0}^{3} a_i\beta_i$ merely amounts to a rotation of the coefficients: $\gamma^2 = a_2\beta_3 + a_1\beta_2 + a_0\beta_1 + a_3\beta_0$. Such a basis, called a *normal basis*, may be useful in some implementations, since it makes squaring an element an extremely simple and fast operation. It is also useful when attempting to solve quadratic equations in fields of characteristic 2 as we shall discuss in Sect. 6.3.

Our observation that an extension of a prime field is (among other things) a vector space over that prime field, leads to another useful fact:

If $\mathbb{F}$ is a finite field, then it contains a prime field $GF(p)$ for some prime p, and therefore is a vector space of some dimension, n say, over that field. This means that every element of $\mathbb{F}$ is of the form $\sum_{i=0}^{n-1} a_i\underline{e}_i$, where the $\underline{e}_i$, $0 \leq i \leq n-1$, are basis vectors, and the a_i are coefficients chosen from $GF(p)$. Thus there are p^n elements in $\mathbb{F}$. This gives us, as we have previously noted, the important

Theorem *Every finite field has p^n elements, for some prime number p and for some positive integer n.*

We now suspect that the n appearing in $GF(p^n)$ may be related to the degree of some polynomial, irreducible over $GF(p)$, which now acquires roots in the extension field.

In the important case where $p = 2$, we can make another interesting observation:

Theorem *In a field of characteristic 2, every element has a unique square root.*

Note that this is quite strange: if the field has characteristic $p > 2$, then only half the elements have square roots, and if a nonzero element of $GF(p)$ (or of an extension field of $GF(p)$) is a square, then it has two square roots. We leave the proof of this as an exercise.

Proof of Theorem If $\mathbb{F}$ has characteristic 2, then the multiplicative group of nonzero elements of F has order $2^n - 1$, so that, for every nonzero $\gamma \in \mathbb{F}$, $\gamma^{2^n-1} = 1$, or $\gamma^{2^n} = \gamma$. Hence if $\zeta = \gamma^{2^{n-1}}$, then $\zeta^2 = \gamma$.

One further observation is worth noting: In our example, we see that the multiplicative group of the nonzero elements of $GF(2^4)$ is a *cyclic* group, generated by α in fact. This is not a coincidence: we shall prove in the next chapter that the multiplicative group of any finite field is cyclic.

Exercises

1. Show that the polynomial $x^4 + x + 1 \in GF(2^2)[x]$ is the product of two quadratic polynomials. (The field $GF(2^2)$ is listed at the end of Sect. 4.1.)
2. Let $\mathbb{F}$ be a field. If $\alpha \in \mathbb{F}$ is a root of the polynomial $f(x) = \sum_{i=0}^{n} a_i x^i \in \mathbb{F}[x]$, show that α^{-1} is a root of $\sum_{i=0}^{n} a_{n-i} x^i$. (Note for future reference that the second polynomial can be written as $x^n f(\frac{1}{x}) = x^n \sum_{i=0}^{n} a_i \frac{1}{x^i}$.)
3. With the representation of $GF(2^4)$ as in the example of this section:
   (a) Compute $(\alpha^4 + \alpha^7)(\alpha^3 + \alpha^{11})$.
   (b) Find the square roots of α^{11} and of α^5.
   (c) α was defined as a root of $x^4 + x + 1$. What equation does α^{14} satisfy?
   (d) Find a quadratic equation with coefficients in $GF(2)$ of which α^5 is a root.
   (e) Show that $\{0, 1, \alpha^5, \alpha^{10}\}$ is a subfield of $GF(2^4)$. Does this field look familiar?
   (f) How many generators does the multiplicative group of $GF(2^4)$ have? What are they?
4. What is the dimension of ${}_{GF(2^2)}GF(2^4)$?
5. Show that $x^3+x+1 \in GF(2)[x]$ is irreducible and hence construct a field with 8 elements. Does this field contain a copy of $GF(2^2)$?
6. Using your construction of $GF(2^3)$, find another irreducible polynomial of degree 3 in $GF(2)[x]$.
   Hint: Consider Exercise 2.
7. Let $\mathbb{F}$ be a field of characteristic $p > 2$. Show that if $\zeta \in \mathbb{F}$ is a square, then it has exactly two square roots.

6

Properties of Finite Fields

In this chapter we shall use the construction of the previous chapter, and discuss some of the most important properties of finite fields. While we shall generally discuss our results in terms of fields of characteristic p, in our examples we shall concentrate on the case $p = 2$, which is, for many applications, and especially those in secret key cryptography, the most important case. For example, in the case of *Rijndael*, the Advanced Encryption Standard, most of the computations take place in the field $GF(2^8)$.[1]

However, in order to keep them simple, we shall take many of our examples from the field $GF(2^4)$, as we defined it in the last section of the previous chapter.

6.1 The Minimal Polynomial of an Element

We have observed that the finite field $GF(p^n)$ has (among other things) the structure of a vector space of dimension n over the prime field $GF(p)$. This means, of course, that any set $\{\underline{e}_0, \underline{e}_1, \ldots, \underline{e}_n\}$ of $n+1$ elements from the space $GF(p^n)$ is linearly dependent, i.e. that there exist "scalars" $a_i \in GF(p)$ such that $\sum_{i=0}^{n} a_i \underline{e}_i = 0$.

In particular, it means that for any $\beta \in GF(p^n)$, the set $\{1, \beta, \beta^2, \ldots, \beta^n\}$ is linearly dependent, i.e. there exist $a_i \in GF(p)$ such that

$$\sum_{i=0}^{n} a_i \beta^i = 0$$

[1]This is a suitable choice, since this means that field elements can be represented by strings of eight bits, i.e. bytes, which fits in well with most computer architectures. Other choices could have been 16-bit or 32-bit words, in which case the fields $GF(2^{16})$ or $GF(2^{32})$ would have been used. One of the operations used in the cipher is inversion, i.e. the mapping $\chi \mapsto \chi^{-1}$, which is most efficiently handled through the use of a look-up table. In the block cipher *Rijndael* this table has $2^8 = 256$ entries of 8 bits each, for a total of 2048 bits. If a larger field, say $GF(2^{16})$, had been used, the table would have needed $2^{16} = 65536$ entries of 16 bits, giving a total of 1 048 576 bits, which might have required too much storage space in some low-end implementations of the cipher.

A.R. Meijer, *Algebra for Cryptologists*, Springer Undergraduate Texts in Mathematics and Technology,
DOI 10.1007/978-3-319-30396-3_6

or, if we define $a(x) = \sum_{i=0}^{n} a_i x^i \in GF(p)[x]$, then

$$a(\beta) = 0.$$

The obnoxious reader of Sect. 5.3.3 would no doubt object to this notation, and would insist that we do the "evaluation" properly, and write (in the notation of that section) "$\psi_\beta(a(x)) = 0$". We shall now ask him to go away and to stop splitting hairs, so that we can return to the important stuff.[2]

We may call such a polynomial an *annihilator* of β. It is almost trivial to confirm that the set of all annihilators of β forms an ideal in the ring $GF(p)[x]$, and since any ideal in this ring is a principal ideal, there must exist a polynomial $m(x) \in GF(p)[x]$ such that

- $m(\beta) = 0$;
- if $a(\beta) = 0$, then $a(x)$ is a multiple of $m(x)$, i.e $m(x)|a(x)$;
- $m(x)$ is irreducible, because if $m(x) = b(x) \cdot c(x)$ is a non-trivial factorisation, then either $b(\beta) = 0$ or $c(\beta) = 0$, contradicting the fact that every annihilator is a multiple of $m(x)$.

Such a polynomial is called a *minimal polynomial of* β. For convenience we shall always insist that our minimal polynomials are monic, which will ensure that they are unique.

Two observations may be made here:

Firstly, since the multiplicative group of nonzero elements of $GF(p^n)$ has order $p^n - 1$, we know from our group theory that $\beta^{p^n-1} = 1$ for every nonzero $\beta \in GF(p^n)$, or, what amounts to the same thing:

$$\beta^{p^n} = \beta \;\; \forall \beta \in GF(p^n)$$

and hence

The minimal polynomial of any element of $GF(p^n)$ *is a divisor of* $x^{p^n} - x$.

Secondly, if $a(x)$ is an annihilator of $\beta \in GF(p^n)$, then, by using the Freshman Theorem, $a(x)$ is also an annihilator of β^p, and therefore also of $\beta^{p^2}, \ldots, \beta^{p^{n-1}}$. In particular, this means that, if $m(x)$ is the minimal polynomial of β, then all of $\beta^p, \beta^{p^2}, \beta^{p^3}, \ldots, \beta^{p^{n-1}}$ will also be roots of $a(x)$, so $a(x)$ is a multiple of the annihilator of β^{p^j} for all $j \in \{0, 1, \ldots, n-1\}$. But we know that the annihilator of β^{p^j} is irreducible, so it must be $m(x)$ itself. In other words:

$$\beta, \beta^p, \beta^{p^2}, \ldots \textit{ all have the same minimal polynomial.}$$

This prompts us to look at the polynomial

$$\chi(x) = \prod_{j=0}^{n-1} (x - \beta^{p^j}),$$

which has all of $\beta, \beta^p, \ldots, \beta^{p^{n-1}}$ as roots.

[2]It has been suggested that I am being unkind to this individual, and that I should rather refer to him as "alert". At least he does seem to be paying attention, and he deserves credit for that.

$\chi(x)$ has an interesting property, viz that its coefficients are the so-called (elementary) "symmetric functions" in the β^{p^i}. If we write $\chi(x) = \sum_{i=0}^{n} a_i x^i$, then (ignoring the + or - signs)

$$a_n = 1,$$

$$a_{n-1} = \beta + \beta^p + \beta^{p^2} + \ldots \beta^{p^{n-1}},$$

$$a_{n-2} = \beta\beta^p + \beta\beta^{p^2} + \ldots + \beta\beta^{p^{n-1}} + \ldots + \beta^{p^{n-2}}\beta^{p^{n-1}},$$

$$\ldots$$

$$a_0 = \beta\beta^p \ldots \beta^{p^{n-1}}.$$

One can easily convince oneself that a_{n-i} is the sum of all possible products of the elements of the set $\{\beta, \beta^p, \beta^{p^2}, \ldots, \beta^{p^{n-1}}\}$ taken i at a time, so a_{n-i} is the sum of $\binom{n}{i} = \frac{n!}{i!(n-i)!}$ terms. Each of these coefficients a_i has the property that $a_i^p = a_i$. We shall later show that the set of elements with the property that they are equal to their own pth powers is precisely the set $GF(p)$. Thus, the polynomial $\chi(x)$ is in fact an element of $GF(p)[x]$. Since it annihilates β and all the $\beta^{p^j}, 0 \leq j \leq n-1$, it must therefore be a multiple of the minimal polynomial of β. In fact it might actually be that minimal polynomial (though it need not be).

Let us apply these ideas to the field $GF(2^4)$ as we defined it in Sect. 5.6:

Example The minimal polynomial of α is a divisor of $(x-\alpha)(x-\alpha^2)(x-\alpha^4)(x-\alpha^8)$. If we multiply these terms out we get the polynomial $x^4 + x + 1$. This should not come as a surprise, really, since this is how we chose α in the first place.

Doing the same for α^3, we have the product $(x-\alpha^3)(x-\alpha^6)(x-\alpha^{12})(x-\alpha^{24}) = (x-\alpha^3)(x-\alpha^6)(x-\alpha^{12})(x-\alpha^9) = x^4 + x^3 + x^2 + x + 1$.

We could do the same for α^7, but we note that $\alpha^{14} = \alpha^{-1}$ has the same minimal polynomial, and use the fact (proven in Exercise 2 of Sect. 5.6) that if $m(x)$ of degree n annihilates α, then $x^n m(x^{-1})$ annihilates α^{-1}. Hence α^{-1} has minimal polynomial $x^4 + x^3 + 1$.

This takes care of the minimal polynomials of 12 of the 16 elements. Of course, 0 has minimal polynomial x and 1 has minimal polynomial $x + 1$.

Finally, if we apply the same trick to α^5, we get $(x-\alpha^5)(x-\alpha^{10})(x-\alpha^{20})(x-\alpha^{40}) = ((x-\alpha^5)(x-\alpha^{10}))^2 = (x^2 + x + 1)^2$, so we conclude that the minimal polynomial of α^5 is $x^2 + x + 1$.

In summary we have:

Element	Minimal polynomial	Trace
0	x	0
1	$x + 1$	0
$\alpha, \alpha^2, \alpha^4, \alpha^8$	$x^4 + x + 1$	0
α^5, α^{10}	$x^2 + x + 1$	0
$\alpha^3, \alpha^6, \alpha^{12}, \alpha^9$	$x^4 + x^3 + x^2 + x + 1$	1
$\alpha^7, \alpha^{14}, \alpha^{13}, \alpha^{11}$	$x^4 + x^3 + 1$	1

The meaning of the last column will be explained in Sect. 6.3. For the moment note that

- All the polynomials in the middle column are irreducible, and every element of $GF(2^4)$ is a root of one of them, so that their degrees add up to 16. On the other hand every element of $GF(2^4)$ is a root of the polynomial $x^{2^4} + x$, which is also of degree 16. It follows that $x^{2^4} + x$ is precisely the product of all the polynomials in the middle column:

$$x^{2^4} + x = x(x+1)(x^4+x+1)(x^2+x+1)$$
$$(x^4+x^3+1)(x^4+x^3+x^2+x+1).$$

- If one were to try constructing the field $GF(2^4)$ using the irreducible polynomial $g(x) = x^4+x^3+x^2+x+1$ and proceeding as before, taking $\beta = x + GF(2)[x]/ < g(x) >$, we find that we don't get all of $GF(2^4)$, but only a cyclic group of order 5:

$$\beta = x+ < g(x) >,$$
$$\beta^2 = x^2+ < g(x) >,$$
$$\beta^3 = x^3+ < g(x) >,$$
$$\beta^4 = x^3+x^2+x+1+ < g(x) >,$$
$$\beta^5 = 1+ < g(x) >= 1.$$

This again should not surprise us: In our previous notation $GF(2^4)\backslash\{0\}$ is the cyclic group of order 15 generated by α. But $\beta = \alpha^3$, and 3 is not relatively prime to 15, so β cannot generate the whole group, but only a subgroup of order $15/\gcd(15,3) = 5$.

This shows the distinction which has to be made between irreducible and *primitive* polynomials.

Definition If $m(x) \in GF(p)[x]$ is an irreducible polynomial such that the coset $\alpha = x + GF(p)[x]/ < m(x) >$ generates the entire multiplicative group of nonzero elements of the field $GF(p)[x]/ < m(x) >$, then $m(x)$ is called a *primitive* polynomial.

Thus a primitive polynomial is irreducible but not every irreducible polynomial is primitive. The polynomial $x^4+x^3+x^2+x+1$ above provides the necessary counterexample.

The reader may ask whether there are reasons why a primitive, rather than just an irreducible, polynomial should be used, when defining an extension field of $GF(p)$. The answer to this question depends on the use that is to be made of the field. As we shall show in the following section, all extensions of a prime field $GF(p)$ by means of an irreducible nth degree polynomial lead to essentially the same field. So if one merely requires a polynomial in order to obtain a basis for calculations in that field, then there is little reason to insist on a primitive polynomial (as long as everyone you deal with knows where the basis came from, or, in other words, knows the polynomial you are using).

The situation is different, however, if you use the polynomial in order to generate a pseudorandom sequence, because an irreducible polynomial which is not primitive may lead to a sequence with a periodicity much less than the maximum of $p^n - 1$ which can be obtained by using a primitive polynomial. We discuss pseudorandom generators, or rather pseudorandom keystream generators, which are not quite the same thing, in the next chapter.

Using a primitive polynomial has the further advantage that one can use a primitive element as basis for a *discrete logarithm*: If α is a primitive element of $GF(p^n)$, then every nonzero element $\chi \in GF(p^n)$ can be written as $\chi = \alpha^k$ and we define $\log_\alpha(\chi) = k$. Multiplication of nonzero elements can then be performed by addition of the logarithms (modulo $p^n - 1$), almost like traditional logarithms.

Exercises

1. Show that if $GF(p^m)$ is a subfield of $GF(p^n)$, then $m|n$. (The converse is also true, as we shall show in the next section.)
2. In Exercise 6 of Sect. 5.6 you constructed the field $GF(2^3)$.
   (a) Find the minimal polynomial of each of its elements.
   (b) Which of these polynomials is or are primitive?
   (c) Factorize $x^8 + x \in GF(2)[x]$ into irreducible factors.
3. Show that every irreducible polynomial in $GF(2)[x]$ of degree 7 is primitive.
4. Using the fact that $x^{2^n} - x$ is the product of all irreducible polynomials in $GF(2)[x]$ of degree d with $d|n$ (with the exception of n itself, of course), prove the following:
   If $\psi(n)$ denotes the number of irreducible polynomials of degree n, then

$$\psi(n) = \frac{1}{n}\left(2^n - \sum_{d|n, d \neq n} d \cdot \psi(d)\right).$$

   Using the fact that $\psi(1) = 2$ (since x and $x + 1$ are the only two irreducible polynomials of degree 1), find $\psi(2), \psi(3), \psi(4), \psi(8)$ and $\psi(p)$, where p is a prime.

6.2 More on the Structure of Finite Fields

We have, more than once, hinted that the multiplicative group of nonzero elements of a finite field is cyclic. It is now high time that we actually proved this result:

Theorem *Let $\mathbb{F}$ be a finite field, $\#(\mathbb{F}) = p^n$, for some prime p and positive integer n. Then the group $\{\mathbb{F}\backslash\{0\}, \cdot\}$ is a cyclic group.*

Proof We know (by definition) that $\{\mathbb{F}\backslash\{0\}, \cdot\}$ is a finite group, so every element has finite order. Let r be the maximum of the orders of the elements, and let α be an element of that order. Clearly, since $r|p^n - 1$, we must have that $r \leq p^n - 1$.

We shall firstly show that if l is the order of any element β in the group, then $l|r$.

For let π be any prime, and write $r = \pi^a r', l = \pi^b l'$, where r' and l' are not divisible by π. Then α^{π^a} has order r', $\beta^{l'}$ has order π^b, and (since these two orders are relatively prime, and using the result of Exercise 1 of Sect. 3.5) $\alpha^{\pi^a}\beta^{l'}$ has order $\pi^b r'$. Hence $b \leq a$, or else r would not be maximal. This is true for every prime π, and it follows that $l|r$.

Hence every $\beta \in F\backslash\{0\}$ satisfies the equation $x^r - 1 = 0$, so that $\prod_{\beta\in F\backslash\{0\}}(x-\beta)$ is a factor of $x^r - 1$. Since the product has degree $p^n - 1$, we must have that $r \geq p^n - 1$. Hence $r = p^n - 1$, and α is a generator of the group.

Definition If α is a generator of the group $\{F\backslash\{0\}, \cdot\}$, where $\mathbb{F}$ is some finite field, then we call α a *primitive element* of $\mathbb{F}$.

A primitive polynomial is therefore the minimal polynomial of a primitive element.

Theorem *Any two finite fields with the same number of elements are isomorphic.*

Here "isomorphism" refers to isomorphism in the ring-theoretic sense: after all, a field is just a special kind of ring. In actual fact, if one has a homomorphism from one field, $\mathbb{F}_1$, say, to another, $\mathbb{F}_2$, then this mapping is either a monomorphism (i.e. one-to-one) or the function that maps everything onto the 0 element of F_2. This follows from the isomorphism theorem and from the fact that a field only has the two trivial ideals, and no others. In the first case, the non-trivial one, it follows from the fact that F_1 and F_2 have the same number of elements that a monomorphism must be an epimorphism (i.e. onto) so it is an isomorphism.[3]

Proof of Theorem Let $\mathbb{F}$ and $\mathbb{G}$ be fields, with $\#(\mathbb{F}) = \#(\mathbb{G}) = p^n$. Let α be a primitive element of $\mathbb{F}$, with minimal polynomial $m(x)$. Since $m(x)|x^{p^n} - x$, and since every element of $\mathbb{G}$ satisfies $x^{p^n} - x = 0$, we have, by the uniqueness of factorization, that there exists an element $\beta \in \mathbb{G}$ which also has minimal polynomial $m(x)$. Define $\psi : \mathbb{F} \longrightarrow \mathbb{G}$, by $\psi(\alpha) = \beta$ or, more precisely, by

$$\psi : \mathbb{F} \longrightarrow \mathbb{G} : \sum_{i=0}^{n-1} a_i\alpha^i \mapsto \sum_{i=0}^{n-1} a_i\beta^i,$$

where the $a_i \in GF(p)$.

Example Let $\mathbb{F}$ be the field $GF(2^4)$ as we defined it in Sect. 5.6, and let $\mathbb{G}$ be the field defined by means of the primitive polynomial $x^4 + x^3 + 1$. With the notation of the example of Sect. 5.6, α is a primitive element of $\mathbb{F}$ and the primitive element β of $\mathbb{G}$ corresponds to our α^7 (or α^{14}, α^{13} or α^{11}, for that matter).

Hence the mapping $\psi : \mathbb{G} \longrightarrow \mathbb{F}$ defined by $\psi(\beta) = \alpha^7$ (its inverse is $\psi^{-1} : \mathbb{F} \longrightarrow \mathbb{G} : \psi^{-1}(\alpha) = \beta^{13}$) is the required isomorphism.

The above theorem justifies our use of the term *the* field $GF(p^n)$, because essentially there is only one such field; all that may differ is the names we give to the elements.

---

[3] Remember Exercise 5 in Sect. 1.4?

Theorem *If $\mathbb{F}$ is a finite field, $\#(\mathbb{F}) = p^n$, then $\mathbb{F}$ contains a subfield with p^m elements, whenever $m|n$.*

Proof If $m|n$, then $p^m - 1|p^n - 1$. (Why?) Consider the set

$$\{\beta|\beta \in \mathbb{F} \text{ and } \beta^{p^m} = \beta\},$$

which is the set of all elements of order $p^m - 1$ in $\mathbb{F}$, or whose order is a divisor of $p^m - 1$, together with the element 0. This is easily seen to be a subfield of $\mathbb{F}$, using the Freshman Theorem to prove that the set is closed under addition. If α is a primitive element of $\mathbb{F}$, then α^r, where $r = \frac{p^n-1}{p^m-1}$, generates the cyclic group of nonzero elements in the subfield, so the subfield contains $p^m - 1$ elements.

It is also easy to see that in this case the subfield $GF(p^m)$ is precisely the set

$$\{\beta \in GF(p^n)|\beta^{p^m} = \beta\}$$

or, in other words, precisely the set of roots of the polynomial $x^{p^m} - x$ in $GF(p^n)$. After all, all these roots lie in $GF(p^n)$ and there are precisely p^m of them. (As an example, consider the set $\{0, 1, \alpha^5, \alpha^{10}\}$ in our representation of $GF(2^4)$; this is (isomorphic to) $GF(2^2)$.)

We know that any field $\mathbb{F}$ of characteristic p contains $GF(p)$ as a subfield. What we have now also proved is that that subfield is precisely the set of elements $\{\zeta \in \mathbb{F}|\zeta^p = \zeta\}$.

Finally, our results show that a polynomial $x^{p^m-1} - 1$ is precisely the product of all irreducible polynomials whose degree is a divisor of $p^m - 1$. For example, we have seen that if $p = 2$, then

$$x^{15} - 1 = (x + 1)(x^2 + x + 1)(x^4 + x + 1)(x^4 + x^3 + 1)(x^4 + x^3 + x^2 + x + 1)$$

and

$$x^3 - 1 = (x + 1)(x^2 + x + 1).$$

Exercises

1. If $\psi : F \longrightarrow G$ is an isomorphism between two fields, show that
   (a) $\psi(0) = 0, \psi(1) = 1$ and $\psi(a^{-1}) = (\psi(a))^{-1}$.
   (b) $\psi^{-1} : G \longrightarrow F$ is an isomorphism.
2. Show that the mapping $\phi : GF(p^n) \longrightarrow GF(p^n) : \zeta \mapsto \zeta^p$ is an isomorphism. (This is called the *Frobenius mapping.*)
3. Show that there are $\frac{\phi(p^n-1)}{n}$ primitive polynomials of degree n with coefficients from $GF(p)$.

6.3 Solving Quadratic Equations over $GF(2^n)$ and the Trace Function

We digress briefly to discuss quadratic equations over $GF(2^n)$. Over a field of characteristic 2 the usual "high school" formula cannot work, since if the equation is $ax^2 + bx + c = 0$, the formula involves division by $2a$ which in the case of characteristic 2 does not make sense.[4]

In the process of doing so, we introduce an important concept, namely that of the *trace* of an element, and take the opportunity to look at linear functions in a novel way.

We have already seen that in the field $GF(2^n)$ every element has a (unique) square root; in fact the square root of $\alpha \in GF(2^n)$ is $\alpha^{2^{n-1}}$, since $(\alpha^{2^{n-1}})^2 = \alpha^{2^n} = \alpha$.

But it seems likely that we may occasionally need to solve more complicated quadratic equations, of the form $x^2+\alpha x+\beta = 0$. In such cases, we note that, first of all, a solution need not exist (e.g. in the field of Sect. 5.6 the equation $x^2+x+\alpha^3 = 0$ has no solution in $GF(2^4)$ as the reader may verify by trial and error, or from the following theory), whereas any equation of the form $x^2 + x + \beta = 0$ which has a solution, will have two: because if γ is a solution, then so is $\gamma + 1$. The latter is a consequence of the fact that $(\gamma + 1)^2 + (\gamma + 1) = \gamma^2 + \gamma$.

In order to be able to separate these cases, and for other purposes as well, we need the following

Definition Let $\alpha \in GF(p^n)$. The *trace* of α, $\text{Tr}(\alpha)$, is defined[5] by

$$\text{Tr}(\alpha) = \alpha + \alpha^p + \alpha^{p^2} + \cdots + \alpha^{p^{n-1}}.$$

Note that by the Freshman Theorem, $(\text{Tr}(\alpha))^p = \alpha^p + \alpha^{p^2} + \cdots + \alpha^{p^{n-1}} + \alpha^{p^n}$, and this equals $\text{Tr}(\alpha)$, because $\alpha^{p^n} = \alpha$. Hence $\text{Tr}(\alpha) \in GF(p)$, because, as we have seen, $GF(p)$ is precisely the subset of elements with this property. In particular, if $p = 2$, then the trace of any element of $GF(2^n)$ is either 1 or 0.

More remarkable is the fact that the trace is a linear function: If $a, b \in GF(p)$ and $\alpha, \beta \in GF(p^n)$, then $\text{Tr}(a\alpha + b\beta) = a\text{Tr}(\alpha) + b\text{Tr}(\beta)$, but this is also easily seen, using the Freshman Theorem and the fact that $a^p = a$ and $b^p = b$.

Now consider the equation $x^2 + x + \beta = 0$ over $GF(2^n)$, which is, of course the same as the equation $x^2 + x = \beta$. If we take the trace of both sides we get that $\text{Tr}(\beta) = \text{Tr}(x + x^2) = \text{Tr}(x) + \text{Tr}(x^2) = \text{Tr}(x) + \text{Tr}(x) = 0$. Hence if $\text{Tr}(\beta) \neq 0$, there can be no solution to the equation.

Finding the solution, if there is one, depends on a concept briefly touched upon in Sect. 5.6: A set $\{\alpha, \alpha^2, \alpha^4, \ldots, \alpha^{2^{n-1}}\} \subset GF(2^n)$ which forms a basis for the field $GF(2^n)$, considered as a vector space over $GF(2)$, is called a *normal basis*.[6] It can be proved that every finite field has a normal basis, but the proof is beyond the scope of this book. If you

[4] If you can remember the derivation of the formula for solving quadratic equations, you can easily convince yourself that it works fine for all fields, except those of characteristic 2.

[5] Strictly speaking we define the trace of $GF(p^n)$ over $GF(p)$. One can define the trace of $GF(p^n)$ over any subfield of $GF(p^n)$ in an analogous manner, but this is the only one we need.

[6] The set $\{1, \alpha, \ldots, \alpha^{n-1}\}$, where α is a root of an irreducible polynomial in $GF(2)[x]$ is called a *polynomial basis*. Up to now, we have always used polynomial bases.

need a normal basis for $GF(2^n)$, the quick and dirty method is to pick an element α, compute $\{\alpha^{2^i} | 0 \leq i \leq n-1\}$ and see whether that gives you a linearly independent set. If not, try another α. For example, in the field $GF(2^4)$, as defined in Sect. 5.6, the set $\{\alpha^{3 \cdot 2^i} | i = 0, 1, 2, 3\} = \{\alpha^3, \alpha^6, \alpha^{12}, \alpha^{24} = \alpha^9\}$ forms a normal basis. If we express x in terms of a normal basis, solving the quadratic equation becomes easy as in the following example. We hasten to point out that while squaring and solving quadratic equations is easy when expressing field elements in terms of the normal basis, multiplication is much easier when using a polynomial basis.

Example Solve: $x^2 + x = 1$ in $GF(2^4)$.

We express x as $x = \sum_{i=1}^{4} x_i \alpha^{3 \cdot 2^i}$, and note that $1 = \sum_{i=0}^{4} \alpha^{3 \cdot 2^i}$. (This is always a property of normal bases over $GF(2)$: The sum of all the vectors in the basis is 1. This sum must lie in GF(2), since it is merely the trace of any one of the elements of the basis, but it can't be 0, because of the demand for linear independence.)

Now $x^2 = x_4\alpha^3 + x_1\alpha^6 + x_2\alpha^{12} + x_3\alpha^9$, so that

$$x^2 + x = (x_1 + x_4)\alpha^3 + (x_2 + x_1)\alpha^6 + (x_3 + x_2)\alpha^{12} + (x_4 + x_3)\alpha^9 = \alpha^3 + \alpha^6 + \alpha^{12} + \alpha^9$$

and we have the linear equations

$$x_1 + x_4 = 1,$$

$$x_1 + x_2 = 1,$$

$$x_2 + x_3 = 1,$$

$$x_3 + x_4 = 1,$$

with solutions $(x_1, x_2, x_3, x_4) = (\delta, 1 + \delta, \delta, 1 + \delta)$, $\delta \in GF(2)$, so either

$$x = \alpha^6 + \alpha^9 = \alpha^5$$

or

$$x = \alpha^3 + \alpha^{12} = \alpha^{10} = 1 + \alpha^5.$$

Finally, on this topic, we note that an equation of the form $x^2 + \beta x = \gamma$ can be brought to the form $z^2 + z = \chi$, by making the substitution $x = \beta z$.

There are two further points to be made about the trace:

Firstly, in the field $GF(2^n)$ there are exactly as many elements with trace 0 as there are with trace equal to 1. In fact, more generally: in $GF(p^n)$, each of the p sets $\{x \in GF(p^n) | \mathrm{Tr}(x) = i\}$ $(0 \leq i \leq p-1)$ contains p^{n-1} elements. This follows from the fact that the equation $\mathrm{Tr}(x) = x + x^p + \cdots + x^{p^{n-1}} = i$, which, as an equation of degree p^{n-1}, has at most p^{n-1} solutions. But every element of $GF(p^n)$ must be a solution of one of these p equations, so for every i there must be exactly p^{n-1} solutions.

Let us, at this point, introduce some terminology. We shall call a function which maps onto the set $\{0, 1\}$ a *Boolean function*, regardless of whether the 0 and the 1 are interpreted as `False` and `True`, as is sometimes the case. If we have a finite set A and a Boolean function f defined on that set, we shall call f *balanced* if $\#\{x \in A | f(x) = 0\} = \#\{x \in A | f(x) = 1\}$.

Secondly, let us define a *linear function* $L : (GF(p))^n \longrightarrow GF(p)$ of n variables $x_0, x_1, \ldots, x_{n-1}$ to be a function of the form $L(x) = \sum_{i=0}^{n-1} a_i x_i$, where the $a_i \in GF(p)$. Clearly there are p^n choices for the a_i, so there are p^n distinct linear functions of n variables.

In the particular case where $p = 2$ we would speak of a linear *Boolean* function.

If we have fixed a basis for $GF(p^n)$, we have an obvious one-to-one correspondence between the elements of $(GF(p))^n$ and those of $GF(p^n)$ and we can think of our linear function as a function from $GF(p^n)$ to $GF(p)$:

$$L(x_{n-1}\alpha^{n-1} + \cdots + x_1\alpha + x_0) = \sum_{i=0}^{n-1} a_i x_i.$$

The remarkable thing is that for some $\omega \in GF(p^n)$

$$L(x_{n-1}, \ldots, x_1, x_0) = \mathrm{Tr}[\omega \cdot (x_{n-1}\alpha^{n-1} + \cdots + x_1\alpha + x_0)].$$

Thus, in a way, any linear function on n variables is a (modified) version of the trace. This conclusion follows from the fact that the functions $\mathrm{Tr}(\omega \cdot x)$ are obviously linear, that for different ω_1 and ω_2 we get different functions $\mathrm{Tr}(\omega_1 \cdot x)$ and $\mathrm{Tr}(\omega_2 \cdot x)$, and that there are p^n choices for ω as well as p^n distinct linear functions, so every linear function must be one of the form $\mathrm{Tr}(\omega \cdot x)$.

Exercises

1. Solve, if possible, the following equations in the field $GF(2^4)$ as defined in Sect. 5.6:
   (a) $x^2 + x = \alpha^2$;
   (b) $\alpha^2 x^2 + x = \alpha^{12}$.
2. Show that, considered as an element of $GF(2^n)$, the trace of the element 1 is 0 if and only if n is even.
3. Show that any linear Boolean function (other than the zero function, which maps every input onto 0) is balanced. In particular, the trace function is balanced.
4. Show that if $\mathrm{Tr}(\alpha x) = 0$ for all $x \in GF(p^n)$ then $\alpha = 0$.
5. Let n be a positive integer, not divisible by the prime p. Show that the elements of $GF(p^n)$ whose trace is zero form a *subspace* of the $GF(p)$-vector space, but do not form a *subfield* of $GF(p^n)$.

6.4 Operations in Finite Fields

Although these notes concentrate on the theoretical and mathematical aspects of Algebra as needed in Cryptology, it is worth our while to devote a little time to the computational side of things. In particular, we need to look at efficient (or moderately efficient) ways of performing the operations in rings of the form $\mathbb{Z}_n$ and in finite fields.

A disclaimer may be appropriate here: the methods we suggest are not the most efficient ones, but they are easy to implement, and, for that matter, easy to understand. There is a very wide ranging literature on the subject of efficient algorithms for performing the arithmetic operations, especially as a consequence of the increasing interest in elliptic curves over finite fields $GF(p)$, where p is a very large prime. (See Sect. 6.6 below.) Moreover, differences about the most efficient algorithms will arise depending on whether the implementation is in software or hardware, and, if the latter, on the relevant processor architecture. As a consequence, as already noted in the *Handbook of Applied Cryptography*,

> Efficiency can be measured in numerous ways; thus, it is difficult to definitively state which algorithm is the best. An algorithm may be efficient in the time it takes to perform a certain algebraic operation, but quite inefficient in the amount of storage it requires. One algorithm may require more code space than another. Depending on the environment in which computations are to be performed, one algorithm may be preferable over another. For example, current chipcard technology provides very limited storage for both precomputed values and program code. For such applications, an algorithm which is less efficient in time but very efficient in memory requirements may be preferred.

The reader interested in such implementation issues may consult Chap. 14 of the *Handbook*, or as a more recent reference, Chaps. 2 and 5 of Hankerson et al.'s *Guide to Elliptic Curve Cryptography*.[7]

6.4.1 Addition in $\mathbb{Z}_n$

Addition in $\mathbb{Z}_n$ is simple: To find $a \bmod n + b \bmod n$, simply add a and b. If the answer is less than n, return $a + b$, otherwise return $a + b - n$. (After all, $a + b < 2n$.)

Addition in a finite field $GF(p^n)$ is almost equally simple: The elements of the field are polynomials of degree less than n, and to add two such polynomials is simply a case of adding (in $\mathbb{Z}_p$, of course) the coefficients of terms of the same degree. If $p = 2$, it is really easy, since addition in that field is the same as XORing them. (C, and most other scientific programming languages, include XOR ($\oplus$) as an operation on bitstrings.)

This leaves us with the operations of multiplication in $GF(p^n)$, and of finding the inverses of nonzero elements in $\mathbb{Z}_n$ and $GF(p^n)$:

6.4.2 Inverses in $\mathbb{Z}_n$

We recall that $a \bmod n$ has an inverse in $\mathbb{Z}_n$ if and only if the greatest common divisor of a and n is 1. One finds the greatest common divisor by means of the Euclidean Algorithm, of course. But in Sect. 2.3 we also showed how the Extended Euclidean Algorithm also returned (at very little extra cost) the constants x and y satisfying

$$xa + yn = \gcd(a, n).$$

[7] Hankerson, D., Menezes, A. and Vanstone, S.: *Guide to Elliptic Curve Cryptography*, Springer-Verlag, 2004.

If $\gcd(a, n) \neq 1$, the Extended Euclidean Algorithm tells us that $a \bmod n$ does not have an inverse; otherwise it tells us that the inverse exists and equals $x \bmod n$.

One might fear that inverses in $GF(p^n)$ would be found in the same way, since the elements of such a field are also residue classes modulo a principal ideal, and the Extended Euclidean Algorithm works just the same in rings of polynomials as it does in $\mathbb{Z}$. One could do things that way, but we shall give a more efficient (and less unpleasant) method, once we have dealt with multiplication.

6.4.3 Multiplication in $GF(p^n)$

For many cryptological applications, especially those in symmetric cryptography, $p = 2$. We let α be a root of an irreducible polynomial $f(x)$ of degree n, and use a polynomial basis for the field $GF(2^n)$, i.e. we identify the bitstring $b_{n-1}b_{n-2}\ldots b_2b_1b_0$ with the field element $B = b_{n-1}\alpha^{n-1} + b_{n-2}\alpha^{n-2} + \cdots + b_2\alpha^2 + b_1\alpha + b_0$. If $f(x) = x_n + \sum_{i=0}^{n-1} a_i x^i$, then multiplying an element by α amounts to shifting the bitstring one place to the left, and, moreover, if $b_{n-1} = 1$ (in other words if the polynomial $B(\alpha)$ has a term of degree $n - 1$) XORing the bitstring $a_{n-1}a_{n-2}\ldots a_2a_1a_0$ (in other words adding $\sum_{i=0}^{n-1} x_i\alpha^i = \alpha^n$).

Once we have implemented multiplication by α, multiplication of $B(\alpha)$ by an expression $a(\alpha) = a_0 + a_1\alpha + \cdots + a_{n-1}\alpha^{n-1}$ is straightforward: form, iteratively, the expressions $B(\alpha), \alpha\cdot B(\alpha), \alpha^2\cdot B(\alpha), \ldots, \alpha^{n-1}\cdot B(\alpha)$ and XOR those $\alpha^i\cdot B(\alpha)$ for which the corresponding $a_i = 1$.

In the cases where $p \neq 2$, the procedures are the same, but one doesn't have the convenience of simple shifting the bitstring, and some calculations modulo p will be required in place of the simple XOR.

6.4.4 Exponentiation

Exponentiation (in a ring or a field) is conveniently performed by means of the "square-and-multiply" technique. Finding a^k can, of course, be done by means of the iteration $a_0 = 1$; $a^i = a \times a^{i-1}$, but that would require k multiplications. The following requires fewer than $2\log_2(k)$ multiplications, and depends on the obvious fact that

$$a^{\sum_{i=0}^{t} k_i 2^i} = \prod_{i=0}^{t} a^{k_i 2^i} = (a^{2^0})^{k_0}(a^{2^1})^{k_1}\cdots(a^{2^t})^{k_t}.$$

For example, to compute $a^{22} = a^{16+4+2}$, we compute a^2, a^4, a^8 and a^{16}, and then multiply out those that we need: $a^{16} \times a^4 \times a^2$. The following program in pseudocode does precisely that, except that, starting with $b = 1$, it multiplies b by the appropriate a^{2^i} as we go along.

Algorithm for exponentiation in $GF(p)$

INPUT: Ring element α, non-negative integer $k = \sum_{i=0}^{t} k_i 2^i$
OUTPUT: α^k

1. Set $b \longleftarrow 1$. If $k = 0$ return 1;
2. Set $A \longleftarrow a$;
3. If $k_0 = 1$ then set $b \longleftarrow a$;
4. For i from 1 to t do
   - Set $A \longleftarrow A^2$;
   - If $k_i = 1$ then set $b \longleftarrow A \cdot b$;
5. Return b.

6.4.5 *Inverses in $GF(p^n)$*

Using the above (efficient) method of exponentiation, as well as the fact that in $GF(p^n)$ one has that $a^{p^n-1} = 1$ for all nonzero $a \in GF(p^n)$, and that therefore the inverse of a nonzero element a is a^{p^n-2}, we have an efficient method of finding inverses. Note in particular that if $p = 2$ then $p^n - 2$ has the binary representation $111\ldots110$.

Exercises

1. Compute $3^{2671} \bmod 769$. (769 is prime.)
2. Compute $3^{2671} \bmod 256$.
3. The polynomial $x^{32} + x^{28} + x^{27} + x + 1$ is irreducible over $GF(2)$. Use this to write a program to perform addition, multiplication, exponentiation and inversion in the field $GF(2^{32})$.

6.5 Factoring Polynomials

We shall, in this section, discuss an algorithm for factoring polynomials over a finite field. This is not strictly needed for any of our purposes, so feel free to skip this section.

The algorithm we describe only works for polynomials which do not contain repeated factors, or, in other words, which are *square-free*. We therefore start with a method for ensuring that we do not have repeated factors, then discuss how to determine whether or not a given polynomial is irreducible, and, finally, arrive at the desired algorithm. The algorithms for these purposes were invented in 1967 by E.W. Berlekamp, and the factoring algorithm is therefore always referred to as *Berlekamp's algorithm*.

6.5.1 *Square-freeness*

In an exercise in Sect. 5.3.4 we defined the *derivative* of a polynomial $f(x) = \sum_{i=0}^{n} a_i x^i$ (where the coefficients a_i come from some field $\mathbb{F}$) as the polynomial $f'(x) = \sum_{i=1}^{n} i a_i x^{i-1}$.

This is, of course, a purely formal derivative; the concept of limit, so fundamental in the Calculus, makes no sense in our (usually finite, and always discrete) environments. We are only concerned with finite fields, so $\mathbb{F} = GF(q)$ where $q = p^n$, for some prime p and some n. Our main interest is, as always, in the case $p = 2$.

If $f'(x) = 0$, then $f(x)$ is a pth power of some polynomial, which can be found by inspection. For example, the polynomial $x^8 + x^4 + 1 \in GF(2)[x]$ has derivative 0, which is because $x^8 + x^4 + 1 = (x^4 + x^2 + 1)^2$. In fact, the term in brackets is still a square, and $x^8 + x^4 + 1 = (x^2 + x + 1)^4$.

Otherwise, consider $g = \gcd(f, f')$. If g has nonzero degree, then we have found a factor of f (which is, after all, our aim), and we feel pleased with ourselves. If we want a complete factorization of f, we can now continue with f/g.

The reason why this works is quite easy to see, if one notes that this formal derivative behaves just like the traditional one. In particular, the rule for differentiating products applies. So, for the case $p = 2$, if, say, $f(x) = (g(x))^2 \cdot h(x)$, then $f'(x) = 0 \cdot h(x) + (g(x))^2 \cdot h'(x)$ and $\gcd(f, f') = (g(x))^2$. The general case ($p \neq 2$) we leave as an easy exercise.

Finally, if g has degree 0 (i.e. g is a constant), then f is square-free, and we can continue with the rest of the programme.

6.5.2 *A Test for Irreducibility*

Assume therefore that f is square-free. We shall write q for p^n, to simplify notation.

Let us assume that $f(x) = \prod_{i=1}^{k} e_i(x)$, where the $e_i(x)$ are irreducible. We consider the ring $R = GF(q)[x]/ < f(x) >$ which is, as we have seen in Sect. 5.5, isomorphic to the ring $\prod_{i=1}^{k} GF(q)[x]/ < e_i(x) >$. We shall call this isomorphism the *CRT map*.

But $GF(q)[x]/ < f(x) >$ can also be considered a vector space over $GF(q)$ with basis $\{1, x, x^2, \ldots, x^{\text{degree}(f)-1}\}$.

We now consider the Frobenius type mapping $Q_f : R \longrightarrow R$, which maps elements of R onto their qth powers. From the Freshman theorem it follows that this is, in fact, a linear mapping:

$$Q_f(\alpha(x) + \beta(x) + < f(x) >) = Q_f(\alpha(x) + < f(x) >) + Q_f(\beta(x) + < f(x) >).$$

Therefore $Q_f - I$ is also a linear mapping, where I is the identity map on R, and we consider the subspace $S = \text{Ker}(Q_f - I)$. Thus $\gamma(x) \in S$ if and only if $(\gamma(x))^q \equiv \gamma(x) \mod f(x)$.

Now comes the crunch: the dimension of S is the number k of irreducible factors of f.

For let $\gamma(x) \in S$, and let the image of $\gamma(x)$ under the CRT map be $(a_1 + < e_1(x) > , \ldots, a_k + < e_k >)$. But $(\gamma(x))^q = \gamma(x)$, so, for each i, $a_i^q + < e_i(x) > = a_i + < e_i(x) >$ or $a_i^q - a_i + < e_i(x) > = 0 + < e_i(x) >$. This implies that $a_i \in GF(q)$. Thus in every term in the product, the image of S occupies the $GF(q)$ part, so that $S \cong (GF(q))^k$. Thus the kernel of $Q_f - I$ has dimension k.

6.5.3 Finding a Factor

The work is now almost done. We prove the following (where the notation is as before)

Lemma *For every* $\gamma(x) \in S$, $f(x) = \prod_{c \in GF(q)} \gcd(f(x), \gamma(x) - c)$.

Proof The right-hand side is a divisor of the left-hand side: clearly every one of the greatest common divisors is a divisor of the left-hand side, and there are no repeated terms of degree greater than 0, since the $\gamma(x) - c$ will be relatively prime for different values of c.

Conversely, we need the fact that for any $\gamma(x)$,

$$\gamma(x)^q - \gamma(x) = \prod_{c \in GF(q)} (\gamma(x) - c).$$

This is just a consequence of the fact that every element of $GF(q)$ is a solution of the equation $y^q - y$, and whether we call the unknown y or $\gamma(x)$ is of no consequence. Thus the fact that $\gamma(x) \in S$, and therefore $(\gamma(x))^q - \gamma(x)$ is divisible by $f(x)$, shows that $f(x)$ is a divisor of the right-hand side.

Now it follows that for some $c \in GF(q)$ we must have that $\gcd(f(x), \gamma(x) - c)$ must have degree greater than 1. We therefore have a non-trivial factor of f.

Example Let $f(x) = x^7 + x^5 + x^4 + x^2 + 1 \in GF(2)[x]$. Considering the map $h(x) \mapsto (h(x))^2 \mod f(x)$ we have

$$\begin{aligned} 1 &\mapsto 1, \\ x &\mapsto x^2, \\ x^2 &\mapsto x^4, \\ x^3 &\mapsto x^6, \\ x^4 &\mapsto x + x^3 + x^5 + x^6, \\ x^5 &\mapsto 1 + x + x^2 + x^4 + x^5 + x^6, \\ x^6 &\mapsto 1 + x, \end{aligned}$$

and therefore the matrix of $Q_f - I$ (w.r.t. to the basis $\{1, x, x^2, x^3, x^4, x^5, x^6\}$) is

$$\begin{pmatrix} 0&0&0&0&0&1&1 \\ 0&1&0&0&1&1&1 \\ 0&1&1&0&0&1&0 \\ 0&0&0&1&1&0&0 \\ 0&0&1&0&1&1&0 \\ 0&0&0&0&1&0&0 \\ 0&0&0&1&1&1&1 \end{pmatrix}.$$

Applying row reduction to this matrix (and throwing away all rows which contain only zeros) we get

$$\begin{pmatrix} 0\,1\,0\,0\,1\,1\,1 \\ 0\,0\,1\,0\,1\,0\,1 \\ 0\,0\,0\,1\,1\,0\,0 \\ 0\,0\,0\,0\,1\,0\,0 \\ 0\,0\,0\,0\,0\,1\,1 \end{pmatrix}.$$

The kernel of $Q_f - I$ therefore has dimension 2, showing that f has 2 irreducible factors. This kernel has basis $\{1, x^2 + x^5 + x^6\}$, and trying the gcd of $f(x)$ and $x^6 + x^5 + x^2$ we find the factor $x^4 + x^3 + 1$ without difficulty.

Exercises

1. Test for square freeness: $x^6 + x^5 + x + 1 \in GF(2)[x]$.
2. Find a factor of $x^9 + x^7 + x^4 + x^3 + x^2 + x + 1 \in GF(2)[x]$.

6.6 The Discrete Logarithm Problem on a Finite Field

When implementing a DLP-based protocol, one needs a large cyclic group. It is tempting to use the multiplicative group of the nonzero elements of a finite field of low characteristic (2 or 3, say) rather than one of those recommended in Sect. 4.4.5. After all, we have proved that the multiplicative group $GF(2^n)\backslash\{0\}$ is cyclic and it can be made large simply by choosing a large enough n. Moreover, exponentiation of elements is computationally not too demanding.

Unfortunately current indications are that such a choice may be unwise: since 2013 enormous progress has been made in solving the DLP over such fields, starting with the work of Joux[8] in December 2012, and continuing with work by numerous others. At the time of writing, Wikipedia[9] gives the record for solving the DLP over binary fields as that of Granger, Kleinjung and Zumbrägel who found logarithms over the field $GF(2^{9234})$. The record for fields of characteristic 3 stands, according to the same source, at $GF(3^{5 \cdot 479})$, by Joux and Pierrot in December 2014, who beat the previous record of $GF(2^{6 \cdot 163})$ set by Menezes et al. in February 2014. The reader will note that in these cases the fields concerned are extensions of prime degree (479 and 163) of smaller fields $GF(3^5)$ and $GF(2^6)$ respectively.[10]

[8] Joux, A.: *A new index calculus algorithm with complexity L(1/4+o(1)) in very small characteristic*, IACR ePrint Archive: Report 2013/095.

[9] Accessed 10 October 2015.

[10] Even in some fields of "moderate" characteristic, according to Wikipedia, the DLP has been solved, such as the solutions for $GF(65537^{25})$ by Durand et al. in October 2005 and for $GF(p^2)$, where p is the 80-digit prime

$$p = 31415926535897932384626433832795028841971693993751058209749445923078164063079607$$

by Barbulescu et al. in June 2014.

While we will not go into the details of these methods of solving the DLP, similar to our omission of any reasonable methods of factoring integers, it is interesting to note that a technique known as *Index Calculus* is usefully employed in both these problems. For a description of this we refer the reader to the *Handbook of Applied Cryptography* by Menezes et al. or Vaudenay's *Classical Introduction*.[11] Joux[12] notes that

> All index calculus algorithms have in common two main algorithmic phases. The first of these phases is the generation of multiplicative relations, which are converted into linear or affine equalities involving the logarithms of the elements which appear in the multiplicative relations. The second phase is the linear algebra phase, which solves the resulting system of equations. For factoring, the linear algebra is performed modulo 2. For discrete logarithms, it is done modulo the order of the relevant group.

Vaudenay states that in fact "Some researchers believe that the two problems have the same intrinsic complexity". The appeal of elliptic curves lies in the fact that no multiplicative relations, in terms of the so-called "factor base", can be found: an analogue for the factor base does not exist. So let's give some indication of what elliptic curves actually are.

6.7 Elliptic Curves over a Finite field

If $\mathbb{F}$ is any field (of characteristic not equal to 2 or 3), an *elliptic curve over* $\mathbb{F}$ is the set of points $(x, y) \in \mathbb{F}^2$ which satisfy an equation of the form $y^2 = x^3 + ax + b$, $(a, b \in \mathbb{F})$ together with a "point at infinity", denoted by $\mathcal{O}$. If the field is of characteristic 2, the equation is of the form $y^2 + y = x^3 + ax + b$. This set of points can be turned into an abelian group.

For the purposes of explanation, we shall start with the case where $\mathbb{F}$ is the field of real numbers, and define how addition and additive inverses are defined. For other fields, the definitions are analogous, but the precise definitions would lead us too far afield.[13]

If P and Q are points on the curve, then

- If $P = \mathcal{O}$ then $-P = \mathcal{O}$ and $P + Q = Q$. In other words, $\mathcal{O}$ is going to be the identity element of the (additive) group we are trying to construct.
- The additive inverse of the point $P = (x, y)$ is the point $-P = (x, -y)$.
- If $Q = -P$ then $P + Q = \mathcal{O}$.
- If P and Q have different x-coordinates, then we construct the secant through P and Q. This line will intersect the curve in a third point. This third point is defined to be $-(P + Q)$.

---

[11] Menezes, A., van Oorschot, P.C. and Vanstone, S.A.: *Handbook of Applied Cryptography*, CRC Press, Boca Raton, 1997;

Vaudenay, S.: *A Classical Introduction to Cryptography*, Springer Science + Business Media Inc., 2006.

[12] Joux, A.: *Faster index calculus for the medium prime case; Application to 1175-bit and 1425-bit finite fields*, IACR ePrint Archive 2012/720.

[13] No pun intended. We recommend the book by Washington for a suitable exposition of the definition and properties of elliptic curves, including elementary applications to asymmetric cryptology: Washington, L.C.: *Elliptic Curves—Number Theory and Cryptography*, Chapman & Hall/CRC Press, Boca Raton, 2003.

- If $P = Q$, we replace in the last construction the secant through P and Q by the tangent to the curve at P. The point where this tangent intersects the curve is then defined to be $-(P+P) = -2P$.
- As usual when additive notation for the group operation, we denote, if $n \in \mathbb{Z}$, and P is a point on the curve, the sum $\underbrace{P+P+\cdots+P}_{n \text{ times}}$ as nP. The DLP thus becomes the problem of finding n if P, Q and the equation $nP = Q$ are given.

It is clear that these operations can be performed in terms of the coordinates, and for fields other than the reals that is what we do: we define the group operation in terms of the coordinates of P and Q. What is not clear is that this definition results in turning the set of points into a group. (The worst part of checking that it works is in verifying that the operation "+" as we defined it is actually associative.) Never mind: it is true.

As usual we are (most) interested in the case where $\mathbb{F}$ is a finite field. In this case a theorem due to Hasse states that the order $\#E$ of the group E of points on the elliptic curve (including the point at infinity) satisfies

$$|q+1-\#E| \leq 2\sqrt{q},$$

where q is the number of elements of the field $\mathbb{F}$.

The discrete logarithm problem for elliptic curves is simply the additive version of what we have seen before:

Let E be an elliptic curve over the finite field $\mathbb{F}$, and let G be a cyclic subgroup of E. Let B be a generator of G, and let $P \in G$. Then the discrete logarithm problem is to determine x such that $xB = P$.

Some precautions have to be taken when using the elliptic curve analogues of (e.g.) the Diffie–Hellman or the Digital Signature algorithms. There are reasons to believe that curves over prime fields ($GF(p)$) are safer than curves over binary fields ($GF(2^n)$). Even so, some curves are definitely better than others, and it is advisable to use either curves which have been selected by international standards organisations, or to use suitable groups of points on randomly generated curves and change them frequently, in order to ensure that an adversary cannot collect enough material to mount an attack, even if the group turns out to have unanticipated weak properties. Within these guidelines it seems to be accepted that in order to achieve security comparable to that provided by n-bit symmetric cryptography, the order of the elliptic curve group should be of the order of 2^{2n}.

7 Applications to Stream Ciphers

In this chapter, after some general observations on stream ciphers and block ciphers and on the fundamental concept of entropy as defined in Information Theory, we apply our ideas of finite fields to linear feedback shift registers (LFSRs), a frequent component of stream cipher designs. We also discuss methods in which LFSRs are used, which brings us to the problems involved in stream cipher design, and then provide a survey of such design methods.

7.1 Introduction

7.1.1 Stream Ciphers vs Block Ciphers

Traditionally secret key encryption algorithms are divided into block ciphers and stream ciphers. *Block ciphers* operate with a fixed (though key dependent) transformation on large blocks of plaintext data, whereas *stream ciphers* operate with a time-varying transformation on individual bits (or bytes, or at any rate, small blocks) of plaintext. The boundary between these two kinds of system is rather vague: there are modes of operation of block ciphers (Output Feedback and Counter Mode, usually referred to as OFB and CTR modes, for example) which are really stream ciphers according to this description, apart from the size of the block which is being encrypted.

Stream ciphers are an important class of encryption algorithms, but there are few fully specified stream ciphers in the open literature, and by and large the theory of block cipher design is much better understood than that of stream ciphers. Stream ciphers have the advantages of speed and ease of implementation over block ciphers when implemented in hardware, and are more convenient when buffering is limited or when bits (or bytes or whatever) need to be processed individually. Padding of plaintext before encryption, which sometimes creates security weaknesses, is limited or completely unnecessary. Against all this are the disadvantages of greater uncertainty about the security of stream ciphers and far less standardisation has taken place.

A.R. Meijer, *Algebra for Cryptologists*, Springer Undergraduate Texts in Mathematics and Technology,
DOI 10.1007/978-3-319-30396-3_7

Consequently the question of whether stream ciphers have a future has been raised; the consensus of opinion seems to be that they will continue to survive in "niche" applications where the advantages are of vital importance, though the general trend over the last few decades has been towards the use of block ciphers. We return to these matters in Sect. 7.9 when we briefly discuss the European Commission's NESSIE and eStream projects.

Stream ciphers may be thought of as having developed as a substitute for the *one time pad*. The one time pad, in which (to take the simple binary case) the plaintext message bitstream is XORed, bit by bit, with a random bit stream, the latter being known to both transmitter and receiver of the message (and only to them!), to create a ciphertext bitstream, only dates from about 1882, and was patented by Vernam in 1917. We shall show in Sect. 7.3 that this is, as long as the (key) bitstream is never re-used, a secure method of encryption, even in the information-theoretic sense.

The name actually arises from early implementations where the key material was distributed in the form of a pad of paper; the sheets were torn off and destroyed as they were used, so making re-use impossible as long as everyone followed the rules.

The problem is, in practice, that the key distribution is extremely demanding: the amount of key data that has to be distributed being equal to the amount of data that needs to be encrypted. Moreover, the key material has to be truly random: given any subsequence of key bits, it should not be possible to guess any preceding or following bits with a non-negligible probability of success greater than that provided by pure guessing. Generating such random sequences is difficult and time consuming.

Because of this, the search started for keystream generators which, on input of a much shorter key, generate a bitstream. Thus, typically, a stream cipher consists of a keystream generator, the output of which "appears random" and is added modulo two to the plaintext bits. The goal in stream cipher design is therefore to produce random-looking sequences of bits, in other words, sequences which are indistinguishable from coin-tossing sequences. Loosely, one may consider a sequence to be pseudorandom if no patterns can be detected in it, no predictions about the next output can be made, and no short description of it can be made. The last requirement is, of course, impossible, since the program generating the key sequence is itself a short description! Nevertheless, performing statistical tests can give an indication of how closely the keystream approximates a genuine random sequence.

7.1.2 Design Principles

Rainer A. Rueppel in his paper *Stream Ciphers* in G. Simmons: *Contemporary Cryptology: The science of information integrity*[1] distinguishes between four different approaches to the construction of stream cipher systems. These are

- *The information-theoretic approach:* The cryptanalyst is assumed to have unlimited time and computing power. Cryptanalysis is the process of determining the plaintext message (or the key) given the ciphertext and the *a priori* probabilities of the various message and keys. The system is considered "broken" when there is what amounts to a unique

[1] IEEE Press, New York, 1992.

solution: one message (or key) occurs with probability essentially 1, while the others have probabilities essentially equal to 0. Because the attacker is assumed to be infinitely powerful, the corresponding notion of security is independent of the complexity of the encryption method. The cipher system is called *perfectly secure* if the plaintext and the ciphertext are statistically independent, or more precisely, if the mutual information between plaintext and ciphertext (in the absence of the key) is zero. We shall discuss these ideas a little further later in this chapter, where we shall also make precise the definition of what exactly we mean by "information". A cipher system is called *ideally secure* if the attacker cannot find a unique solution, no matter how much ciphertext he is allowed to observe.

- *The system-theoretic approach:* The objective for the designer in this approach is to create new and difficult unknown problems for the cryptanalyst—a new such problem for each new system. (Breaking such a new system is much less "glamorous" than breaking, say, the discrete logarithm problem!) Thus the designer attempts to construct a system on which none of the known cryptanalytic techniques, such as statistical analysis (frequency counts, etc.), correlation attacks, etc. will work. This is, in all likelihood, the main design approach followed in practice.
- *The complexity-theoretic approach:* In this approach all computations are parametrized by a security parameter—usually the key length—and only cryptanalytic algorithms whose complexity is polynomial in this security parameter are assumed to be feasible. The cryptanalyst is taken to have "won" the contest if he or she can either predict a bit of the keystream or can distinguish the keystream from a truly random sequence. The designer will attempt to make his generator computationally equivalent to a known computationally infeasible problem: a keystream generator is called *perfect* if it is

  (a) unpredictable; or
  (b) indistinguishable from a random sequence

  by all polynomial time statistical tests.
- *Randomized stream ciphers:* The designer may instead try to ensure that the cryptanalytic effort has an infeasible size, rather than an infeasible time complexity. The objective is to increase the number of bits that the analyst has to consider, while still keeping the key size small. For example: a large random string may be made publicly available, while the key indicates the part or parts of this string which is or are actually used in the encryption process.

We shall briefly consider the information-theoretic approach, but otherwise concentrate on the system-theoretic side of things.

7.1.3 Terminology

In the following we introduce the terminology of Rueppel *op. cit.* to describe stream ciphers and their modes of operation:

Let $\mathcal{X}, \mathcal{Y}, \mathcal{Z}$ denote, respectively, the plaintext alphabet, the ciphertext alphabet and the keystream alphabet. Let $\mathcal{K}$ denote the key space, and $\mathcal{S}$ the internal state space of the generator. Let x_i, y_i, z_i and s_i denote, respectively, the plaintext symbol, the ciphertext symbol, the keystream symbol and the internal state at time i. (We note here that, almost invariably, the key is drawn from the key space according to the uniform probability distribution.)

A general stream encryption algorithm may be described by means of the equations

$$s_{i+1} = F(k, s_i, x_i),$$
$$y_i = f(k, s_i, x_i),$$

where F is the next-state function and f is the output function, and $k \in \mathcal{K}$. Usually

$$y_i = x_i + f(k, s_i).$$

(It can be shown that this form is both necessary and sufficient for ensuring that the decryption process take place without delay.)

The sequence

$$\{z_i = f(k, s_i) : i \geq 1\}$$

is called the *keystream.*

7.1.3.1 Synchronous Stream Ciphers

For a synchronous stream cipher (also called keystream generator or running-key generator) the equations above are of the form

$$s_{i+1} = F(k, s_i),$$
$$z_i = f(k, s_i).$$

The initial state s_0 may be a function of the key k and possibly some other random variable, such as an initialisation value (*IV*), as well. The purpose of a keystream generator G is to expand a short random key k into a long pseudorandom string[2]

$$G : \mathcal{K} \to \mathcal{Z}^l$$
$$z^l = G(k).$$

[2]In this respect one may think of a keystream generator as a pseudorandom sequence generator. But this is only partially true. For a pseudorandom generator, e.g. one used in Monte Carlo methods, it does not matter whether the sequence is predictable or whether, given a state, the previous or subsequent states of the generator can be found. But for a keystream generator this should be impossible. Ideally, knowing a small part of the message, e.g. "Dear Sir", should not enable an attacker to find out anything more of its contents.

Synchronous stream ciphers can further be specified according to the mode in which they are operated:

- *Counter Mode:*

$$s_{i+1} = F(s_i),$$
$$z_i = f(k, s_i),$$

  i.e. the next-state function does not depend on the key, but steps through (at least most of) the state space. F could, for example, represent an ordinary counter or a Linear Feedback Shift Register (LFSR), about which much more later in this chapter. We shall show that while the outputs of LFSRs have excellent statistical properties, LFSRs are cryptologically very weak when used on their own and therefore any cryptographic strength the cipher may have will have to come from the output function f.
- *Output Feedback Mode:* Here we have

$$s_{i+1} = F(k, s_i),$$
$$z_i = f(s_i),$$

  i.e. the output function does not depend on the key. Very often the output function produces a single bit which is simply XORed with the plaintext bit x_i.

These two modes of operation are also recognised modes of operation of block ciphers, which, as we observed earlier, essentially means that we use the block cipher algorithm as a (stream cipher) keystream generator.

Note that a synchronous stream cipher has no error propagation: an error in a ciphertext bit simply produces a single bit error in the plaintext. This is clearly a desirable property. But it comes at a price: a synchronous stream cipher demands perfect synchronization between sender and receiver. If a single bit is lost, or an extra bit is erroneously introduced, the receiver will only produce garbled data from that point onwards. In particular "late entry" is impossible using these techniques. This presents a problem in some applications: while stream ciphers are usually economical in the complexity of their implementation, and can also be very fast, their usability is limited to good communication channels where bit losses (or false bit insertion) are unlikely. The self-synchronising ciphers, which we'll look at next, provide a solution to problems of this kind, but—again— at a price!

7.1.3.2 Self-Synchronising Stream Ciphers

What we need is a method for the system to resynchronise itself when the vitally important synchronicity of sender and receiver is lost. The most common method of achieving such self-synchronization is the *cipher feedback mode:*

$$s_i = F(y_{i-1}, y_{i-2}, \dots, y_{i-N}),$$
$$z_i = f(k, s_i).$$

Thus the state of the stream cipher is determined by the previous N ciphertext symbols. Note that whereas the encryptor uses feedback, the corresponding decryptor has to employ feedforward. No feedback other than that of the ciphertext can be employed in the encryptor without losing the self-synchronising property. For example, the state of a finite state machine cannot be used as input into the function F. This implies in particular that the use of LFSRs (see Sect. 7.4) in self-synchronising ciphers is excluded. The cryptographic strength[3] resides in the output function: an attacker is at all times aware of the state of the encryptor or decryptor. Note that a self-synchronising cipher has only limited error propagation. If an error occurs in transmission, the decryptor will be in an incorrect state for N units of time, viz until N correct symbols have been received. This is obviously an attractive feature of such ciphers. Because of the self-synchronising property, ciphers based on this principle allow for "late entry".

Again, cipher feedback (CFB) is a recognised mode of operation of block ciphers (recognised in the sense that it appears as an option in international standards) but we are uncertain about the extent of its usage.

Synchronous stream ciphers can be implemented with a more limited self-synchronising feature, by stipulating that when some pre-agreed condition is met, both encryptor and decryptor re-synchronize. This appears to be the most common method of restoring or obtaining synchronisation between transmitter and receiver; remarkably little research has been reported in the open literature about the security of such implementations.[4]

7.1.3.3 Stream Ciphers with Memory

In a paper presented at SAC (Selected Areas in Cryptography) 2000, Jovan Goliĉ discussed what he called SCM: "Stream ciphers with memory".[5]

For these

$$s_{t+1} = F(k, s_t, x_t),$$
$$y_t = x_t + f(k, s_t).$$

"The next state function depends on the current plaintext symbol in such a way that both the encryption and decryption transforms have infinite input memory. This type [of stream cipher] is typically either not mentioned or just overlooked in the open literature on stream ciphers." Goliĉ claims that the PKZIP stream cipher is of this type.

---

[3]If any! We warn the reader here that there appears to be consensus in the cryptological community that self-synchronising stream ciphers are inherently weak. The eStream project (see Sect. 7.9) failed to find an acceptable one, and there is currently no standardisation body such as ISO/IEC or IEEE which has approved a self-synchronising cipher, other than block ciphers in CFB mode. In such cases the self-synchronising property does not hold if block boundaries are lost.

[4]Two exceptions are the paper by Daemen, Govaerts and Vandewalle: *Resynchronization weaknesses in synchronous stream ciphers*, in the Proceedings of Eurocrypt '93 (LNCS 765, Springer-Verlag, 1994) and, 10 years later, Armknecht, F., Lano, J. and Preneel, B.: *Extending the resynchronization attack*, Proc. SAC 2004, LNCS 3357, Springer-Verlag 2005.

[5]Proc. SAC 2000, LNCS 2012, pp. 233–247, Springer-Verlag 2001.

The SCM type of stream cipher is sensitive to both substitution and synchronization errors. Because of this, however, they may be useful for ensuring message-integrity, rather than message confidentiality. Errors may be dealt with using error-correction techniques, and their number kept within reasonable bounds through frequent resynchronisation.

7.2 Some Notes on Entropy

We have made the point that in an ideal world the ciphertext should not leak any information about the plaintext (except, grudgingly, perhaps the length of the plaintext message). But what exactly do we mean by "information"? What better way to answer that question than to turn to Information Theory? Which is what we'll do in this section.

7.2.1 Entropy = Uncertainty

To start with an example which, on the face of it, has nothing to do with cryptography or with communications even: Joe is playing poker. As a not quite honest player, he is anxious to have a peek at what is in his opponent's hand, because he is uncertain about how the cards in his hand compare with those of his opponent. If he could be certain about his opponent's hand, he would know how to bet (or to throw in his hand). In other words: getting this *information* would remove *uncertainty*.

This was, in essence, Shannon's great insight: information means the removal or lessening of uncertainty. Now the question is: what is uncertainty and/or how do we measure it? This is where *entropy* comes into our discussion. The term, and in fact the concept, was introduced by Shannon.[6]

The term "entropy" was taken from thermodynamics, where it can be interpreted as disorder (and as a numerical measure of disorder). By a very rough analogy, disorder can be taken as the absence of information: the more disorderly a system is, the more information is required to describe it: a randomly dealt hand in bridge takes more words to describe than a hand which contains the 13 Diamonds.

Leaving aside these vague ideas, let us consider Shannon's definition and we shall then show that the implications of this definition agree with our common sense views of the properties of uncertainty.

---

[6]Shannon, Claude E.:*A mathematical theory of communication*, Bell System Tech. J. **27** (1948), pp. 379–423 and 623–656. This paper represents the start of the mathematical and engineering discipline of Information Theory, of which Shannon is seen as the "father".

The following exposition is heavily based on the excellent exposition in the book by D. Welsh: *Codes and Cryptography*; Oxford University Press, 1988.

If you will pardon a very personal note: these two, reading Shannon's paper and some years later the book by Welsh, are what got me hooked on Cryptology.

Definition Let Z be a random variable which assumes the values $\{a_0, a, \ldots .a_{n-1}\}$ with probabilities $\Pr(a_i) = p_i$ for $i = 0, 1, \ldots, n-1$ (where $\sum_{i=0}^{n-1} p_i = 1$), then the entropy of Z is

$$H(Z) = -\sum_{i=0}^{n-1} p_i \log_2(p_i).$$

Strictly speaking $H(Z)$ is clearly dimensionless (i.e. it's a purely numerical expression), but it is sometimes convenient to express this as measuring uncertainty in bits. For example, suppose that a cryptographic key K consists of 128 randomly selected bits (determined, say, by flipping a coin 128 times). There are 2^{128} possible outcomes, each with probability $\frac{1}{2^{128}}$, and the entropy is

$$H(K) = -2^{128} \times \frac{1}{2^{128}} \times \log_2(\frac{1}{2^{128}}) = -\log_2(\frac{1}{2^{128}}) = 128$$

and it seems natural to refer to this uncertainty as being 128 *bits*.

Since in the definition, the function H depends only on the probability distribution of Z, we shall in the following express H as a function of the probabilities. We note the following properties:

1. $H(p_0, p_1, \ldots, p_{n-1}) = H(p_{\pi(0)}, H(p_{\pi(1)}, \ldots, p_{\pi(n-1)})$ for any permutation π of the set $\{0, 1, \ldots, n-1\}$. This seems an obvious requirement: rearranging the outcomes does not increase or decrease the uncertainty.
2. $H(p_0, p_1, \ldots, p_{n-1}) = 0$ if and only if one of the $p_i = 1$. If one of the outcomes is certain, there is no uncertainty! Note, by the way, that we take the value of $p\log(p)$ to be 0 if $p = 0$.
3. $H(p_0, p_1, \ldots .p_{n-1}, 0) = H(p_0, p_1, \ldots, p_{n-1})$. A zero probability, in case there are only a finite number of outcomes, implies impossibility. So our statement says that if one includes an impossibility in one's considerations, the uncertainty remains the same.
4. For any n

$$H(p_1, \ldots, p_n) \leq \log_2(n),$$

with equality if and only if $p_1 = p_2 = \cdots = p_n (= \frac{1}{n})$.

This states, for example, that there is less uncertainty about the outcome of flipping a biased coin ($\Pr(Heads) = \frac{3}{4}, \Pr(Tails) = \frac{1}{4}$, say) than if the coin is unbiased.

Proof Recall that the logarithmic function is a convex function, i.e. that

$$\ln x \leq x - 1$$

for all $x > 0$, with equality if and only if $x = 1$. Hence, if $(q_1, \ldots, q_n)$ is any probability vector, then

$$\ln\left(\frac{q_k}{p_k}\right) \leq \frac{q_k}{p_k} - 1,$$

with equality if and only if $q_k = p_k$. Hence

$$\sum_i p_i \ln\left(\frac{q_i}{p_i}\right) \leq \sum_k q_k - \sum_k p_k = 0$$

so that

$$\sum_i p_i \ln(q_i) \leq \sum_i p_i \ln(p_i).$$

If we put $q_i = \frac{1}{n}$ we get

$$H(p_1, \ldots, p_n) = -\sum_i p_i \ln(p_i) \leq \ln(n),$$

with equality as stated.[7]

5. In proving the previous statement, we also proved the following:
If $\sum_{i=0}^{n-1} p_i = \sum_{i=0}^{n-1} q_i = 1$, then the minimum of the function $G(q_0, \ldots q_{n-1}) = -\sum_{i=0} p_i \log(q_i)$ is achieved when $q_i = p_i$ for all i.
6. This result can in turn be used to prove that if X and Y are random variables, taking only finitely many values, then

$$H(X, Y) \leq H(X) + H(Y),$$

with equality if and only if X and Y are independent.

Example If the pairs $(0,0), (0,1), (1,0), (1,1)$ have the probability distribution $\frac{3}{8}, \frac{1}{8}, \frac{1}{8}, \frac{3}{8}$, then the entropy is approximately 1.81 bits. But the entropy of each of the two components is 1 (0 and 1 are equally likely as first component, and also as second component), so (with the obvious notation) $H(X, Y) \approx 1.81 < 2 = 1 + 1 = H(X) + H(Y)$. The difference arises from the fact that $\Pr(X = Y) = \frac{3}{4}$, so X and Y are not independent. Clearly this means that knowing something about X means that we know something about Y: in fact, if we know the first component, we have a 75 % chance of guessing the second component correctly, instead of just 50 %. Thus X "leaks" information about Y (and conversely, in fact). We shall soon make idea this more precise.

The proof of the statement is left as an exercise.

7. $H(\frac{1}{n}, \frac{1}{n}, \ldots \frac{1}{n}) = H(\frac{1}{n}, \frac{1}{n}, \ldots \frac{1}{n}, 0) \leq H(\frac{1}{n+1}, \frac{1}{n+1}, \ldots, \frac{1}{n+1}, \frac{1}{n+1})$, by 4. and 3. above. There is more uncertainty about the outcome of 6 rolls of a die, than about 5 rolls. And there is more uncertainty about a 64-bit key than about a 56-bit one.

[7] We have used the natural logarithm ln in the proof. This makes no difference to the conclusion: recall that $\log_2(x) = \frac{\ln(x)}{\ln(2)}$.

8.

$$H\left(\frac{1}{mn},\frac{1}{mn},\ldots,\frac{1}{mn}\right)=H\left(\frac{1}{m},\ldots,\frac{1}{m}\right)+H\left(\frac{1}{n},\ldots,\frac{1}{n}\right).$$

The uncertainty about a 128-bit key is the sum of the uncertainties about the first 32 bits and the uncertainty about the remaining 96 bits. But if someone tells you that the first 32 bits are all zeros, you have some valuable information!

The result as stated follows from 6. and from the fact that all the outcomes are independent (as is implied by their probabilities).

9. If $p = p_1 + p_2 + \cdots + p_m$ and $q = q_1 + q_2 + \cdots + q_n$, where each p_i and q_j is positive and $p + q = 1$, then

$$H(p_1,\ldots,p_m,q_1,\ldots,q_n)=H(p,q)+pH\left(\frac{p_1}{p},\ldots,\frac{p_m}{p}\right)+qH\left(\frac{q_1}{q},\ldots,\frac{q_n}{q}\right).$$

If Bob foolishly picks his password as the name of one of his four dogs and three cats, the total uncertainty in the outcome is the uncertainty about the species being canine or feline plus the weighted sum of the uncertainties given that the selected animal is respectively dog or cat.

This is a little tedious to prove, and we shall omit doing so.

Exercises

1. The input source to a noisy communication channel is a random variable X over the four symbols a, b, c, d. The output from this channel is a random variable Y over these same four symbols. The joint distribution of these two random variables is as follows:

$y\backslash x$	a	b	c	d
a	1/8	1/16	1/16	1/4
b	1/16	1/8	1/16	0
c	1/32	1/32	1/16	0
d	1/32	1/32	1/16	0

Compute $H(X), H(Y), H(X,Y)$.

2. Prove assertion 6 above. *Hint:* Put $\Pr(X = a_i) = p_i$ for $i = 1,\ldots,k$, $\Pr(Y = b_j) = q_j$ for $j = 1,\ldots,l$, $\Pr(X = a_i, Y = b_j) = r_{ij}$, and use assertion 5.
3. Two fair dice are thrown. Let X be the outcome of the first and Y the outcome of the second and let $Z = X + Y$. Show that

$$H(Z) < H(X) + H(Y).$$

4. Show that for any random variable X, $H(X, X^2) = H(X)$.

5. A random variable X has the binomial distribution with parameters n and p, i.e. for $0 \le k \le n$

$$\Pr(X = k) = \binom{n}{k} p^k q^{n-k},$$

where $0 < p < 1$ and $q = 1 - p$. Show that

$$H(X) \le -n(p \log p + q \log q).$$

7.2.2 *Conditional Entropy*

We continue to consider only random variables which can only take on a finite number of values. In the same way that one defines *conditional probabilities*, we now define the vitally important (for our purposes, anyway) concept of conditional entropy. Recall that the conditional probability of an event a given an event b is given by $\Pr(X = a|Y = b) = \frac{\Pr(X=a \text{ and } Y=b)}{\Pr(Y=b)}$. In defining *conditional entropy*, we unsurprisingly use conditional probabilities in the place of (common or garden) probabilities. This gives us:

The *conditional entropy* of X given A is

$$H(X|A) = -\sum_{k=1}^{m} P(X = x_k|A)\log(P(X = x_k|A)).$$

Similarly, if Y is any other random variable, taking values y_k with $1 \le k \le m$ then we define the conditional entropy of X given Y by

$$H(X|Y) = \sum_j H(X|Y = b_j)P(Y = y_j).$$

One can think of $H(X|Y)$ as the uncertainty of X given a particular value of Y, averaged over the range of values that Y can take.

Some consequences, in agreement with one's intuitive conceptions, which will follow from our next theorem are

- $H(X|X) = 0$. (There is no uncertainty about X if X is given!)
- $H(X|Y) \le H(X)$. This is in accordance with our interpretation of entropy as uncertainty: if further *information* in the form of Y becomes available, the uncertainty about X must be reduced or, at worst remain the same. This gives an indication of how "information" may be defined.
- $H(X|Y) = H(X)$ if X and Y are independent. (Y has no effect on X, so knowing what it is does not change the uncertainty about X.)

Which brings us to the important

Theorem *For any pair of random variables X and Y that take only a finite set of values*

$$H(X, Y) = H(Y) + H(X|Y).$$

Proof The proof is purely computational, based on the definitions:

$$\begin{aligned}
H(X,Y) &= -\sum_i \sum_j P(X = x_i, Y = y_j)\log(P(X = x_i, y = y_j)) \\
&= -\sum_i \sum_j P(X = x_i, Y = y_j)\log(P(X = x_i|Y = y_j)P(Y = y_j)) \\
&= -\sum \sum p_{ij} \log(P(X = x_i|Y = y_j)) - \sum \sum p_{ij}\log(P(Y = y_j)) \\
&= -\sum \sum P(X = x_i|Y = y_j)P(Y = y_j)\log(P(X = x_i|Y = y_j)) + H(Y) \\
&= -\sum_j P(Y = y_j) \sum_i P(X = x_i|Y = y_j)\log(P(X = x_i|Y = y_j)) + H(Y) \\
&= \sum_j P(Y = y_j)H(X|Y = y_j) + H(Y) \\
&= H(X|Y) + H(Y).
\end{aligned}$$

7.2.3 Information

We are now ready to define information, and the definition should not surprise you: With X and Y as before, the *information* conveyed by Y about X is defined as:

$$I(X|Y) = H(X) - H(X|Y).$$

In other words, the information conveyed by Y about X is the amount of uncertainty about X which is removed by knowledge of Y.

Clearly, following from the observations made in the discussion earlier, we have

- $I(X|X) = H(X)$;
- $I(X|Y) = 0$ if only if X and Y are independent.

It may also be noted that $I(X|Y) = I(Y|X)$. It is therefore customary to use the notation $I(X, Y)$ and refer to this concept as the *mutual* information.

Exercises

1. With X and Y as in Exercise 1 of Sect. 7.2.1. What is $I(X, Y)$?
2. A die (fair and six-sided) is cast twice. How much information is contained in each of the following sentences?

   (a) One of the dice shows a 6.
   (b) The sum of the outcomes is 6.

7.2.4 Redundancy

If a message contains a single bit of information, it is, in theory, possible to encode that message as a single bit. For example, sending a single 0 or 1 as a reply to a simple "no" or "yes" question is a perfectly clear response (provided, of course, that the encoding 1 = "yes", 0 = "no" has been agreed upon beforehand). In a similar way if a message contains n bits of information, it should, in theory, be possible to encode that information in a binary string of length exactly n. In practice, this is rather hard to do.

For example, there are 26 letters in the alphabet available, if we ignore such details as upper and lower case, spaces and punctuation marks. Since $\log_2(26) \approx 4.7$, it should therefore, in an ideal world, be possible to write messages in such a way that every letter conveys some 4.7 bits of information. Instead, a rough estimate of written English is that every letter conveys, on average, about 1.2 bits. An outstanding example of such wastage is the fact that the letter "q" is *always* followed by the letter "u" in English, so that the second letter does nothing useful whatsoever.[8]

Working in bits, the *redundancy* in a message M is the difference between the message length (in bits) and the amount of information in the message.

There are two aspects about redundancy worth noting: On the positive side, redundancy allows for the correction of errors in transmission; this is the fundamental principle of error-correcting codes.[9] By extending the message with some further data, completely determined by the data, and therefore completely redundant in the information-theoretic sense, the recipient is enabled to determine what the data is, even if some errors were introduced in transmission. By way of example, the reader should have no difficulty in reading t th emd fo tihs sntnce.

On the negative side, redundancy can be exploited by a cryptanalyst. If every diplomatic message contains the sentence "We have the honour to inform Your Excellency", the cryptanalyst has some data on which he or she can base a known-plaintext attack. More generally, the frequency of characters is uneven; "e" is by far the most common letter in written English. This makes frequency analysis of simple substitution ciphers possible. For such reasons, it is highly unwise to apply error-correcting coding *before* encryption: the coding introduces known relationships between the message bits, which can be exploited. In any case, after encrypting the encoded message, applying the error correcting mechanism to the received ciphertext will fail. The fact that the bits of codewords before encryption satisfy certain linear equations does not mean that the ciphertext strings will satisfy those equations—in fact, it would be a very poor cipher if they did. Moreover, decrypting corrupted ciphertext strings will lead to text which contains too many errors to allow the error correction to function properly. So encryption after encoding completely undermines the whole purpose of using error correcting codes.

There are data compression techniques available (such as those which "zip" files) which remove some of the redundancy in data. Applying such techniques before encryption seems

---

[8]"Texting", which turns 'See you tomorrow" into "C u 2morro", shows the reduction of such redundancy.

[9]To which we return, for completely different reasons, in Sect. 9.3.

to be a good idea: not only is there less to encrypt, but the statistics of the (compressed) plaintext are much harder to exploit. Interestingly, Kelsey discovered a kind of side-channel attack exploiting the use of data compression algorithms in well-known protocols. The (quite minimal) information provided about their inputs by the size of their outputs can be exploited to break the encryption, thus contradicting our "common sense" belief to a limited extent.[10]

There are some unexpected relationships in this connection. For example, one would expect that encrypted data would be indistinguishable from purely random data; in fact, that is the aim of encryption. It should therefore be impossible to compress an encrypted file. But a group from the University of California, Berkeley, managed to do so nevertheless.[11] (Not surprisingly, such a reversal of the natural order of compression followed by encryption leads to great inefficiencies, and for practical purposes is useless. There appears to have been little work following up on this surprising result.)

7.2.4.1 A Note on Algorithmic Information Theory

Chaitin has described *algorithmic information theory* (AIT) as "the result of putting Shannon's information theory and Turing's computability theory into a cocktail shaker and shaking vigorously. The basic idea is to measure the complexity of an object by the size in bits of the smallest program for computing it."[12]

A more or less theoretically inclined computer scientist can probably do better than the description that follows; the reader whose interest is highly concentrated on cryptology may safely skip it as we shall not need it at any stage.

But for those who remain: If X is a bitstring, we denote by $H_K(X)$ the length of the shortest program on a (given) universal Turing machine (UTM) that generates X. It can then be shown that the function $H_K(X)$ has properties similar to those of the (Shannon) entropy that we have been discussing. $H_K(X)$ is called the Kolmogorov or Kolmogorov–Chaitlin complexity of X.

Kolmogorov complexity and entropy are not the same thing, however, even if they are related: Kolmogorov complexity deals with individual strings, whereas Shannon entropy deals with the statistical behaviour of information sources. The connection between the two is suggested by the fact that if a source S has entropy $H(S)$, then the *average* word length of an optimal binary *uniquely decodable code* will lie between $H(S)$ and $H(S) + 1$.

A uniquely decodable code[13] is one in which the code words may have different lengths, but no codeword is the same as the first bits of any other codeword (or, in other words, once a codeword has been received, it is immediately identifiable as the encoding of its source word).

---

[10] Kelsey, J.: *Compression and information leakage of plaintext*; Proc. FSE 2002, LNCS 2365, Springer-Verlag, 263–276. A little earlier in Benedetto, D., Caglioti, E. and Loreto, V.: *Language trees and zipping*, published on-line as ArXiv.cond-mat/0108530v2.pdf, the authors showed how to determine, from the zipped version of a document, the language in which the original was written.

[11] Johnson, M., Ishwar, P., Prabhakaran, V., Schonberg, D. and Ramchandran, K.: On compressing encrypted data, IEEE Trans. Signal Processing **52** (2004), pp. 2992–3006.

[12] Webpage of the Centre for Discrete Mathematics and Theoretical Computer Science, University of Auckland, New Zealand. Accessed 24 October 2015.

[13] We need to warn the unwary that a *code* is not the same as a *cipher*, contrary to popular usage. In coding data there is no intention of preserving secrecy or authenticity. See also the footnote in Sect. 7.6.3. There is more about codes in Sect. 9.5.

In such a code the encoding of a string can be taken as representing a method of generating that string.

The theorem quoted is, not unexpectedly, due to Shannon.

By way of example, consider the source $S = \{a, b, c, d\}$ with probabilities

$$P(a) = 0.4,$$
$$P(b) = 0.3,$$
$$P(b) = 0.2,$$
$$P(d) = 0.1.$$

Then $H(S) = 1.28$. Under the uniquely decodable encoding

$$a \mapsto 0,$$
$$b \mapsto 10,$$
$$c \mapsto 110,$$
$$d \mapsto 111,$$

the average word length is $1.9 \in [H(S), H(S) + 1]$.

7.3 Perfect Secrecy

Having defined the concept of *mutual information*, we are now in a position to investigate the relationship between plaintext (X) and ciphertext (Y).

Let again $\mathcal{X}$, $\mathcal{Y}$ and $\mathcal{K}$ denote the plaintext, ciphertext and key spaces respectively. Shannon in the second of his two most famous papers[14] defined the concepts of

- *key equivocation*: $H(K|Y)$, the uncertainty about the key when ciphertext is given; and
- *message equivocation*: $H(X|Y)$, the uncertainty about the plaintext when ciphertext is given.

The relationship between these two is given by the following

Theorem *The key equivocation is related to the message equivocation by*

$$H(K|Y) = H(X|Y) + H(K|X, Y).$$

Proof The proof depends only on repeated applications of the identity

$$H(A|B) = H(A, B) - H(B).$$

[14] Shannon, C.E.: *Communication Theory of Secrecy Systems*, Bell System Techn. J. **28**, pp. 657–715 (1949).

This gives, on the one hand

$$\begin{aligned} H(X|Y) &= H(X,Y) - H(Y) \\ &= H(X,K,Y) - H(K|X,Y) - H(Y) \end{aligned}$$

and on the other

$$\begin{aligned} H(K|Y) &= H(K,Y) - H(Y) \\ &= H(X,K,Y) - H(X|K,Y) - H(Y). \end{aligned}$$

But clearly

$$H(X|K,Y) = 0,$$

since if both key and ciphertext are known, there is no uncertainty about the plaintext. Hence

$$H(K|Y) = H(X,K,Y) - H(Y) = H(X|Y) + H(K|X,Y),$$

which was to be proved.

Note that this theorem immediately implies that $H(K|Y) \geq H(X|Y)$, or, in words, the key equivocation is at least as big as the message equivocation.

Let us now consider a practical matter: the cryptographer wants to ensure that anyone intercepting an encrypted message will not be able to extract any information from it; in other words, his or her desire is to have $I(X,Y) = 0$.

So the question is: What does $I(X,Y) = 0$ actually imply?

As we have seen, the condition $I(X,Y) = 0$ is equivalent to the condition that X and Y are statistically independent. This, in turn, is equivalent to the condition that $H(Y) = H(Y|X)$. A cryptosystem with this property is said to have *perfect secrecy*.

Perfect secrecy is hard to accomplish in practice, as the following theorem shows:

Theorem *A necessary condition for a cryptosystem to have perfect secrecy is that* $H(K) \geq H(X)$.

Proof

$$\begin{aligned} H(K) &= H(K,Y) - H(Y|K) \\ &= H(X,Y,K) - H(X|Y,K) - H(Y|K) \\ &= H(X,Y,K) - H(Y|K) \\ &\geq H(X,Y) - H(Y|K) \\ &= H(X) + H(Y|X) - H(Y|K) \\ &= H(X) + H(Y) - H(Y|K) \end{aligned}$$

because X and Y are statistically independent. Therefore

$$H(K) \geq H(X)$$

because $H(Y|K) \leq H(Y)$.

This result is usually quoted as saying that for perfect secrecy the key must be at least as long as the message. In fact it says (more precisely) that the entropy of the key space must be at least as great as the entropy of the message space. This is the principle behind the one time pad, which, while providing perfect secrecy, is highly impractical in all but the applications requiring *very* high security, such as, in the 1960s, the hotline used for fax transmissions between Moscow and Washington.

But the ambition of every cipher designer is to achieve a situation in which the ciphertext is for practical purposes *indistinguishable* from a truly random stream of characters (bits, usually), so that the entropy of the plaintext, given the ciphertext, is, again for practical purposes, equal to the number of bits in the message.

Exercises

1. Suppose that the message space consists of all n-digit decimal integers, and that encipherment is by a Caesar cipher in which the key K is a single decimal digit, with addition modulo 10. (For example: if $n = 3$, $X = 450$, and $K = 7$ then $Y = (4+7) \bmod 10\ (5+7) \bmod 10\ (0+7) \bmod 10 = 127$.)

   Under the assumption that all values of X and all values of K are equally likely, find $H(X|C)$ and $H(K|C)$.
2. Prove that for any cryptosystem

$$H(K, C) = H(M) + H(K).$$

7.4 Linear Feedback Shift Registers

After that somewhat theoretical approach to cryptography, let us return to more practical matters. How does one generate a "good" keystream (whatever "good" may mean)?

Feedback shift registers are very fast generators of binary sequences, which can, moreover, be conveniently implemented in hardware. A (general) feedback shift register of *length n* consists of n memory cells, each holding one bit, and the state of the device is described by the sequence $\{s_0, s_1, \ldots, s_{n-1}\}$ of the contents of these registers. The state transition function is of the form

$$\{s_0, s_1, \ldots, s_{n-1}\} \longmapsto \{s_1, s_2, \ldots, s_{n-1}, f(s_0, s_1, \ldots, s_{n-1})\}$$

where f is a mapping from $\{0, 1\}^n$ to $\{0, 1\}$, i.e. a Boolean function.[15] Thus the Boolean function f is evaluated, the state vector is shifted one space to the left, and its new last component is the output of f.

Boolean functions are, usually, easy to implement in hardware. At time t the shift register will output s_t and its state will change from $\{s_t, s_{t+1}, \ldots, s_{t+n-1}\}$ to $\{s_{t+1}, s_{t+2}, \ldots, s_{t+n}\}$, where $s_{t+n} = f(s_t, s_{t+1}, \ldots, s_{t+n-1})$. In the simplest (and hopelessly insecure!) application of this to stream ciphers, the tth plaintext bit x_t is simply XORed with s_t to produce the tth ciphertext bit $y_t = x_t \bigoplus s_t$.

In the case where f is a linear function, say

$$\begin{aligned} f(s_t, s_{t+1}, \ldots, s_{t+n-1}) &= c_0 s_t + c_1 s_{t+1} + \cdots + c_{n-1} s_{t+n-1} \\ &= \sum_{i=0}^{n-1} c_i s_{t+i} \end{aligned}$$

(with addition modulo 2, of course), we speak of a *Linear Feedback Shift Register* or LFSR. Equivalently, we may write

$$\sum_{i=0}^{n} c_i s_{t+i} = 0$$

where $c_n = 1$ by definition.

We shall refer to the polynomial

$$f(x) = \sum_{i=0}^{n} c_i x^i$$

as the *connection polynomial* of the LFSR.

For the time being, we shall assume that $c_0 = 1$, since otherwise the output sequence will yield $s_0, \ldots,$ before degrading to the output of an LFSR of length less than n. We also note that, with a linear connection polynomial, the LFSR will output a constant sequence of 0s should it ever be in the all-zero state.

Example Suppose that the next state function f is given by

$$s_{t+4} = s_t + s_{t+1}$$

as in the figure below, and that the initial state is $s_0 = 1, s_1 = s_2 = s_3 = 0$.

[15] More generally, the s_i could come from any set, such as a field $\mathbb{F}$, in which case the function f would be a function mapping $\mathbb{F}^n$ to $\mathbb{F}$. In that case the cells would hold elements of $\mathbb{F}$ as well. We discuss only the case where $\mathbb{F} = GF(2)$.

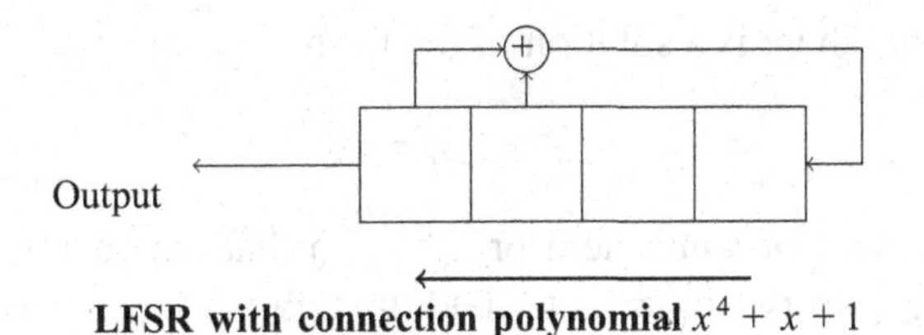

LFSR with connection polynomial $x^4 + x + 1$

Then (writing the subsequent states in the order $\{s_t, s_{t+1}, s_{t+2}, s_{t+3}\}$) the LFSR passes through the states

1000
0001
0010
0100
1001
0011
0110
1101
1010
0101
1011
0111
1111
1110
1100

before repeating itself. We now ask ourselves whether there is a neat closed formula for the output sequence 100010011010111 . . .

In the theory of (linear) difference[16] equations, there is a standard trick for finding a solution. Suppose we are given a difference equation

$$s_{t+n} = \sum_{i=0}^{n-1} c_i s_{t+i}.$$

[16]Sic. Not to be confused with linear differential equations, even though there are some beautiful analogies between the two theories.

We then assume that there is a solution of the form

$$s_t = \alpha^t$$

where α is an element of some field or other, so that exponentiation makes sense. In our case we shall take it to be some finite field, though we haven't yet established which one. We shall try the smallest $GF(2^n)$ in which the connection polynomial has a root. (In the case of difference equations, the field would probably be $\mathbb{R}$ or $\mathbb{C}$.) We substitute this into the equation, and solve for α. Observe that if α^t is a solution, then so is $A\alpha^t$ for any constant A and we can use A later to fit the solution to the initial conditions, if these are given.

Example Continued

In the example above, we are faced with the equation

$$s_{t+4} = s_t + s_{t+1}.$$

Assuming that $s_t = \alpha^t$, and substituting this into the equation, we get

$$\alpha^{t+4} = \alpha^{t+1} + \alpha^t$$

or

$$\alpha^4 = \alpha + 1.$$

But this is precisely the equation which α, in our example of $GF(2^4)$ as an extension of $GF(2)$ (Sect. 5.6), was designed to satisfy! In fact there are four roots: $\alpha, \alpha^2, \alpha^4$ and α^8:

$$x^4 + x + 1 = (x - \alpha)(x - \alpha^2)(x = \alpha^4)(x - \alpha^8).$$

So the most general solution to the equation $s_{t+4} = s_{t+1} + s_t$ would be

$$s_t = A\alpha^t + B\alpha^{2t} + C\alpha^{4t} + D\alpha^{8t}$$

for some constants A, B, C, D, which we will determine later in order to satisfy the initial conditions $s_0 = 1, s_1 = s_2 = s_3 = 0$.

However, the reader will now raise the perfectly valid objection that each of the components of the state of the LFSR, i.e. each of the s_t, is supposed to be either 1 or 0, and not an element of some large field. This is where the trace function, with its linearity, comes to the rescue. For if

$$\alpha^{t+4} = \alpha^{t+1} + \alpha^t$$

then also

$$\text{Tr}(\alpha^{t+4}) = \text{Tr}(\alpha^{t+1}) + \text{Tr}(\alpha^t),$$

so that $s_t = \text{Tr}(\alpha^t)$, or, more generally, $s_t = \text{Tr}(A\alpha^t)$, is a solution.

We can now adjust A to fit the initial conditions $s_0 = 1, s_1 = s_2 = s_3 = 0$. A is an element of the field $GF(2^4)$, so we can write it as $A = a_0 + a_1\alpha + a_2\alpha^2 + a_3\alpha^3$, where the $a_i \in \{0, 1\}$. Substituting this into the initial conditions we get

$$a_0\mathrm{Tr}(1) + a_1\mathrm{Tr}(\alpha) + a_2\mathrm{Tr}(\alpha^2) + a_3\mathrm{Tr}(\alpha^3) = 1,$$
$$a_0\mathrm{Tr}(\alpha) + a_1\mathrm{Tr}(\alpha^2) + a_2\mathrm{Tr}(\alpha^3) + a_3\mathrm{Tr}(\alpha^4) = 0,$$
$$a_0\mathrm{Tr}(\alpha^2) + a_1\mathrm{Tr}(\alpha^3) + a_2\mathrm{Tr}(\alpha^4) + a_3\mathrm{Tr}(\alpha^5) = 0,$$
$$a_0\mathrm{Tr}(\alpha^3) + a_1\mathrm{Tr}(\alpha^4) + a_2\mathrm{Tr}(\alpha^5) + a_3\mathrm{Tr}(\alpha^6) = 0$$

or $a_3 = 1, a_2 = 0, a_1 = 0, a_0 + a_3 = 0$, so $a_0 = 1$. Hence $A = 1 + \alpha^3 = \alpha^{14}$ and the complete solution is

$$s_t = \mathrm{Tr}(\alpha^{t+14}).$$

We can draw some interesting conclusions from this example.[17]

Definition A *pseudo-noise* or PN-sequence is an output sequence from an n-stage LFSR with period $2^n - 1$.

We shall refer to a subsequence $s_a, s_{a+1}, \ldots, s_{a+N}$ (for some a and N) of the output sequence which consists entirely of 1s as a *block*, and a similar subsequence which consists entirely of 0s as a *gap*.

- If the connection polynomial is primitive, then the LFSR will run through all the powers of a root α of that polynomial, and hence will generate a PN-sequence. (Unless it starts in the all-zero state, of course.)
- If the connection polynomial is irreducible but not primitive, the sequence generated will have a period which is a divisor of $2^n - 1$: In this case a root of the connection polynomial will be some element β of the field $GF(2^n)$, whose order will be a divisor of the order of the multiplicative group of nonzero elements of $GF(2^n)$, i.e. of $2^n - 1$.
- A PN-sequence satisfies the following conditions, known as Golomb's randomness postulates.[18]
  - **G1:** $\mathrm{Tr}(\alpha^k)$ will be output exactly once in each period. There are $2^{n/2}$ values of k for which the trace is 1 and $2^{n/2} - 1$ values for which the trace will be 0 (because 0 will never occur). The output sequence is therefore "almost" balanced.
  - **G2:** We note that every one of the possible $2^n - 1$ nonzero states will occur once per period. Looking at these states, we note that there are 2^{n-k-2} states in which the leftmost

[17] In general one should not "prove" mathematical results from examples. On the other hand, a mathematical proof is not very helpful when considering applications, unless it helps clarify matters. In this case, I don't believe that rigorous proofs would contribute to understanding, and I have recourse to the fact that we are only interested in applying all this stuff. But please don't tell anyone.

[18] These go all the way back to 1967: Golomb, S.W.: *Shift Register Sequences*, Holden-Day, San Francisco, 1967.

$k+2$ coordinates are of the form $1000\dots01$ and the same number with the leftmost $k+2$ coordinates of the form $0111\dots10$. Therefore gaps and blocks of length $k \leq n-2$ will occur exactly 2^{n-k-2} times per period. The state $0111\dots11$ will occur once per period. Its successor is $111\dots11$ (Why can't it be $111\dots10$?), which is followed by $111\dots10$ (Why not by $111\dots11$?). Hence there is one block of length n and there are no blocks of length $n-1$. In the same way there is no gap of length n and there are two gaps of length $n-1$.

- **G3:** A shifted version of $\{s_k | k \geq 0\}$, i.e. of $\{\text{Tr}(\alpha^k) | k \geq 0\}$ is $\{\text{Tr}(\alpha^{k+l}) | k \geq 0\}$. Now the number of agreements between these two sequences is the number of zeros in the sequence $\{\text{Tr}(\alpha^k + \alpha^{k+l}) | k \geq 0\}$ which is itself only a shifted version of the original sequence.[19] Hence the number A of agreements is $2^{n/2} - 1$ and the number D of disagreements is $2^{n/2}$: $A - D = -1$.

- Note that it also follows from our discussion that in an LFSR (with a connection polynomial in which the constant term is nonzero) every state has a unique predecessor; in fact if the state and the connection polynomial are known, there is no difficulty in finding all the previous states and outputs.

All this (except possibly the last observation) is most satisfactory: With a primitive connection polynomial we get an output sequence which looks pretty random, and without too much effort we get a long period (for if $n = 128$, say, we get a period of more than 3×10^{38}). It is therefore very sad that *PN-sequences are very unsafe*.

For if $2n$ consecutive bits are known (for example, through a known plaintext),[20] then one can easily determine the coefficients of the connection polynomial and hence all previous and all future outputs. This follows from the fact that $s_{t+n} = \sum_{i=0}^{n-1} c_i s_{t+i}$. Now if $\{s_t, s_{t+1}, s_{t+2}, \dots, s_{t+2n-1}\}$ are known, then we have a system of equations which may be written in matrix form as

$$\mathbf{A}(c_0, c_1, \dots, c_{n-1})^T = (s_{t+n}, s_{t+n+1}, \dots, s_{t+2n-1})^T,$$

where the superscript 'T' denotes the transpose of a vector, and where **A** is the matrix

$$\begin{pmatrix} s_t & s_{t+1} & \cdots & s_{t+n-1} \\ s_{t+1} & s_{t+2} & \cdots & s_{t+n} \\ \cdots & \cdots & \cdots & \cdots \\ s_{t+n-1} & s_{t+n} & \cdots & s_{t+2n-2} \end{pmatrix}.$$

Note that $2n$ output bits are required to determine the connection polynomial uniquely.

---

[19] $1 + \alpha^l = \alpha^m$ for some m, because α is a primitive element.

[20] A known plaintext attack, as the name suggests, is an attack in which the cryptanalyst tries to find the key used, having available some plaintext and the corresponding ciphertext. This is a much more common event than one might at first imagine, because of the ubiquity of common headers in files and messages, such as `#include <stdio.h>` in C-code, and "Dear Sir" in letters. A common piece of plaintext used by the cryptanalysts at Bletchley Park reading the German Enigma traffic during World War II was "Nothing to report", or, more probably, "Nichts zu melden".

Exercises

1. Find the connection polynomial of the LFSR of length 3, which generates the sequence

$$\ldots 1\,0\,1\,0\,0\,1 \ldots$$

and find the next ten output bits.
2. Show that no connection polynomial of degree 3 can generate the sequence

$$\ldots 1\,1\,1\,0\,1\,1\,1 \ldots$$

3. Find the general solution of the difference equation $s_{t+5} = s_{t+4} + s_t$. (*Hint:* The factorisation of $x^5 + x^4 + 1 \in GF(2)[x]$ is $(x^2 + x + 1)(x^3 + x + 1)$. You will have to work in $GF(2^6)$ in order to get a solution; your general solution should contain two arbitrary constants.) What is the period of a sequence generated by an LFSR with this as connection polynomial?
4. If $f(x) \in GF(2)[x]$ is a primitive polynomial, what can you say about the period of an LFSR which has $f(x)^2$ as connection polynomial? (*Hint:* Try it with, say, $x^4 + x^2 + 1$ or $x^8 + x^2 + 1$.)

7.5 LFSRs: Further Theory

We may start a further theoretical discussion with the observation that an LFSR represents a type of finite state machine: An abstract finite state machine is defined by a number of states (with a designated initial state) and a state update function which represents the transition from one state to another. In a deterministic finite state machine, the next state is uniquely determined by the current state and input events. In the case of an LFSR the input may be regarded as the ticking of a clock: when the clock ticks, the LFSR shifts to its next state.

It is obvious that a finite state machine when it keeps running must eventually reach a state in which it has been previously; in the case of a deterministic finite state machine, when a state is repeated, the entire sequence of states will then repeat, because of the uniqueness of the next state.

Thus the output of any LFSR will eventually be periodic: denoting the state of the LFSR by $\mathbf{s}_t$ (or, if preferred, writing $\mathbf{s}_t = (s_t, s_{t+1}, s_{t+2}, \ldots, s_{t+n-1})$, where n is the length of the LFSR) there will exist a T and an N such that

$$\mathbf{s}_{t+T+i} = \mathbf{s}_{t+i} \ \forall\, i \geq N.$$

The smallest value of T with this property, would, of course, be considered to be the period of the LFSR. Note that if T is the period, then, for all t large enough (i.e. large enough so that the non-periodic part of the sequence has been worked out of the system) one has that

$$\mathbf{s}_{T+t} \oplus \mathbf{s}_t = \mathbf{0}.$$

Now suppose that $s_0, s_1, s_2, \ldots$ is any periodic binary sequence. Consider the set of all polynomials (of arbitrary, but finite, degree) $\sum_i a_i x^i \in GF(2)[x]$ with the property that $\sum_i a_i s_i = 0$. This set is nonempty, since, by the previous observation $x^T + 1$ belongs to it. It is not hard to see that the set is, in fact, an ideal, and, because $GF(2)[x]$ is a principal ideal ring, there exists a unique polynomial which generates it. We may call this polynomial the minimal polynomial of the sequence. If $f(x) = \sum_i^n a_i x^i$ belongs to the ideal, then we must have that $f(x)$ is a multiple of the minimal polynomial, but we also know that $\sum_i^n a_i s_i = 0$, i.e. that

$$s_n = a_{n-1}s_{n-1} + a_{n-2}a_{n-1} + \cdots + a_0 s_0,$$

so $f(x)$ may be considered a connection polynomial for an LFSR generating the sequence.

Example Consider the sequence

$$00010\ 01101\ 01111 \text{ recurring.}$$

This satisfies both the equations

$$s_{t+7} = s_{t+5} + s_{t+3} + s_{t+2} + s_t$$

with initial state 0001001, and

$$s_{t+6} = s_{t+5} + s_{t+4} + s_{t+3} + s_t$$

with initial state 000100. The gcd of the two polynomials is $x^4 + x + 1$ (which is irreducible) so we could also generate the sequence using this as connection polynomial, which must indeed be the minimal polynomial.

We note the following theorem:

Theorem *Let* $\mathbf{s}$ *be the output of an LFSR of length L with connection polynomial f. Then f is the minimal polynomial of* $\mathbf{s}$ *if and only if L is the smallest value of k for which the first* $k+1$ *state vectors*

$$\mathbf{s}^{(0)} = (s_0, s_1, s_2, \ldots, s_{L-1}),$$
$$\mathbf{s}^{(1)} = (s_1, s_2, s_3, \ldots, s_L),$$
$$\ldots$$
$$\mathbf{s}^k = (s_k, s_{k-1}, \ldots, s_{L+k-1})$$

are linearly dependent.

For the purpose of the proof of this theorem (and for other purposes, such as getting a better mental grip on things, as well), we note here that the state transition may be described by means of a matrix multiplication:

If the vector $\mathbf{s}_t = (s_t, s_{t+1}, \ldots, s_{t+L-1})$ describes the state of the LFSR at time t, then

$$\mathbf{s}_{t+1} = \mathbf{s}_t\mathbf{A},$$

where

$$\mathbf{A} = \begin{pmatrix} 0\,0\,0 \ldots 0 & a_0 \\ 1\,0\,0 \ldots 0 & a_1 \\ 0\,1\,0 \ldots 0 & a_2 \\ \vdots\;\vdots\;\vdots & \vdots \\ 0\,0\,0 \ldots 1 & a_{L-1} \end{pmatrix}$$

is the *companion matrix* of the connection polynomial $\sum_{i=0}^{L} a_i x^i$.

Proof of Theorem Suppose the state vectors $\mathbf{s}_0, \ldots, \mathbf{s}_{k-1}$, where $k < L$, are linearly dependent, say

$$\lambda_0 \mathbf{s}^{(0)} + \lambda_1 \mathbf{s}^{(1)} + \cdots + \lambda_{k-1}\mathbf{s}^{(k-1)} = \mathbf{0},$$

where $k < L$. Then multiplying this equation repeatedly by the matrix $\mathbf{A}$ one gets

$$\lambda_0 \mathbf{s}^{(t)} + \lambda_1 \mathbf{s}^{(t+1)} + \cdots + \lambda_{k-1}\mathbf{s}^{(t+k-1)} = \mathbf{0}\ \forall\, t \geq 0.$$

In particular, considering, say, the first component of each vector in this equation, we have that

$$\sum_{i=0}^{k-1} \lambda_{i+t} s_{i+t} = 0,$$

which means we could have generated the sequence by using the connection polynomial $\sum_{i=0}^{k-1} \lambda_i x^i$ of a lower degree, contradicting the fact that f was supposed to be the minimal polynomial.

Conversely, it is easily seen that if the minimal polynomial has degree less than L, then the vectors $\mathbf{s}_0, \mathbf{s}_1, \ldots, \mathbf{s}_{L-1}$ cannot be linearly independent.

The value L defined by the preceding theorem is called the *linear complexity* of the sequence $\mathbf{s} = s_0 s_1 s_2 \ldots$, and as should be clear from the above, represents the length of the shortest LFSR that can generate the sequence. The theorem actually shows how this may be found, and one may also determine (from the linear dependence) the connection polynomial of the LFSR required. The following exercises put this into practice. We warn, however, that this is a very inefficient method, especially if the sequence has high linear complexity. A better method, called the Berlekamp–Massey algorithm, is discussed in Sect. 7.8.

Exercises Use the method suggested by the theorem to find the linear complexity of the following sequences:

10101 00001 10010

and

10010 11000 00101 11001

and sketch LFSRs which will, with suitable initialisation, generate these sequences.

7.6 Galois Field Counters

7.6.1 Shift Registers in Feedforward Mode

There is an equivalent method of generating PN-sequences which sometimes has practical advantages when it comes to implementation. This is known as the method of Galois Field Counters, where one essentially uses "feedforward" rather than feedback. This technique may be useful in cases where block ciphers (with block length = n) are used in what is called *Counter Mode*. In this mode the input of the block cipher consists of the state of the counter, and the output is simply XORed bitwise onto blocks of plaintext to produce ciphertext.

Let again

$$f(x) = \sum_{i=0}^{n} c_i x^i$$

be a primitive polynomial. Again we use a shift register of length n. (Let us, for the sake of avoiding argument, label the individual registers from left to right as $s_{n-1}, s_{n-2}, \ldots, s_0$, and assume that shifting takes place to the left. This is the opposite of the order in which we labelled the cells of the LFSR.) We now identify shifting with multiplication by α, where α is a root of the polynomial. We regard the contents of s_i as the coefficient of α^i. Then the state transition function is the function

$$F(s_{n-1}, s_{n-2}, \ldots, s_0) = (s'_{n-1}, s'_{n-2}, \ldots, s'_0),$$
$$s'_i = s_{i-1} + c_i s_{n-1} \qquad \forall i = 1, \ldots, n-1,$$
$$s'_0 = 0 + c_0 s_{n-1}.$$

This amounts to saying that if α^{n-1} does not appear in the field element, then multiplication by α is simply a shift (the coefficient of 1 becomes the coefficient of α, the coefficient of α becomes the coefficient of α^2, etc.). But if α^{n-1} occurs (in other words, when $s_{n-1} \neq 0$), then, apart from the shift, one also has to take account of the fact that

$$\alpha^n = \sum_{i=0}^{n-1} c_i \alpha^i$$

and adjust the coefficients of the α^i's accordingly. It is clear that the resulting output sequence (obtained by, for example, taking z_t as the coefficient of α^{n-1} after t such multiplications by α) has exactly the same properties as one obtained from an LFSR with the same connection polynomial, since in both constructions we essentially run through all the powers of some primitive element α.

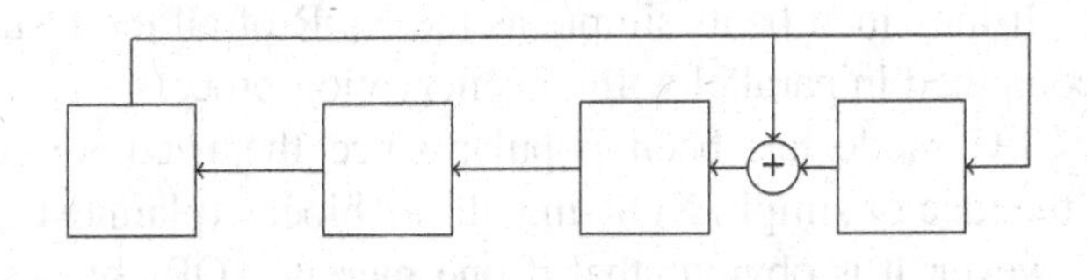

A feedforward shift register representing multiplication by $\alpha \in GF(2^4)$. Here $\alpha^4 = \alpha + 1$, the leftmost register contains the coefficient of α^3, the next the coefficient of α^2, etc.

7.6.2 *Finite Field Multiplication*

It is clear that this description corresponds with our implementation of finite field arithmetic as we described in Sect. 6.4.3: the contents of the shift register is interpreted as an element of $GF(2^n)$ and a shift of the register then corresponds with multiplication by α, where α is a root of the connection polynomial. From this it follows, of course, that if the connection polynomial is irreducible but not primitive, the period of the Galois counter will not be $2^n - 1$, but only $2^m - 1$, where m is a proper factor of n.

If the connection polynomial of the LFSR is reducible, the situation gets a bit more complicated, as you will have found in Exercises 3 and 4 of Sect. 7.4.

7.6.3 *Galois Counters and Authenticated Encryption*

Contrary to intuition and to a belief which is still widely held among users of cryptography, encryption of plaintext does not actually give any guarantee of the authenticity of the plaintext or ciphertext received. Authentication requires a further operation on either plaintext or ciphertext, usually involving hashing and a digital signature (by means of the RSA mechanism or something similar) or else the use of a "Message authentication Code" or

MAC.[21] A MAC may be considered to be a keyed hash function and the receiver will only accept a message if it has been correctly signed or if the correct (secret) key has been used for computing the MAC (which is sent along as a final block, a "tag", with the encrypted message).

These forms of combining secrecy with authentication lead to a degree of inefficiency, since two passes through the data are now required: once to encrypt the data and once more in order to hash-and-sign or compute the MAC. In recent years various methods have been proposed to increase the efficiency: some of them depend on using the outputs of a Galois Field Counter (with secret initial state) which are then XORed with the blocks of plaintext, before encryption. Encryption takes place by means of a block cipher of adequate strength, usually in CipherBlock Chaining (CBC) mode.[22]

The MAC itself may then be as simple as the XOR of all the resulting ciphertext blocks, which can be computed in parallel with the encryption process.

The use of CBC mode has been popular since the need for authentication was first identified, and the idea of simply XOR-ing all the blocks (plaintext or ciphertext) goes back just as long. However, it is obvious that if one merely XORs blocks then an adversary can rearrange the blocks in any order he or she wishes without this affecting the "tag", so the receiver will accept the altered message as authentic. By using the outputs of a Galois Field Counter as a "whitener" through XORing them with the message blocks before encryption, the position of a block in the message is fixed by the value of the whitener.

7.7 Filter and Combining Functions

Our interest in LFSRs springs from the fact, already noted in Sect. 7.4, that the sequences they generate have excellent statistical properties. In fact, if our only intention was to generate "random looking" binary sequences, we could hardly do better than just take the output of an

[21]This terminology is now so well established that there is no reasonable hope of changing it. It is nevertheless unfortunate: *codes* are employed in communication in order to reduce bandwidth or, more commonly nowadays, to enable the receiver to correct random errors introduced in the transmission process. However, when it comes to authentication, what is needed is protection against changes in the message deliberately introduced by an intelligent adversary, which won't be random at all. Thus a MAC is not a code in the technical sense.

[22]CBC is one of the recognised modes of operation of block ciphers. Let the message consist of n-bit blocks $P_1, P_2, \ldots, P_m$, and denote the encryption function (with block length n) using key K by E_K. The encryption of the message is the sequence of ciphertext blocks $C_0, C_1, \ldots, C_m$, where C_0 is some initial value, and

$$C_i = E_K(C_{i-1} \oplus P_i) \text{ for } i = 1, \ldots, m.$$

Decryption is obvious:

$$P_i = D_K(C_i) \oplus C_{i-1}.$$

Note that if P_i is modified, then all ciphertext blocks from the ith one onwards are changed. This diffusion property probably explains its popularity in designs of MACs.

LFSR. If we wish to encrypt a message, and let us assume that it is in the form of a binary string, we can simply take the output bits generated by the LFSR and XOR them with the plaintext bits, to obtain something which looks like a random string of bits. In other words, we treat the "random looking" output string of the LFSR as if it is a one-time pad.

This is a very tempting approach, because LFSRs have several desirable properties:

- They are easy to implement, especially in hardware.
- Such hardware implementations are fast.
- They can produce sequences with long periods.
- The sequences produced have good statistical properties.
- As already noted, they can be readily analysed—which also brings disadvantages!

The only problem with this is that, while such a cryptographic system may keep your neighbour's teenage son without access to your correspondence, we have already seen in Sect. 7.4 that PN-sequences are very unsafe.

In order to rescue something from the wreckage, we need to introduce nonlinearity in some form or other, by some means or other, since it is the linearity of all the operations concerned that makes finding the initial loading of the LFSR easy to find, as the discussion in Sects. 7.4 and 7.5 has shown. Once the state of the LFSR is known at any instant, one can simply run the LFSR forward or backward to find its state at any other time.

We shall discuss various methods to obtain nonlinearity, using two distinct approaches. On the one hand we may take the state vector of a single LFSR and use it as the input into a highly nonlinear Boolean function to form the keystream. The relevant function would then be called a *filter function*. On the other hand we might consider several LFSRs running simultaneously and use a highly nonlinear *combining function* which takes the outputs of these LFSRs as input and returns keystream bits.

In practice, it turns out that every construction of stream ciphers seems to have its advantages as well as its weaknesses and disadvantages, and also that apart from nonlinearity other properties of the functions used are required and need to be taken into account. Frequently these other properties are incompatible to some degree and compromises have to be made. It is probably mainly for these reasons that stream ciphers are gradually disappearing from the scene, remaining in use, if at all, in situations where either cost is a significant factor or where very high encryption speeds are required.

But first we need to deal with another matter.

7.8 Linear Complexity

7.8.1 Introduction

Having agreed that linearity in the generation of a bitstream is undesirable, we discuss in this section a method of determining how "far" such a stream differs from a sequence generated by an LFSR. This idea is formalized in the concept of linear complexity, which we touched upon in our earlier discussion of LFSRs and which we now define properly below.

We shall use the following notation:

s denotes an infinite sequence with terms $s_0, s_1, s_2, \ldots$
s^n denotes a sequence of length n with terms $s_0, s_1, s_2, \ldots, s_{n-1}$.
We shall also need the notations

$$s(x) = s_0 + s_1x + s_2x^2 + \ldots$$

and

$$s^n(x) = s_0 + s_1x + s_2x^2 + \ldots s_{n-1}x^{n-1}.$$

We shall say that an LFSR generates a sequence s (or s^n) if there is some initial state for which the output of the LFSR will be s (respectively s^n).

Definition The *linear complexity* $L(s)$ of an infinite binary sequence s is

- 0 if s is the all-zero sequence $0000\ldots$
- ∞ if no LFSR can generate s
- L, if the length of an LFSR of minimal length which can generate s is L.

In the same way the linear complexity of a finite binary sequence s^n is defined as the length of a shortest LFSR that can generate a sequence having s^n as its first n terms.

We have seen that this could also be defined as the maximum number of linearly independent vectors among the "state vectors"

$$\mathbf{s}_0 = s_0s_1s_2\ldots$$

$$\mathbf{s}_1 = s_1s_2s_3\ldots$$

$$\mathbf{s}_2 = s_2s_3s_4\ldots$$

$$\ldots$$

Some properties of linear complexity:

1. If s^n has linear complexity $L(s^n)$ then $0 \le L(s^n) \le n$.
2. $L(s^n) = 0$ if and only if s^n is the all-zero sequence.
3. $L(s^n) = n$ if and only if $s^n = 000\ldots01$.
4. If s is periodic with period p then $L(s) \le p$.
5. $L(s||t) \ge L(s)$, where "$||$" denotes concatenation.
6. $L(s \bigoplus t) \le L(s) + L(t)$.

Proofs 1, 3. If s^{n-1} is not the all-zero sequence, we can load a LFSR of length $n-1$ with s^{n-1} as its initial state. By suitably choosing the coefficients in the connection polynomial, we can ensure that the next output bit assumes the correct value. Hence $L(s^n) \le n-1$. On the other hand, if s^{n-1} consists entirely of zeros and $s_n = 1$, then no choice of connection polynomial can give us s_n, so the only way to produce this output is to load an LFSR of length n with 000...01.

Items 2, 4 and 5 are self-evident.

But to prove item 6 needs a lot more work. It's worth doing, however, because it proves that combining the outputs of several LFSRs by simple XORing does little to improve the security.

7.8.2 The Berlekamp–Massey Algorithm

We shall eventually arrive at the Berlekamp–Massey algorithm, which is an efficient method of determining the linear complexity of a finite sequence.[23]

First of all note that any LFSR can be described as an ordered pair $< f(x), L >$, where $f(x)$ is the connection polynomial and L is the length of the LFSR. Note that the degree of the connection polynomial need not be equal to the length of the shift register, This allows for LFSRs whose output only eventually becomes periodic. Take, for example, the sequence 00010101010... This can be described by

$$s_0 = 0,$$
$$s_1 = 0,$$
$$s_2 = 0,$$
$$s_3 = 1,$$
$$s_{n+2} = s_n \quad \forall n \geq 2,$$

and the shortest LFSR that can generate this sequence consequently has length 4, even though the connection polynomial only has degree 2. In fact, the connection polynomial is $f(x) = 1 + x^2$:

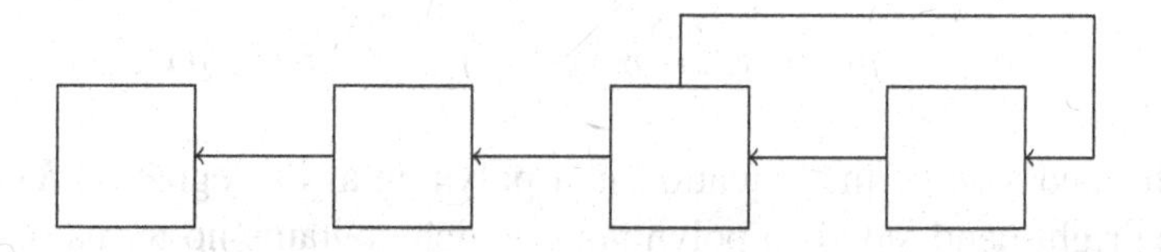

Now consider the LFSR $< f(x), L >$ with output $s_0, s_1, s_2, \ldots$, put $s(x) = s_0 + s_1x + s_2x^2 + \cdots + s_nx^n + \ldots$, and consider the product $p(x) = s(x)f(x)$.

Then even though $s(x)$ is an infinite series, $p(x)$ is just a polynomial, of degree less than L: For, in terms of the s_i, we have, from the connection polynomial, that

$$\sum_{j=0}^{L} d_j s_{n-L-j} = 0 \qquad \forall n \geq L.$$

[23] In what follows we shall largely follow the exposition of J.L. Massey in a set of notes used in lectures offered at the University of Pretoria in 1991.

But the left-hand side of this equation is precisely the coefficient of x^n in the product $s(x)f(x)$ if $n \geq L$.

The following theorem *can* be proved, but we'll take it on trust:

Theorem *The LFSR* $< f, L >$ *generates the sequence* $s_0, s_1, \ldots$ *if and only if*

$$s(x) = \frac{p(x)}{f(x)},$$

where $p(x)$ *is a polynomial of degree less than* L. *Moreover, if this condition is satisfied for polynomials* $f(x)$ *and* $p(x)$, *with* $f(0) = 1$ *and* $gcd(f(x), p(x)) = 1$, *then* $< f(x), L >$ *is the shortest LFSR that generates* $s_0, s_1, s_2, \ldots$, *where* $L = max\{$ *degree of* $f(x)$, $1 +$ *degree of* $p(x)\}$.

Consider, for example, the sequence 00010 101010... again. Here $s(x) = x^3 + x^5 + x^7 + \cdots = \frac{x^3}{1+x^2}$ and, by the foregoing, $< 1 + x^2, 4 >$ is indeed the shortest LFSR which can generate the sequence. In this case $L = 1 + \deg(p(x)$.

On the other hand, the sequence 110110110... gives us $s(x) = 1 + x + x^3 + x^4 + \cdots = (1 + x)(1 + x^2 + x^4 + \ldots) = \frac{1+x}{1+x^2}$ and $L =$ the degree of the numerator.

We next show that if an infinite sequence s has finite (and nonzero) linear complexity K, then this is also the linear complexity of any subsequence s^n, provided that $n \geq 2K$:

Suppose that $< f_1(x), K_1 >$ and $< f_2(x), K_2 >$, with $K_1, K_2 \leq K$, both generate s^n. Then there exist polynomials $p_1(x)$ and $p_2(x)$ of degrees less than K_1, K_2 respectively, and determined by s^{K_1} and s^{K_2} respectively, such that

$$\frac{p_1(x)}{f_1(x)} - \frac{p_2(x)}{f_2(x)} = x^n \Delta,$$

where Δ is some (possibly unknown) infinite series, i.e.

$$p_1(x)f_2(x) - p_2(x)f_1(x) = x^n \Delta f_1(x)f_2(x).$$

But the left-hand side of this equation is a polynomial of degree $< K_1 + K_2 \leq 2K \leq n$, whereas the right-hand side is a polynomial which contains no terms of degree less than n. Hence $\Delta = 0$. It follows that *if* $n \geq 2K$ *then every LFSR of length* K*, or less, that generates* s^n *will generate* s.

We are now in a position to prove number 6 of the properties listed on page 156, which, we recall, states the following:

$$L(s^n \oplus t^n) \leq L(s^n) + L(t^n).$$

Proof If $L(s^n) + L(t^n) \geq n$ then there is nothing to prove, since the linear complexity of $s^n \bigoplus t^n$ is at most n. So we assume that $L(s^n) + L(t^n) < n$. Let $< f_1(x), L_1 >$ and $< f_2(x), L_2 >$ generate s^n and t^n, respectively. Put $s^n(x) = s_0 + s_1 x + \cdots + s_{n-1}x^{n-1}$ and define $t^n(x)$ in the same way. Then

$$\frac{p_1(x)}{f_1(x)} = s^n(x) + x^n\Delta_1,$$

$$\frac{p_2(x)}{f_2(x)} = t^n(x) + x^n\Delta_2,$$

so that

$$\frac{p_1(x)}{f_1(x)} + \frac{p_2(x)}{f_2(x)} = \frac{p_1(x)f_2(x) + p_2(x)f_1(x)}{f_1(x)f_2(x)}$$
$$= s^n(x) + t^n(x) + x^n(\Delta_1 + \Delta_2),$$

which means that $< f_1(x)f_2(x), L_1 + L_2 >$ generates $s^n \bigoplus t^n$.

As an immediate consequence of this result, we have the following:
If $< f(x), L >$ generates s^n, but not s^{n+1}, then $L(s^{n+1}) \geq n + 1 - L$.

Proof Let t^{n+1} differ from s^{n+1} only in the last bit, so that $s^{n+1} \bigoplus t^{n+1} = 00\ldots01$. Then $L(s^{n+1} \bigoplus t^{n+1}) = n+1 \leq L(s^{n+1}) + L(t^{n+1}) \leq L(s^{n+1}) + L$, since t^{n+1} is generated by the same LFSR as s^n. That's it.

Example $01100110 = s^8$ and $011001100 = s^9$ are generated by $< 1 + x + x^2 + x^3, 3 >$, but $011001101 = t^9$ is generated by $< 1 + x + x^2 + x^3 + x^4 + x^5 + x^6, 6 >$.

In order to arrive at the Berlekamp–Massey algorithm we shall need the following result:

Lemma *Suppose that* $< 1 + d_1x + \cdots + d_Lx^L, L >$ *generates* s^n *but possibly not* s^{n+1} *(which means that necessarily* $L \leq n$*). Then*

$$\frac{p(x)}{f(x)} = s^{n+1}(x) + \delta_n x^n + x^{n+1}\Delta,$$

where δ_n *is the "next discrepancy", defined by*

$$\delta_n = s_n + d_1 s_{n-1} + \cdots + d_L s_{n-L}$$

and $< 1 + d_1x + \cdots + d_Lx^L, L >$ *also generates* s^{n+1} *if and only if* $\delta_n = 0$.

Proof The coefficient of x^n in $f(x)s^{n+1}(x)$ is $s_n + d_1 s_{n-1} + \cdots + d_l s_{n-L} = \delta_n$. It therefore follows that $< f(x), L >$ correctly produces the last bit s_n of s^{n+1} if and only if $\delta_n = 0$.

We also need the following, dubbed by Massey the "Two-Wrongs-Make-a-Right Lemma":

Lemma *Suppose that* $m < n$ *and that* $< f^*(x), L^* >$ *generates* s^m *but not* s^{m+1}*, and that* $< f(x), L >$ *generates* s^n *but not* s^{n+1}.

Then $< f(x) - \frac{\delta_n}{\delta_m^*} x^{n-m} f^*(x), \max\{L, L^* + n - m\} >$ *generates* s^{n+1}.

Proof We use the notation $\Delta(x)$ for various Δs for series whose precise form is of no consequence.

By the previous lemma

$$\frac{p^*(x)}{f^*(x)} = s^{m+1}(x) - \delta_m^* x^m + x^{m+1}\Delta^*(x),$$

$$\frac{p(x)}{f(x)} = s^{n+1}(x) - \delta_n x^n + x^{n+1}\Delta(x).$$

The first of these implies that

$$\frac{p^*(x)}{f^*(x)} = s^{n+1}(x) - \delta_m^* x^m + x^{m+1}\Delta_1(x),$$

absorbing the error into the $x^{m+1}\Delta_1(x)$ term, so that

$$p^*(x) = f^*(x)s^{n+1}(x) - \delta_m^* x^m + x^{m+1}\Delta_2(x)$$

by multiplication, remembering that $f^*(0) = 1$, and absorbing the other terms into $x^{m+1}\Delta_2$.

From the second we have, similarly

$$p(x) = f(x)s^{n+1}(x) - \delta_n x^n + x^{n+1}\Delta_3(x)$$

for the same reason: $f(0) = 1$.

Hence

$$p(x) - \frac{\delta_n}{\delta_m^*} x^{n-m} p^*(x) = \left[f(x) - \frac{\delta_n}{\delta_m^*} x^{n-m} f^*(x)\right] s^{n+1}(x) + x^{n+1}\Delta^\dagger(x).$$

Comparing the term on the left-hand side and the first term on the right-hand side, we note that the former is of degree $< \max\{L, L^* + n - m\}$, while the term within the square brackets on the right is of degree $\leq \max\{L, L^* + n - m\}$ and has constant term 1.

Thus, if we put $p^\dagger(x) =$ the left-hand side, and $f^\dagger(x) =$ the term in square brackets on the right, we have

$$p^\dagger(x) = f^\dagger(x)s^{n+1}(x) + x^{n+1}\Delta^\dagger(x)$$

or

$$\frac{p^{\dagger}(x)}{f^{\dagger}(x)} = s^{n+1}(x) + x^{n+1}\Delta^{\sharp}(x),$$

as required.

The following theorem is the fundamental one for the understanding of the Berlekamp–Massey algorithm, which we shall state afterwards:

Theorem ("Linear Complexity Change Theorem") *If* $< f(x), L(s^n) >$ *generates* s^n *but not* s^{n+1} *then*

$$L(s^{n+1}) = \max\{L(s^n), n + 1 - L(s^n)\}.$$

Proof By induction on n.

If $s^n = 000\ldots0$ then $L = 0$, $s^{n+1} = 000\ldots01$, so that $L(s^{n+1}) = n + 1$, as required.

If $s^n \neq 000\ldots0$, let m be the sequence length for which

$$L(s^m) < L(s^{m+1}) = L(s^n).$$

The induction hypothesis gives us that

$$L(s^n) = m + 1 - L(s^m).$$

Let $< f^*(x), L^* >$ (where $L^* = L(s^m)$) be an LFSR that generates s^m but not s^{m+1}.

Then, by the "Two-wrongs-make-a-right" lemma

$$\begin{aligned} L(s^{n+1}) &\leq \max\{L(s^n), L^* + n - m\} \\ &= \max\{L(s^n), n + 1 - L(s^n)\}. \end{aligned}$$

Of course, $L(s^{n+1}) \geq L(s^n)$, and by the observation on page 156

$$L(s^{n+1}) \geq n + 1 - L(s^n),$$

so that

$$L(s^{n+1}) = \max\{L, L^* + n - m\}.$$

This brings us to the concept of the *linear complexity profile*:

Definition Let $s = s_0s_1s_2\ldots$ be a binary sequence, and let s^n be defined as before by $s^n = s_0s_1s_2\ldots s_n$. Let L_N denote the linear complexity of s^N. Then the sequence $L_1, L_2, L_3, \ldots$ is called the linear complexity profile of s.

Now we are ready and able to state

The Berlekamp–Massey Algorithm

INPUT: a binary sequence $s_0 s_1 s_2 \dots s_{n-1}$ of length n.
OUTPUT: the linear complexity $L(s^n)$

1. *Initialisation:* Put
   $f(x) = 1$
   $L = 0$
   $m = -1$
   $B(x) = 1$
   $N = 0$
2. While $N < n$:
   - Compute the next discrepancy $d = s_N + \sum_{i=1}^{L} d_i s_{N-i}$
   - if $d = 1$ then
     Put $Temp(x) = f(x)$
     Put $f(x) = f(x) + x^{N-m} B(x)$
     If $L \leq N/2$ put $L = N + 1 - L, m = N, B(x) = Temp(x)$
   - Put $N = N + 1$
3. Return L.

Example For the sequence 1101 1000 0100 1101 0010 0011 0001 1100 we get the following results from the Berlekamp–Massey algorithm (the reader is cordially invited to check these):

n	L	$f(x)$
0	1	$x+1$
1	1	$x+1$
2	1	x
3	2	x^2+1
4	2	x^2+x+1
5	2	x^2+x+1
6	4	$x^4+x^3+x^2+x$
7	4	x^4
8	4	x^4
9	5	x^5+x^2+x+1
10	5	x^5+x^2+x+1
11	5	x^5+x^2+x+1
12	5	x^5+x^2+x+1
13	5	x^5+x^2+x+1
14	9	$x^9+x^6+x^5$
15	9	$x^9+x^6+x^5$
16	9	$x^9+x^6+x^5$
17	9	$x^9+x^5+x^3+x^2+x$
18	9	$x^9+x^5+x^3+x^2+x$
19	10	$x^{10}+x^6+x^5+x^4+x^3+x+1$
20	10	$x^{10}+x^6+x^5+x^4+x^3+x+1$
21	10	$x^{10}+x^6+x^5+x^4+x^3+x+1$
22	12	$x^{12}+x^9+x^8+x^7+x^6+1$
23	12	$x^{12}+x^{11}+x^9+x^8+x^5+x^4+x^2$
24	12	$x^{12}+x^{11}+x^9+x^8+x^5+x^4+x^2$
25	12	$x^{12}+x^{11}+x^9+x^8+x^5+x^4+x^2$
26	14	$x^{14}+x^{13}+x^{12}+x^{11}+x^8+x^6+x^2+x+1$
27	14	$x^{14}+x^{10}+x^9+x^8+x^7+x^6+x^4+x^3+1$
28	14	$x^{14}+x^{10}+x^9+x^8+x^7+x^6+x^4+x^3+1$
29	15	$x^{15}+x^{12}+x^7+x^6+x^5+x^4+x^3+x^2+1$
30	15	$x^{15}+x^{12}+x^7+x^6+x^5+x^4+x^3+x^2+1$

7.8.3 *The Linear Complexity Profile*

Let us first of all make the (obvious?) observation that for cryptological applications a high linear complexity on its own is not sufficient. Take for example, the sequence $\{s_i\}_{i=0}^{999}$ defined by

$$s_i = 1 \quad \text{if} \quad 0 \leq i < 500,$$
$$s_i = 0 \quad \text{if} \quad 500 \leq i < 1000.$$

This has linear complexity 501 (the connection polynomial is x^{500}), which looks pretty satisfactory, but its linear complexity profile shows that the sequence is indeed as "unrandom" as one would expect from such a silly sequence. There is only one jump in the profile.

Nevertheless, applying the Berlekamp–Massey algorithm may be useful in assessing the "randomness" (whatever that means) of a given sequence. If the sequence has high linear complexity in the Berlekamp–Massey sense and a "good" linear complexity profile, it may be (or thought to resemble) a purely random sequence.

A disadvantage of the algorithm, however, is its complexity: the number of operations required increases as the square of the length of the sequence.

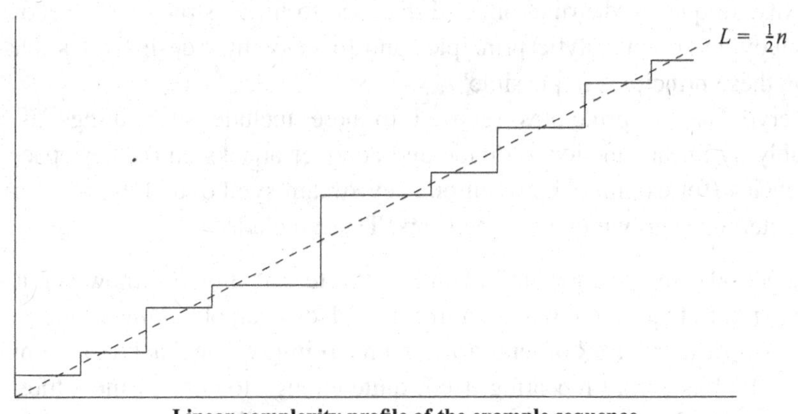

Linear complexity profile of the example sequence

We state without proof some further properties of linear complexity and linear complexity profiles.

- Let s^n be chosen randomly from the set of all binary sequences of length n. The expected linear complexity of s^n is

$$E(L(s^n)) = \frac{n}{2} + \frac{4 + B(n)}{18} - \frac{1}{2^n}\left(\frac{n}{3} + \frac{2}{9}\right).$$

  Here $B(n)$ is the parity of n, i.e. $B(n) = 0$ if n is even, and $B(n) = 1$ if n is odd. The variance (i.e. the square of the standard deviation) is approximately $\frac{86}{81}$ for moderately large n.

- One can plot the linear complexity of s^n against n and should obtain a set of points which closely follow the line $L = n/2$. (We know that we have an increasing function, and that a jump in the linear complexity can only occur from below this line, to a point symmetrically positioned above the line as illustrated by the results of the randomly selected binary sequence tabulated in the previous subsection.) This provides a statistical test for randomness: if the linear complexity profile deviates too far from the line $L = n/2$ the sequence being considered is (probably) not random (e.g. the silly sequence we have just considered). On the other hand, if the correlation is too good, one should also be suspicious.
- It can be shown that the sequence s_n defined by $s_n = 1$ if $n = 2^j - 1$ for some j, and $s_n = 0$ otherwise, follows the line $l = n/2$ as closely as possible. But one would hardly consider this to be a random sequence!

7.9 On the Design of Stream Ciphers

Rueppel[24] describes the system-theoretic approach to stream ciphers as having two distinct aspects. Firstly, there is the objective of developing "methods and building blocks that have provable properties with respect to certain system-theoretic measures such as period, linear complexity, frequency distribution, and distance to linear structures". Secondly, we should aim at studying "cryptanalytic principles and [developing] design rules that render attacks based on these principles impossible".

The cryptanalytic principles referred to here include such things as approximation (preferably by linear functions), divide-and-conquer attacks on the key space, and statistical dependencies (for example, between one key stream symbol and the next). A set of general design criteria has grown up over the years. These include:

- Long period. Since the generator is a finite state machine, we know that it will eventually have to return to a state it has been in before. However, obtaining a long period is the least of our worries: an LFSR of length 128 with a primitive feedback polynomial will generate some 10^{38} bits before repeating itself, quite enough to encrypt most files. Unfortunately, knowing just 256 consecutive ones of those bits will enable an attacker to find all the others.
- Large linear complexity, "good" linear complexity profile. If the output sequence has a relatively low linear complexity, L say, it may be considered as simply the output of an equivalent LFSR of that length, and as we know, would be "broken" if the attacker possesses just $2L$ bits of keystream.
- Statistical criteria, such as good distribution of k-tuples of keystream bits. In fact, recent criteria for stream ciphers demand that the keystream should be indistinguishable from

[24] In his article on stream ciphers in the book edited by Gustavus Simmons, which we referenced in footnote 1 of this chapter.

a "genuine" random bit sequence, by means of any statistical tests which run in time polynomial in the length of the sequence. This is a tough requirement.
- Confusion: every keystream bit must be a function of many or preferably all of the key bits;
- Diffusion: redundancies in substructures must be spread evenly over the output;
- Non-linearity criteria, for example distance between the Boolean function(s) used and affine functions.

This list of course needs updating as new attacks become known. The reader will raise the objection to this approach that we are, in MBA-speak, using a *reactive* as opposed to a *pro-active* one. This, however, seems to be unavoidable. Just as the medical profession cannot discover cures for diseases which do not yet exist, so the stream cipher designer can only attempt to make his or her cipher secure against attacks which have succeeded against previous designs.[25]

It may also be worthwhile remembering that what the stream cipher designer is attempting to do is, under a strict interpretation, impossible. He or she is trying to create a one-time pad, i.e. a bit string of unlimited length with unbounded entropy. All he or she has to use for this purpose is a key string with (on a good day) n bits of entropy, if the key length is n. Clearly, since entropy cannot be manufactured out of nothing, this cannot be done. A more realistic interpretation of the designer's job is to require from him or her that finding the key cannot be done by any method more efficient than an exhaustive search through the entire key space. Let us hope that this is achievable.

In the years 2000–2003 the NESSIE (New European Schemes for Signatures, Integrity and Encryption) project funded by the European Commission was run, with as its objective establishing a new set of cryptographic primitives (such as ciphers, signature schemes, message authentication codes, etc.) which would provide better security and confidence levels than the ones then in use. While to some extent modelled on the Advanced Encryption Standard "competition", run by the U.S. National Institute for Standards and Technology (NIST) from 1997 to 2000, NESSIE's scope was much wider. A call for proposals was issued in February 2000 and a total of 39 were received, of which five for stream ciphers and twelve for block ciphers. Sadly, none of the proposed stream ciphers were judged to be satisfactory.

NESSIE was followed by another research initiative, eCRYPT, one of whose target areas was specifically symmetric encryption, and, as a result of the failure of NESSIE to find an acceptable stream cipher, the eStream project was established to concentrate on these. Proposals were requested for, essentially, two classes of stream ciphers:

1. Stream ciphers for software implementation with high throughput, with a key length of 128 bits;
2. Stream ciphers for hardware implementation with limited resources (such as memory, gate count, power consumption, etc.), and a key length of 80 bits.

[25]The designer who breaks his own cipher at the time of designing it, though not completely unknown, remains an exception.

The latter category may be considered to be that of the "niche" applications, to which we referred.

Thirty-four proposals were received, of which sixteen made it into the final eStream portfolio, eight in each of the two categories. Details and specifications of the survivors may be found in the book *New Stream Cipher Designs*.[26] It may be interesting to consider the criteria according to which they were judged. In a way they are rather more bland than those of Rueppel:

1. Security. In fact this trumps all other requirements. It might be argued that Rueppel's requirements are (some of) the means to achieve this.
2. Performance, when compared with the AES cipher in one of the possible "stream cipher modes" (as referred to in Sect. 7.1).
3. Performance, when compared to the other submissions.
4. Justification of the design and its analysis, as given in the submitted documentation.
5. Simplicity and flexibility.
6. Completeness and clarity of the submitted documentation.

It should be noted that in the discussion of the following sections we restrict ourselves to designs which are essentially based on LFSRs. While this is a rather traditional approach,[27] it is worth noting that several of the eStream finalists include LFSRs in their design, as do the GSM telephone encryption algorithms A5/1 (which has been broken) and A5/2 (which is even weaker), and E0, used in Bluetooth, which is also weak. But the LFSR based members of the SOBER family are considerably stronger and SOBER-128 remains unbroken, as far as this author is aware. SNOW3G, chosen as the 3GPP (3rd Generation Partnership Project, representing collaboration of various communications associations) encryption mechanism is also LFSR-based.[28]

Of the constructions that we deal with in the following sections, the first two, namely combination and filter generators, are the most important, and certainly the most widely studied.[29]

[26]Robshaw, M. and Billet, O. (eds.): *New Stream Cipher Designs*, LNCS 4986, Springer-Verlag, 2008.

[27]"Stream cipher design has traditionally focussed on bit-based linear feedback registers (LFSRs), as these are well studied and produce sequences which satisfy common statistical criteria." Dawson, E., Henrickson, M. and Simpson, L.: *The Dragon stream cipher: Design, analysis and implementation issues*; in Robshaw and Billet, *op. cit.*, pp. 20–38.

[28]The reader may also wish to read Alex Biryukov's article *Block Ciphers and Stream Ciphers: The State of the Art*, now somewhat dated, obtainable at https://eprint.iacr.org/2004/094.ps.

[29]In their analysis of stream cipher designs (Braeken, A. and Lano, J.: *On the (im)possibility of practical and secure nonlinear filters and combiners*, Proc. Selected Areas in Cryptography 2005, LNCS 3897, Springer-Verlag, pp. 159–174.) Braeken and Lano suggest that this is a good reason to prefer ciphers based on these constructions when considering dedicated stream ciphers. They also come to the conclusion that using nonlinear filters is more practical than using combiners.

7.10 Combination Generators

In a combination generator, the outputs of several independent LFSRs are combined by means of a (nonlinear, of course, because otherwise there is little point in this exercise) function f, called the *combining function*.

We note here the following:

Any Boolean function f of m Boolean variables $x_0, x_1, \ldots, x_{m-1}$ can be written as the sum (modulo 2) of distinct mth order products of the variables, i.e.

$$f(x_0, x_1, \ldots, x_{m-1}) = a_0 + \sum_i a_i x_i + \sum_{i_1,i_2} a_{i_1 i_2} x^{i_1} x^{i_2} + \ldots$$
$$+ \sum_{i_1,i_2,\ldots,i_{m-1}} a_{i_1 i_2 \ldots i_{m-1}} x^{i_1} x^{i_2} \ldots x^{i_{m-1}} + a_{0,1,\ldots,m-1} x_0 x_1 \ldots x_{m-1},$$

where the as are constants belonging to the set $\{0, 1\}$. This expression of the function is known as its *algebraic normal form* or ANF. In the next chapter we shall describe an efficient way of determining the algebraic normal form of a function whose values are only known from a table or listing.

Example The function f of four Boolean variables whose values are as given in the following table

$x_3x_2x_1x_0$	$f(x)$	$x_3x_2x_1x_0$	$f(x)$
0000	1	1000	1
0001	0	1001	1
0010	1	1010	0
0011	1	1011	0
0100	0	1100	0
0101	0	1101	0
0110	1	1110	0
0111	1	1111	1

has the ANF

$$f(x_3, x_2, x_1, x_0) = 1 + x_0 + x_0x_1 + x_2 + x_0x_2$$
$$+ x_1x_2 + x_0x_1x_2 + x_0x_3$$
$$+ x_1x_3 + x_0x_1x_3 + x_0x_2x_3.$$

This function (which is balanced, by the way, having equal numbers of 0s and 1s in its output) therefore has degree 3, this being the degree of the term(s) with highest degree and nonzero coefficient.[30]

[30]This is frequently in the cryptological literature referred to as the "algebraic degree" or even the "(algebraic or nonlinear) order". The word "algebraic" appears to be unnecessary, redundant and superfluous, to coin a phrase. There also appears to be no reason to use the overworked word "order" when mathematicians have happily agreed on the term "degree".

For the moment we merely point out the following interesting theorem:

Theorem[31] *Suppose that n maximum-length LFSRs, whose lengths* $L_0, L_1, \ldots, L_{n-1}$ *are pairwise distinct, with* $L_i > 2$ *for all i, are combined by a function* $f(x_0, \ldots, x_{n-1})$. *Then the linear complexity of the resulting keystream is* $f(L_0, L_1, \ldots, L_{n-1})$ *(evaluated as an integer-valued function).*

For example, taking the function of the example above to combine four LFSRs of lengths 3, 4, 5 and 6, would result in a keystream with linear complexity 320. This is considerably better than what we could have got by simply XORing the outputs.

An early proposal for a combination generator was a design by Geffe, which sadly acquired its fame for a wrong reason: its use of a poor combining function.

The Geffe generator used three maximum-length LFSRs, of pairwise relatively prime lengths L_0, L_1 and L_2, and the combining function

$$f(x_2, x_1, x_0) = x_0x_1 \oplus x_1x_2 \oplus x_2.$$

This is cryptographically a poor choice, because the output sequence leaks information about the states of $LFSR_0$ and $LFSR_2$. This can be seen as follows:

Let $x_0(t), x_1(t), x_2(t)$ be the outputs of $LFSR_0$, $LFSR_1$ and $LFSR_2$ respectively, and let $z(t)$ be the keystream output. Then the *correlation probability* between $x_0(t)$ and $z(t)$ is

$$\begin{aligned} \Pr\{z(t) = x_0(t)\} &= \Pr\{x_1 = 1\} \\ &+ \Pr\{x_1 = 0 \quad \text{and} \quad x_2 = x_0\} \\ &= 1/2 + 1/4 = 3/4. \end{aligned}$$

Similarly, $\Pr(z(t) = x_2(t)) = 3/4$. If a sufficiently long string of keystream bits is available, then the initial state of $LFSR_0$ can be determined by going through the 2^{L_0} possible initial states of $LFSR_0$ and selecting the one which produces a sequence with a 75 % correlation with the keystream. Similarly the initial state of $LFSR_2$ can be found, after which a further search will reveal the initial state of $LFSR_1$. The complexity of the attack therefore drops from $2^{L_0+L_1+L_2}$ to $2^{L_0} + 2^{L_1} + 2^{L_2}$. This is an example of a *divide-and-conquer* attack: clearly an attack of this sort will succeed (in the sense of being more efficient than a brute force attack through all the possible initial states) whenever the keystream leaks information about the states of the LFSRs, i.e. whenever there is a correlation probability different from $\frac{1}{2}$ between the keystream and an LFSR. Such so-called *correlation attacks* form an important class of attacks on stream ciphers, and consequently it is important to choose a combining function

[31]This theorem, which we shall not prove, is due to E.L. Key, and appeared in an article entitled *An analysis of the structure and complexity of nonlinear binary sequence generators*, IEEE Trans. Info. Th. **IT-22**, (1976).

which does not "leak" information in this way, or, at the very least, confines leakage to a minimum. This desideratum leads to the following concept:

Definition A Boolean function F of n independent identically distributed variables $x_0, x_1, \ldots, x_{n-1}$ is called *t-correlation immune* if for any choice of $x_{i_0}, \ldots, x_{i_{t-1}}$, we have the mutual information

$$I(x_{i_0}, x_{i_1}, \ldots, x_{i_{t-1}}, Y) = 0$$

where

$$Y = F(x_0, \ldots, x_{n-1}).$$

Another way of saying this is:

$$\Pr(Y = 1) = \Pr(Y = 1 | x_{i_0}, \ldots, x_{i_{t-1}}).$$

A *balanced* t-correlation immune function is also called *t-resilient.*

We shall in the next chapter prove that there is necessarily an unfortunate trade-off between correlation-immunity and degree, first proved by Siegenthaler[32]: If n is the number of variables, t the order of correlation-immunity and d the degree of the function, then

$$d + t \leq n.$$

If the function is balanced, i.e. if it is t-resilient, with $1 \leq t \leq n - 2$, then

$$d + t \leq n - 1.$$

This is very unfortunate, since one of the desirable properties of a combining function would have been that of high degree, for the simple reason that Boolean equations of high degree are regarded as very difficult to solve.[33]

Examples The function

$$f_1(x_3, x_2, x_1, x_0) = x_0x_1 + x_2x_3$$

[32]T. Siegenthaler: *Correlation-immunity of nonlinear combining functions for cryptographic applications*, IEEE Transactions in Information Theory, **30**, 1984, pp. 776–780. It is an interesting observation by M. Liu, P. Lu and G.L. Mullen in their paper *Correlation-immune functions over finite fields*, IEEE Transactions on Information Theory, **44**, 1998, pp. 1273–1276, that this trade-off is only a problem if one works in the finite fields $GF(2)$ or $GF(3)$.

[33]There is a technique which uses so-called "Gröbner bases" for solving Boolean equations of high degree. However, it does not appear to have lived up to the original expectations. Using Gröbner bases still seems to lead to very high complexity. But we shall return to the stated assumption when we discuss "algebraic immunity" of Boolean functions.

has degree 2 and is 2-correlation-immune, as may be verified: we shall, when discussing Boolean functions in the next chapter, find a fairly quick way, due to Massey and Xiao, of establishing this. Since $n = 4$, this function satisfies Siegenthaler's inequality with equality. On the other hand the function

$$f_2(x_3, x_2, x_1, x_0) = x_0 + x_1 + x_2x_3$$

has degree 2 and is balanced, but it is only 1-correlation-immune. The function

$$f_3(x_3, x_2, x_1, x_0) = x_0 + x_1 + x_2 + x_3$$

is 3-correlation-immune and balanced, but only has degree 1.

There are quite a few properties that we would like to have in Boolean functions for cryptographic applications; correlation-immunity is only one of them. The trade-off between degree and correlation immunity that we see here is only one of many such.

7.11 Filter Generators

A filter generator is a keystream generator consisting of a single LFSR and a function—nonlinear, for what should by now be obvious reasons—whose inputs are taken from some shift register stages to produce the output. The key determines the initial state of the LFSR and/or the nonlinear function. A generalisation of this construction would be to use an LFSR and a suitable function on the contents to produce a block of keystream material: this is the idea employed in the counter mode (CTR) of a block cipher. In the latter case it is the filtering function which is key-dependent, the initial state (IV) probably being sent in clear.

In the kind of design that we are currently considering, i.e. where the output of the filtering function consists of just a single bit, it is tempting to use some form of *tree function*. In such a structure one repeatedly uses a function

$$F : GF(2)^N \longrightarrow GF(2)^m$$

where $N > m$, in parallel in each of a series of "layers". Such a structure is also suitable for self-synchronising mode. Unfortunately, it has been shown to be cryptographically weak.[34] A variation on this theme might be to use a "hash-like" function on the contents of an LFSR, to produce a short block (or a single bit) of output.

It is traditional in the (older) cryptographic literature to use the key only to determine the initial state of the LFSR or LFSRs, especially in the case of combination generators. There is, however, no real reason for this restriction. In fact, in the case of self-synchronising

[34] See R.J. Anderson: *Tree functions and cipher systems*, Cryptologia **15**, 1991, pp. 194–202; and W. Millan, E.P. Dawson and L.J. O'Connor: *Fast attacks on tree-structured ciphers*, Electronics Letters **30** (1994), pp. 941–943.

stream ciphers one does not even have a choice: it *must* be the output function which is key-dependent, since an attacker intercepting ciphertext knows exactly what is going into the register. It may therefore be argued that the theory of self-synchronising ciphers is much closer to that of block ciphers than to stream ciphers. In practice this approach is complicated by the fact that self-synchronising stream ciphers are frequently employed in equipment with severe hardware constraints, whereas most block ciphers make considerable demands on space and computational power. For reasons such as these, the history of attempts of devising good self-synchronising stream ciphers is filled almost exclusively with failures; it seems that self-synchronisation inevitably leads to weakness. There are currently no self-synchronising ciphers approved by any standardisation body.

In both combination and filter generators the LFSR or LFSRs essentially serves or serve to obtain good local statistics and long period, whereas the nonlinear function is used to provide "confusion" and large linear complexity. Rueppel has shown that if in a filter generator the LFSR has length L and the filtering function has degree k then the output sequence has linear complexity approximately $\binom{L}{k}$. Thus it is easy to obtain high linear complexity by using this kind of construction; the period will (with high probability) remain the same as that of the LFSR. On the other hand, there will, as we shall show in the next chapter, always be a nonzero correlation between the output sequence and *some* linear combination of the input bits into the filtering function. Since any linear combination of input bits satisfies the same recursion as the LFSR itself, this opens up a possible way of attack, which may need replacing the original initial value by an equivalent one. In the easiest case, "easy" from the attackers' point of view, one might have a high correlation between the output bit and one particular stage in the LFSR; this will allow the attacker to find the initial state of the LFSR. However, by the nature of LFSRs "we can always discount a linear function [of the stages of the LFSR] by moving to a different phase of the underlying shift register" as Anderson puts it.[35] This forms the basic idea of a type of attack discovered by Siegenthaler[36] and later improved by Meier and Staffelbach.[37] These attacks are particularly efficient if the connection polynomial of the underlying LFSR has few nonzero terms, or if a decimation of the keystream has such a "low weight" connection polynomial.

The solution here may lie in selecting a filtering function such that these correlation coefficients take on approximately the same value for all the possible linear combinations of input bits, and with a large number of input bits. These are two of the recommendations made by Golić.[38] As far as the first of these is concerned, it would suggest a move towards the so-called bent functions, about which we shall also have something to say in the next

[35]Anderson, R.J.: *Searching for the optimum correlation attack*; Proc. Fast Software Encryption 2005, LNCS 1008, Springer-Verlag 2005.

[36]T. Siegenthaler: *Correlation immunity of nonlinear combining functions for cryptographic applications*, IEEE Trans. Info. Th. **IT-30**, 1984, pp. 776–780.

[37]Meier, W. and Staffelbach, O.: *Fast correlation attacks on certain stream ciphers*, J. Cryptology **1**, 1989, pp. 159–176.

[38]J.D. Golić: *On the security of nonlinear filter generators*; Proc. Fast Software Encryption '96, LNCS 1039, Springer-Verlag 1996, pp. 173–188.

chapter, but Anderson[39] claims that bent functions perform badly when measured against a modified (dual) version of Siegenthaler's attack.[40]

7.12 Clock Controlled Generators

In the LFSR-based generators discussed so far, the LFSRs are "clocked" regularly, in the sense that the movement of data in the LFSRs are all controlled by the same clock and keystream bits are produced according to the rate of that clock. A way of introducing non-linearity alternative to the two discussed so far, is to use irregular clocking, or alternatively, to discard some of the data in an LFSR according to the output of another (or the same, see below) LFSR. We shall discuss four constructions based on this idea.

7.12.1 The Stop-and-Go Generator

This generator[41] is, as far as this author knows, the oldest of the designs of clock-controlled keystream generators using LFSRs. It uses two LFSRs, of lengths L_1 and L_2. The output sequence is found in the following simple way (variations on the basic theme are easily thought up):

- Clock $LFSR_1$ once to get x_1;
- If $x_1 = 0$ take $z =$ output of $LFSR_2$;
- else clock $LFSR_2$ and take z equal to its output.

As given, it is easy to obtain information about $LFSR_1$: if two successive key bits are the same, then with probability $\frac{1}{2}$ the output of $LFSR_1$ was 0; if they are different, then with

[39] *Loc. cit., supra.*

[40] The following quotation from a paper by Alex Biryukov: *Block ciphers and stream ciphers: The state of the art* (https://eprint.iacr.org/2004/094.ps), may be relevant here:

> Prior to design of DES stream ciphers [were] ruling the world of encryption, either rotor machines (like Hagelin or Enigma), or secret military hardware-based designs using LFSRs all belonged to this class. Appearance of fast block ciphers has caused a shift of interest, due to convenience of use of block ciphers in various protocols, including stream-like behavior which can be obtained via modes of operation in counter, OFB or CBC, as well as due to a shift from hardware to software designs.... Still in cases where there is a need to encrypt large quantities of fast streaming data one would like to use a stream cipher. Popular trends in design of stream ciphers is to turn to block-wise stream ciphers (i.e. output is a block of bits, either a byte or 32-bits instead of a single bit) like RC4, SNOW 2.0, SCREAM, oriented towards fast software implementation. Stream ciphers which use parts of block-cipher like rounds intermixed with more traditional LFSR-like structure (MUGI, SCREAM).[sic]

[41] Due to Beth and Piper: Beth, T. and Piper, F.: *The stop-and-go generator*; Proc. Eurocrypt '84, LNCS 209, Springer-Verlag.

probability 1 its output was 1. This problem is easily overcome: stipulate that $LFSR_2$ will always be clocked either once or twice depending on the output of $LFSR_1$.

Beth and Piper prove that if $\gcd(2^{L_1} - 1, 2^{L_2} - 1) = 1$, then the linear complexity of the output sequence is $(2^{L_1} - 1)L_2$. Interestingly, the designers do not make any claims about the security of the stop-and-go generator: on the contrary they suggest

- "In order to guarantee a good statistical behaviour of the Stop-and-Go-sequence it is suggested that the output sequence [be] finally XOR-gated with another sequence."
- "The statistical behaviour of [the output sequence] itself—though theoretically quite good in special cases—is so that a cryptanalytic attack would be promising in spite of the extremely high linear [complexity] of the sequence."

7.12.2 Alternating Step Generator

In this, a generalization of the previous design, three LFSRs of suitable lengths $l_i, i = 1, 2, 3$, are used, and the keystream bit is determined as follows:

- Clock $LFSR_1$ and let t be its output;
- If $t = 0$ clock $LFSR_2$ and take its output z_2; do not clock $LFSR_3$ but repeat its previous output z_3;
- else clock $LFSR_3$ and take its output z_3, do not clock $LFSR_2$ but repeat its output previous z_2;
- The keystream bit is $z_2 \oplus z_3$.

The resulting keystream sequence has good statistical properties. If, as seems sensible, the lengths are pairwise relatively prime, and approximately equal, $L_i \approx l, i = 1, 2, 3$, then the best known attack, according to HAC,[42] is a divide-and-conquer attack on $LFSR_1$, of complexity of the order of 2^l.

7.12.3 Shrinking Generator

In this design, originally proposed by Coppersmith, Krawczyk and Mansour,[43] two LFSRs are used, the one determining whether or not an output bit of the other shall or shall not be used as a keystream bit. Thus the output is a *shrunken* or *irregularly decimated* subsequence of the output of the second LFSR. Thus:

- Clock both LFSRs;
- If the output of $LFSR_1 = 1$ use the output of $LFSR_2$ as a keystream bit;
- else discard the output of $LFSR_2$.

---

[42] A.J. Menezes, P.C. van Oorschot and S.A. Vanstone: *Handbook of Applied Cryptography*, CRC Press, 1996. I am not aware of any more recent results about this cipher.

[43] Coppersmith, D., Krawczyk, H. and Mansour, Y.: *The shrinking generator*, Proc. Eurocrypt '93, LNCS 765, Springer-Verlag 1994.

The designers prove *inter alia* that if T_1 and T_2 are the periods of the two LFSRs (assumed to be maximal length LFSRs, i.e. with primitive feedback polynomials, and with $\gcd(T_1, T_2) = 1$) then the output sequence has period T_1T_2. Under the same assumptions the linear complexity LC of the output sequence satisfies

$$L_1 \cdot 2^{L_2-2} < LC \leq L_1 \cdot 2^{L_2-1},$$

where L_1, L_2 are the lengths of $LFSR_1$ and $LFSR_2$, respectively. It appears to have satisfactory security: Under the same assumptions as before, and assuming moreover that the secret key is used (only) to initialise the two LFSRs, the effective key length, under the attacks considered by the designers, equals the length of $LFSR_2$. The best currently known attack on the shrinking generator, assuming that the connection polynomials are known to the attacker, is a mere "distinguishing attack", which distinguishes the output sequence from a random sequence if the feedback polynomial (or a multiple thereof) has low weight (but degree as high as 10^4) when 2^{32} output bits are known.[44]

However, despite the fact that it has remained resistant to cryptanalysis, the fact that the keystream is irregularly generated (nothing will come out if $LFSR_1$ hits a gap of some length) is a severe disadvantage in many applications. The designers consider this problem in their original paper. We quote:

> [The output] rate is on *average* 1 bit for each 2 pulses of the clock governing the LFSRs. This problem has two aspects. One is the reduced throughput relative to the LFSRs' speed, the other the irregularity of the output. We show here that this apparently practical weakness can be overcome at a moderate price in hardware implementation …
>
> "The problem of irregular output rate can be serious in real-time applications where repeated delays are not acceptable. … The solution is to use a short buffer for the [$LFSR_2$] output, intended to gather bits from the [$LFSR_2$] output when they abound in order to compensate for sections of the sequence where the rate output is reduced."

The authors then quote results from a paper by Kessler and Krawczyk to claim that "even with short buffers (e.g. 16 or 24 bits) and with a speed of the LFSRs of above twice the necessary throughput from [$LFSR_2$] the probability to have a byte of pseudorandom bits not ready in time is very small. (Examples are a probability of 5×10^{-3} for buffer size 16 and speedup factor of 9/4, or a probability of $3 \cdot 10^{-7}$ for a buffer of size 24 and speedup factor of 10/4. These probabilities decrease exponentially with increasing buffer size and speedup factors.)"

A variation on this theme is the (a, b)-shrinking generator, in which $LFSR_2$ is clocked a times if the output of $LFSR_1$ is a 0, and b times otherwise.[45] Little if any cryptanalysis appears to have been done on this kind of cipher, but one might expect the security to be similar to that of the standard shrinking generator. An interesting feature is that a and b can be adjusted

[44] Ekdahl, P., Meier, W. and Johansson, T.: *Predicting the shrinking generator with fixed connections*, Proc. Eurocrypt 2003, Springer-Verlag.

[45] This proposal emanates from a paper, available on the Internet, *The (a, b)-shrinking generator* by A.A. Kanso of the King Fahd University of Petroleum and Minerals, Haid, Saudi Arabia. Dr. Kanso's Ph.D. thesis, in which this design also appears, is available on the website of the Royal Holloway, University of London, Information Security Group.

(provided that they remain relatively prime) while keeping the connection polynomials of the two LFSRs the same.

7.12.4 Self-Shrinking Generator

This is, in a way, a variation on the previous design due to Meier and Staffelbach.[46] Only a single LFSR is used, but its output bits are considered in pairs:

- All pairs of the form (0,0) and (0,1) are discarded;
- Of the pairs (1,0) and (1,1) the second entry is taken as the keystream bit.

The designers point out that a self-shrinking generator can be implemented as a special case of the shrinking generator, and conversely, that a shrinking generator can be implemented as a self-shrinking generator. Nevertheless, the self-shrinking generator has turned out to be more resistant to cryptanalysis than its shrinking parent: intuitively this may appear obvious, since in the self-shrinking case there is no possibility of a divide-and-conquer attack.

Let A denote the LFSR and $(a_0, a_1, \ldots, a_{L-1})$ its contents, and let z_0 be the first output bit. Now observe that

$$\Pr((a_0, a_1) = (0,0)|z_0) = \frac{1}{4},$$

$$\Pr((a_0, a_1) = (0,1)|z_0) = \frac{1}{4},$$

$$\Pr((a_0, a_1) = (1,z_0)|z_0) = \frac{1}{2},$$

so that the entropy of the first bit pair, given z_0 is

$$H = -\frac{1}{4}\log\left(\frac{1}{4}\right) - \frac{1}{4}\log\left(\frac{1}{4}\right) - \frac{1}{2}\log\left(\frac{1}{2}\right) = \frac{3}{2}$$

and the total of an initial state of $\frac{L}{2}$ such pairs is therefore $0.75L$, implying that an optimal exhaustive search for the initial state will have complexity $\mathcal{O}(2^{0.75L})$. Mihaljević presented a faster attack that, however, needs an unrealistic amount of known keystream.[47] The best attack currently known is due to Zenner et al.[48] It will find the key, of length L, say, in at most $\mathcal{O}(2^{0.694L})$ steps.

---

[46] W. Meier and O. Staffelbach: *The self-shrinking generator*, Proc. Eurocrypt '94, LNCS 950, Springer-Verlag, 1995.

[47] Mihaljević, M.J.: *A faster cryptanalysis of the self-shrinking generator*, Proc. ACISP '96, LNCS 172, Springer-Verlag.

[48] Zenner, E., Krause, M. and Lucks, S.: *Improved cryptanalysis of the self-shrinking generator*; Proc. ACISP 2001, LNCS 2119, Springer-Verlag, pp. 21–35.

The designers proved in their original paper that if the LFSR is maximum length of length L, then the period of the output key sequence is a divisor of $2^L - 1$, and is at least $2^{\lfloor L/2 \rfloor}$. The output sequence is balanced. Its linear complexity is greater than $2^{\lfloor L/2 \rfloor - 1}$, and in fact appears to be close to 2^{L-1}. With L large enough, and the connection polynomial of the LFSR *kept secret*, no practical attack on this design appears to be known, as indicated above. Again, however, we do have the problem that the keystream is not being generated at a uniform rate.

7.12.5 Bitsearch Generators

The most recent version of "self-shrinking" generator was the BitSearch Generator proposed by Gouget and Sibert[49] in 2004. We assume that a binary sequence $\{s_i\}_{i \geq 0}$, such as, once again, the output of an LFSR, is given.

The authors then describe the decimation process as follows:

> The principle of the BSG consists in searching for some bit along the input sequence, and to output 0 if the search ended immediately (that is, if the first bit read during the search was the good one), and 1 otherwise. Consider a window that is located before the first bit on the input sequence. The window moves on to read the first bit of the sequence, and then moves along the sequence until it encounters this bit again. If the window has read only two bits (i.e., the first bit read by the window was followed by the same bit), then the BSG outputs 0, otherwise it outputs 1. The window then reads the next bit following its position, then moves along the input sequence to find it, and so on.

In pseudocode this can be described as follows:

Input: sequence $\{s_i\}_{i \geq 0}$
Output: sequence $\{z_j\} j \geq 0$
Initialisation: Set $i = j = -1$

`Repeat:`

1. $i \leftarrow i + 1$
2. $j \leftarrow j + 1$
3. $b \leftarrow s_i$
4. $i \leftarrow i + 1$
5. `if` $s_i = b$ `then` $z_j \leftarrow 0$ `else` $z_j \leftarrow 1$
6. `while` $s_i \neq b, i \leftarrow i + 1$

The designers note that the rate of output of this generator is (on average) 1/3 of the rate of output of the LFSR. They also show that different initialisations of the LFSR will give different outputs—different in the sense that they are not shifted versions of each other, as one might have expected from the use of an LFSR. Experimentally they observe two such possible output sequences, of lengths $(2^L - 1)/3$ and $2(2^L - 1)/3$ respectively, where L is the length of the shift register.

---

[49]Gouget, A. and Sibert, H.: *The Bitsearch generator*, Proc. State of the Art in Stream Ciphers Workshop 2004, 60–68.

All clock controlled keystream generators suffer from the inconvenience of having an irregular rate of output. While this can be overcome in practice by buffering the output, this still leaves the problem that they may be vulnerable to so-called *side-channel attacks* which measure timing and/or power consumption.

8 Boolean Functions

Following from our previous chapter in which we noted some properties that we require in Boolean functions which are to be used as combining functions and filter functions, we now look at Boolean functions more closely. We start with an efficient way of determining the Algebraic Normal Form of a Boolean function, given its outputs for all possible inputs (and conversely) and then proceed with the Walsh–Hadamard transform and its applications to the kind of problems that we have identified. We end the chapter with a brief introduction to the Discrete Fourier Transform, where our knowledge of finite fields is required once again.

8.1 Introduction

We shall deal only with Boolean functions of Boolean variables, i.e. all our functions are defined on sets $\{0, 1\}^n$ (for some finite n) and with values in the set $\{0, 1\}$. But sometimes it will be useful to regard the domain and/or the range of such a function as a vector space over the field $GF(2)$.

Regardless of whether one considers such a function as just a mapping from one set to another, or as a mapping from one vector space to another, the domain will in our discussions always have a finite number of elements. This means that it is quite feasible to list the elements of the set and their images under the function. In fact, if we can agree on some standard way of ordering the elements, it is sufficient merely to list the images.

A natural way—in fact, surely the most sensible way—of ordering the elements of the domain is to list them in their "numerical" order.[1] If the domain is the set $\{0, 1\}^n$, list its

[1]But there may be times when it might be useful to think of the domain of the function as the finite field $GF(2^n)$, in which case some other ordering like $\{0, 1, \alpha, \alpha^2, \dots, \alpha^{2^n-2}\}$ might be more convenient.

A.R. Meijer, *Algebra for Cryptologists*, Springer Undergraduate Texts in Mathematics and Technology,
DOI 10.1007/978-3-319-30396-3_8

elements in the order

$$00\ldots00,$$
$$00\ldots01,$$
$$00\ldots10,$$
$$00\ldots11,$$
$$\ldots\ldots$$
$$11\ldots10,$$
$$11\ldots11.$$

If we do this, the meaning of the list or sequence (sometimes thought of, and referred to, as its *truth table*)

$$0110\ 1011\ 0011\ 0000$$

is unambiguous, and, as the reader is invited to verify, this sequence corresponds to the function

$$f : \{0,1\}^4 \longrightarrow GF(2):$$
$$(x_3, x_2, x_1, x_0) \mapsto x_0 + x_1 + x_2 + x_1x_2 + x_0x_1x_2$$
$$+x_0x_3 + x_2x_3 + x_0x_1x_2x_3.$$

Representing functions in this way, it becomes clear that there is really only one meaningful way in which the *distance*[2] between two Boolean functions should be defined: If f and g are two such functions, we define the distance between them as $\#\{x|f(x) \neq g(x)\}$; in other words, the number of places in which their sequences differ from each other. It is not hard to show that the distance d defined this way satisfies the usual requirements of a metric: For any functions f, g and h (defined on $\{0,1\}^n$ for one and the same n)

- $d(f,g) \geq 0$ and $d(f,g) = 0$ if and only if $f = g$.
- $d(f,g) = d(g,f)$.
- $d(f,h) \leq d(f,g) + d(g,h)$.

There is another simple interpretation of the distance between two Boolean functions f and g: From addition modulo 2, we obviously have that if $f(x_i) = g(x_i)$, then $(f+g)(x_i) = 0$, and $(f+g)(x_i) = 1$ otherwise. So the number of values of x for which $f+g$ takes the value 1 is precisely the distance between them.

[2]Properly called the Hamming distance, after Richard W. Hamming (1915–1998), American mathematician, and one of the giants in coding theory.

It is customary to call $\#\{x|h(x) \neq 0\}$ the *weight* of a function h, and the previous observation can therefore be rephrased as

$$d(f, g) = \text{weight}(f + g).$$

This is a special case of a more general concept: The weight $W_H(x)$ (or $w_H(\mathbf{x})$) of a binary vector $\mathbf{x}$ is the number of 1s occurring.[3] Thus, using hexadecimal notation, $w_H(\texttt{0xa}) = 2$, $w_H(\texttt{0xffffffff}) = 32$.

There are two further properties of Boolean functions which, while just as obvious as the ones we have mentioned, are also worth noting explicitly. The first is that a Boolean variable a satisfies the convenient property that $a^2 = a$, and hence that $a^n = a$ for all $n > 0$. Let us define the *degree* of a monomial as the number of variables appearing in it. Then any monomial expression involving such variables will never involve powers of the variables higher than 1. If we are dealing with n variables, there are therefore $\binom{n}{s}$ terms of degree s, since there are $\binom{n}{s}$ ways of choosing the variables to be included in that monomial, and furthermore, we need never consider monomials of degree greater than n, since such a term must include one or more variables more than once and can therefore be simplified.

If we define a Boolean *polynomial* as a sum of monomials, then we see that the number of monomials of n variables is

$$\binom{n}{0} + \binom{n}{1} + \binom{n}{2} + \cdots + \binom{n}{n} = 2^n$$

and the number of polynomials is therefore 2^{2^n}, since for each monomial we have the choice of including it or not.

But now, if we look at our representing functions by sequences, we have, in the case of n variables, 2^n entries in our table, and each entry is either 0 or 1, so we have 2^{2^n} possible listings.

We arrive at the pleasant conclusion that every Boolean function is a polynomial. In the next section we show how to move from the listing to the "algebraic" description, and conversely.

For the sake of completeness we mention the following definitions, which will not surprise the reader: A Boolean function is *linear* if it only contains terms of degree 1, and *affine* if all the terms are of degree 1 or 0.

Exercises

1. If f and g are two Boolean functions of n variables, and we define the weight $W_H(x)$ of a function z as before, show that

$$d(f, g) = W_H(f) + W_H(g) - 2 \cdot W_H(f \times g).$$

[3] More properly this is called the Hamming weight of $\mathbf{x}$.

2. Show that every balanced Boolean function[4] of two variables is linear (*i.e.* of degree 1).
3. How many Boolean functions of three variables are there? How many are balanced? Of the latter, how many are linear?

8.2 The Algebraic Normal Form

Suppose that we are given a function f of n variables in the form of a string of length 2^n, and we wish to find the algebraic normal form (ANF) of f. This is the "polynomial" form of the function, a sum of monomials, i.e. an expression of the form

$$f(x) = a + \sum_{i_1} a_{i_1} x_{i_1} + \sum_{i_1} \sum_{i_2} a_{i_1 i_2} x_{i_1} x_{i_2} + \dots$$
$$+ \sum_{i_1} \sum_{i_2} \cdots \sum_{i_{n-1}} a_{i_1 i_2 \dots i_{n-1}} x_{i-1} \dots x_{i_{n-1}} + \dots\dots$$
$$+ a_{0,1,\dots,n-1} x_0 x_1 \dots x_{n-1}$$

where all the a (regardless of the number of subscripts) are either 0 or 1.

Substituting the values of the function into this expression (and using the obvious notation in which we again identify the string $c_{n-1}c_{n-2}\dots c_1c_0$ with the integer $\sum_{i=0}^{n-1} c_i 2^i$) we get

$$\begin{aligned} f(0) &= a, \\ f(1) &= a + a_0, \\ f(2) &= a \quad + a_1, \\ f(3) &= a + a_0 + a_1 + a_{01}, \\ &\dots \\ f(2^n - 1) &= a + a_0 + a_1 + a_{01} + \dots + a_{01\dots(n-1)}, \end{aligned}$$

or, in matrix form

$$[a, a_0, a_1, \dots, a_{01\dots(n-1)}]\mathbf{A}_n$$
$$= [f(0), f(1), \dots, f(2^n - 1)],$$

[4]Recall that a Boolean function f is balanced if $\#\{x|f(x) = 0\} = \#\{x|f(x) = 1\}$.

where $\mathbf{A}_n$ is the $2^n \times 2^n$ matrix

$$\begin{pmatrix} 1 & 1 & 1 & 1 & \dots & 1 & 1 \\ 0 & 1 & 0 & 1 & \dots & 0 & 1 \\ 0 & 0 & 1 & 1 & \dots & 1 & 1 \\ 0 & 0 & 0 & 1 & \dots & 0 & 1 \\ \dots & \dots & \dots & \dots & \dots & \dots & \dots \\ \dots & \dots & \dots & \dots & \dots & \dots & \dots \\ 0 & 0 & 0 & 0 & \dots & 1 & 1 \\ 0 & 0 & 0 & 0 & \dots & 0 & 1 \end{pmatrix}.$$

It is clear that the matrix $\mathbf{A}_n$ has a great deal of structure. In fact, if we put

$$\mathbf{A}_1 = \begin{pmatrix} 1 & 1 \\ 0 & 1 \end{pmatrix}$$

and recall the definition of the *Kronecker* (or *tensor*) product of two matrices **A** and **B** as

$$\mathbf{A} \otimes \mathbf{B} = \begin{pmatrix} a_{00}\mathbf{B} & a_{01}\mathbf{B} & \dots & a_{0s}\mathbf{B} \\ a_{10}\mathbf{B} & a_{11}\mathbf{B} & \dots & a_{1s}\mathbf{B} \\ \dots & \dots & \dots & \dots \\ a_{t0}\mathbf{B} & a_{t1}\mathbf{B} & \dots & a_{ts}\mathbf{B} \end{pmatrix},$$

then we see that

$$\mathbf{A}_n = \mathbf{A}_1 \otimes \mathbf{A}_{n-1}.$$

Note that the matrix $\mathbf{A}_1$ is its own inverse, and, by the properties of the Kronecker product, $\mathbf{A}_n$ is also self-inverse, for every $n \geq 1$.[5]

The recursive structure of the matrices $\mathbf{A}_n$ leads to an efficient way of multiplying a vector of length 2^n by the matrix $\mathbf{A}_n$, of which we only demonstrate the first three steps (i.e. we

---

[5]It can be shown from the definition of the Kronecker product that

$$(\mathbf{A} \otimes \mathbf{B})(\mathbf{C} \otimes \mathbf{D}) = \mathbf{AC} \otimes \mathbf{BD},$$

assuming that these products are defined.

It is also easy to see that, considering identity matrices,

$$\mathbf{I}_n \otimes \mathbf{I}_m = \mathbf{I}_{mn}.$$

Consequently $(\mathbf{A} \otimes \mathbf{B})^{-1} = \mathbf{A}^{-1} \otimes \mathbf{B}^{-1}$.

consider as an example the case $n = 3$):

$$\begin{array}{llll}
a & a & a & a \\
b & a+b & a+b & a+b \\
c & c & a+c & a+c \\
d & c+d & a+b+c+d & a+b+c+d \\
e & e & e & a+e \\
f & e+f & e+f & a+b+e+f \\
g & g & e+g & a+c+e+g \\
h & g+h & e+f+g+h & a+b+c+d+e+f+g+h
\end{array}$$

Thus in each step we have the pattern

$$\begin{array}{ccc}
x & \longrightarrow & x \\
 & \searrow & \\
y & \longrightarrow & x \oplus y
\end{array}$$

with adjacent entries in the first step, entries 2 apart in the next, entries 4 apart in the next, entries 8 apart in the next, and so on.

Using this technique, we see, by way of example, that the ANF of the function

$$[0101 \quad 1010 \quad 0101 \quad 0011]$$

is

$$x_0 + x_2 + x_3x_2 + x_3x_2x_0 + x_3x_2x_1.$$

In the same way, using the fact that the matrices are all self-inverse, we can use the ANF of the function to determine the string which gives the values of the function. Thus it is easy to verify that the string corresponding to the function

$$x_0 + x_0x_2 + x_1x_3 + x_0x_1x_2x_3$$

is

$$[0101 \quad 0000 \quad 0110 \quad 0010].$$

Exercises

1. Write the function whose values are

   0011 1100 1111 0000

   as a polynomial.

2. Use the method above to find the value table of the function

$$x_0 + x_0x_1 + x_0x_1x_2 + x_0x_1x_2x_3 + x_4.$$

3. If f is a balanced function on n variables, explain why its degree is at most $n-1$.
4. More generally than the previous question: If we define the weight of a Boolean function f of n variables to be $\#\{x|f(x) = 1\}$, show that every function whose ANF does not contain the term $x_0x_1 \ldots x_{n-1}$ has even weight.

8.3 The Walsh Transform

The Walsh transform presents a very useful technique for investigating several properties of Boolean functions which are important from the point of view of Cryptology.

8.3.1 Hadamard Matrices

We start our discussion with something which does not immediately seem to be relevant to the promised subject of properties of Boolean functions.

A *Hadamard matrix* H of order n is an $n \times n$ matrix with all entries either 1 or -1, such that

$$\mathrm{HH}^t = n\mathrm{I}_n,$$

where I_n is the $n \times n$ identity matrix and where the superscript t denotes the transpose of a matrix.[6] It can be proved quite easily[7] that if a Hadamard matrix of order n exists, then $n = 1$ or $n = 2$ or n is a multiple of 4. There are various methods of constructing Hadamard matrices, but we shall for the moment only need one of them, constructing the so-called *Sylvester* matrices.

Put

$$\mathbf{H}_1 = \begin{pmatrix} 1 & 1 \\ 1 & -1 \end{pmatrix}.$$

[6]This is not the most general definition of Hadamard matrices, but it is the one we shall use for the moment. A more general definition will be useful when we consider the so-called MDS matrices in Sect. 9.5.

[7]See, for example, the book by F.A. McWilliams and N.J.A. Sloane: *The Theory of Error-Correcting Codes*, North-Holland, 1977, p. 44.

Clearly $\mathbf{H_1}$ is a Hadamard matrix. The same will be true (using a fundamental property of the Kronecker product again) of the matrices $\mathbf{H_n}$ defined recursively by

$$\mathbf{H}_n = \mathbf{H}_1 \otimes \mathbf{H}_{n-1}.$$

Note that $\mathbf{H_n}$ is a $2^n \times 2^n$ matrix.

Proof By induction on n.

$$\begin{aligned}
&(\mathrm{H}_1 \otimes \mathrm{H}_{n-1})(\mathbf{H}_1 \otimes \mathrm{H}_{n-1})^t \\
&= \mathrm{H}_1\mathrm{H}_1^t \otimes \mathrm{H}_{n-1}\mathrm{H}_{n-1}^t \\
&= 2\mathrm{I}_2 \otimes 2^{n-1}\mathrm{I}_{2^{n-1}} \\
&= 2^n\mathrm{I}_{2^n}.
\end{aligned}$$

8.3.2 Definition of the Walsh Transform

Now let, in the following, f be a Boolean function of n variables. We shall find it more useful to deal with an equivalent function, denoted by $\hat{f}$, which is real-valued and defined by

$$\hat{f}(x) = (-1)^{f(x)}.$$

As before, we can represent $\hat{f}$ by a sequence of length 2^n consisting of 1s and -1s.

We now define the *Walsh*-transform[8]

$$\hat{F}(\omega) = \sum_x (-1)^{f(x) \oplus \omega \cdot x}.$$

Here $\omega \cdot x$ is the standard scalar or "dot" product[9] of the vector ω and the vector x. The important thing to notice is that $\omega \cdot x$ is a linear function of x.[10] If we look at the right-hand side of the equation above, we note that if x satisfies $f(x) = \omega \cdot x$ then x contributes 1 to the sum, whereas if $f(x) \neq \omega \cdot x$ a -1 is contributed. Thus

$$\hat{F}(\omega) = \mathrm{A} - \mathrm{D}$$

[8] Also called, among other names, the *Walsh–Hadamard*-transform. The conversion of f to $\hat{f}$ is required because the Walsh transform is defined in general for application to sequences of real or complex numbers of length 2^n for some positive integer n. We may, occasionally, abuse terminology and refer to the result of applying the transform to $\hat{f}$, as "the Walsh transform of f".

[9] With the arithmetic being carried out in $GF(2)$, so that $1 + 1 = 0$, since we are dealing with vectors over $GF(2)$.

[10] In analogy with Fourier analysis, f and $\hat{f}$ are sometimes referred to as being in the "time domain" and $\hat{F}$ as being in the "frequency domain". In our context this terminology does not seem to be helpful, but we will bow to custom in pointing out that $\hat{F}$ is called the (Walsh-) spectrum of f.

where $A = \#\{x|f(x) = \omega \cdot x\}$ is the number of agreements between $f(x)$ and the linear function $\omega \cdot x$ and D is the number of disagreements. Equivalently, if we denote by l_ω the linear function

$$l_\omega(x) = \omega \cdot x$$

then, since A + D = 2^n, we can express the (Hamming-) distance between l_ω and f by

$$d(l_\omega, f) = \frac{1}{2}(2^n - \hat{F}(\omega)).$$

This seems to imply that we are on the way to deriving a convenient way of establishing the distance between a given Boolean function and any linear function of the same number of variables. We need the following

Observation In the matrix $\mathbf{H}_n$, the cth row (and the cth column too, since the matrix is symmetrical) is the string representing the function $\hat{l}_c$, if we identify, in the usual way, integers and binary vectors. (Note that we number rows and columns from 0.)

Proof By induction on n. The statement is true for the case $n = 2$, as may easily be verified by inspection. To prove the inductive step we use the recursive definition of $\mathbf{H}_n$:

$$\mathbf{H}_n = \mathbf{H}_1 \otimes \mathbf{H}_{n-1}.$$

Now if $c < 2^{n-1}$ then the first half of the cth row and its second half are identical, both corresponding to the function $\hat{l}_c$ of $n - 1$ variables, i.e. to the function $\hat{l}_c$ of n variables in which the most significant input bit is set to 0. If $c \geq 2^{n-1}$ then the first half is again the same (corresponding to the fact that in this part the "most significant variable" is 0), and the second half is -1 times the first half, corresponding to the fact that now the most significant bit of c plays a role, and in this half the most significant bit of the input x is equal to 1.

Using this observation, we obtain the significant equation

$$[\hat{f}]\mathbf{H}_n = [\hat{F}].$$

Proof The ωth entry of the row matrix $\hat{F}$ is the sum of the products of $[\hat{f}]$ and the ωth column = the ωth row of $\mathbf{H}_n$. But this is, as we have just seen, the matrix $\hat{l}_\omega$. Hence the ωth entry in the product on the left is

$$\sum_x \hat{f}(x)\hat{l}_\omega(x) = \sum_x (-1)^{f(x)}(-1)^{\omega \cdot x}$$

$$= \sum_x (-1)^{f(x) \oplus \omega \cdot x} = \hat{F}(\omega).$$

Example Consider the function f, defined by $f(x_2,x_1,x_0) = x_2 + x_1x_0$. Then $[f(x)] = [0,0,0,1,1,1,1,0]$ and $[\hat{f}(x)] = [1,1,1,-1,-1,-1,-1,1]$ and $[\hat{F}(\omega)] = [\hat{f}(x)]\,\mathbf{H}_3 = [0,0,0,0,4,4,4,-4]$. Thus, for example, if we consider the 0 constant function, we have that

$$A - D = 0$$

$$A + D = 8$$

so that $A = D = 4$, confirming the obvious fact that f is balanced. On the other hand, if we look at the linear function $x_2 + x_1 = l_6$, we see that

$$A - D = 4$$

so that $A = 6$ and $D = 2$, a fact which can easily be checked. In this case, therefore, we have a positive correlation between the function f and the linear function l_6.

We shall shortly pursue this matter further.

However, we first make a computational observation, similar to the observation we made regarding an efficient way of finding the ANF of a function:

Just as the ANF is calculated by repeated application of the matrix

$$\begin{pmatrix} 1 & 1 \\ 0 & 1 \end{pmatrix},$$

so the Walsh–Hadamard transform is calculated by repeated application of the matrix

$$\begin{pmatrix} 1 & 1 \\ 1 & -1 \end{pmatrix}$$

on adjacent entries, then on entries 2 apart, then on entries 4 apart, then on entries 8 apart, and so on. Applying this to the sequence *abcdefgh*, for example, we get:

$$\begin{array}{llll}
a & a+b & (a+b)+(c+d) & a+b+c+d+e+f+g+h \\
b & a-b & (a-b)+(c-d) & a-b+c-d+e-f+g-h \\
c & c+d & (a+b)-(c+d) & a+b-c-d+e+f-g-h \\
d & c-d & (a-b)-(c-d) & a-b-c+d+e-f-g+h \\
e & e+f & (e+f)+(g+h) & a+b+c+d-e-f-g-h \\
f & e-f & (e-f)+(g-h) & a-b-c+d-e+f-g+h \\
g & g+h & (e+f)-(g+h) & a+b-c-d-e+f+g+h \\
h & g-h & (e+f)-(g-h) & a-b-c+d-e-f+g-h
\end{array}$$

This amounts to repeated applications of the pattern

$$\begin{array}{ccc} x \longrightarrow & & x+y \\ & \nearrow & \\ & \searrow & \\ y \longrightarrow & & x-y \end{array}$$

but remember that the '+' and '-' here indicate addition and subtraction of ordinary integers!

8.3.3 Correlation with Linear Functions

For each of the linear Boolean functions l_ω we have, of course, a corresponding real-valued function $\hat{l}_\omega$. Note that these functions $\hat{l}_\omega$ are mutually orthogonal,[11] as is apparent from the shape of the Hadamard matrix, or can be proved properly.

If we divide each of these vectors by its length $2^{\frac{n}{2}}$, we get an orthonormal basis for the set of all possible $\hat{F}$s since

$$\begin{aligned} \hat{F}(\omega) &= \sum_x (-1)^{f(x)\oplus\omega\cdot x} \\ &= \sum_x (-1)^{f(x)}(-1)^{l_\omega(x)} \\ &= \sum_x \hat{f}(x)\hat{l}_\omega(x). \end{aligned}$$

Thus every Walsh transform $\hat{F}(\omega)$ (ω fixed) of a function f is the dot-product of $\hat{f}$ and $\hat{l}_\omega$. But conversely we also have that

$$\begin{aligned} \sum_\chi \hat{F}(\chi)\hat{l}_\chi(y) &= \sum_\chi \sum_x \hat{f}(x)\hat{l}_\chi(x)\hat{l}_\chi(y) \\ &= \sum_x \hat{f}(x) \sum_\chi \hat{l}_\chi(x)\hat{l}_\chi(y) \\ &= \sum_\chi \sum_x \hat{f}(x)\delta_{xy}, \end{aligned}$$

[11] That is, $\hat{l}_\psi \cdot \hat{l}_\omega = 0$ if $\psi \neq \omega$.

where δ_{xy} is the "Kronecker delta"[12] (since, for a fixed y, $l_\chi(x)$ and $l_\chi(y)$ will agree and disagree the same number of times, unless $x = y$)

$$= \sum_\chi \hat{f}(y)$$

$$= 2^n \hat{f}(y),$$

so that

$$\hat{f}(x) = \frac{1}{2^n} \sum_\omega \hat{F}(\omega) \hat{l}_\omega(x).$$

This may be called the inverse Walsh transform.

This leads us to the important

Parseval's Theorem

$$\sum_\omega \hat{F}(\omega)^2 = 2^{2n}.$$

Proof

$$2^n = \hat{f} \cdot \hat{f} = \frac{1}{2^{2n}} \sum_\omega \sum_\chi \hat{F}(\omega) \hat{F}(\chi) \hat{l}_\omega \cdot \hat{l}_\chi$$

$$= \frac{1}{2^{2n}} \sum_\omega \hat{F}(\omega) \hat{F}(\omega) \hat{l}_\omega \cdot \hat{l}_\omega$$

$$= \frac{1}{2^{2n}} \sum_\omega \hat{F}(\omega) \hat{F}(\omega) \times 2^n$$

$$= \frac{1}{2^{2n}} 2^n \sum_\omega \hat{F}(\omega)^2.$$

What are the implications of this? Recall that $\hat{F}(\omega)$ represents the number of agreements minus the number of disagreements between the functions f and l_ω, or a measure of the *correlation* between these two functions. We define the *correlation coefficient* $c(f, g)$ between two functions f and g, each of n variables as

$$c(f, g) = \frac{A - D}{2^n},$$

[12] $\delta_{xy} = 1$ if $x = y$, 0 otherwise.

where $A = \#\{x|f(x) = g(x)\}$ is the number of inputs for which f and g agree and $D = \#\{x|f(x) \neq g(x)\}$ is the number of inputs for which they disagree. Then we have noted in the previous section that

$$c(f, l_\omega) = \frac{\hat{F}(\omega)}{2^n}.$$

As an example: the fact that f is balanced is equivalent to the fact that $\hat{F}(0) = 0$.

The correlation coefficient may be thought of as measuring the amount of information that the values of f "leak" about the values of g (or conversely). Put this way, it is not surprising that the concept has some relevance to the field of cryptography, the object of which is traditionally to find ways of preventing information leakage.

Looking at Parseval's theorem we note that it implies that every Boolean function must correlate to some linear function of the input variables, since the sum of the squares of the correlation coefficients is a constant (viz 2^{2n}). In fact, the larger the number of linear functions with which the function f has zero correlation, the larger the correlation with the remaining linear functions. In the most extreme case, if one wants a zero correlation with all linear functions of $n-1$ variables then one must have a correlation of ± 1 with the function $\sum_i x_i$. In fact the average value of $(\hat{F}(\omega))^2$ equals 2^n regardless, so if some of these values are 0, the others must assume larger (absolute) values.

If n is even we may consider functions for which $(\hat{F}(\omega))^2 = 2^n$ for all ω, so that $\hat{F}(\omega) = \pm 2^{\frac{n}{2}}$ for all ω. This implies that f correlates with all linear functions to the same extent.[13] Such functions are called *bent* functions and will be discussed later in this chapter. Since $\hat{F}(0) \neq 0$ for a bent function, bent functions cannot be balanced.

8.3.4 Correlation Immunity

We can actually state a further, important, result, due to Xiao and Massey. Recall from the previous chapter that a function f is *t-correlation-immune* if $Z = f(X)$ is statistically independent of any set $\{X_{i_0}, \ldots, X_{i_{t-1}}\}$ of t or fewer input variables. This means that the values of the function do not leak any information about individual input bits, but only about linear combinations of more than t input variables. This is, obviously, an improvement in the cryptographic strength. Note that $t > 0$, so we only consider nonempty sets of input variables. The case where we consider no input variables at all is the case where we consider just the balancedness of f. Xiao and Massey[14] obtained a characterization of correlation-immunity in terms of the Walsh transform.

---

[13] The correlation coefficient can be quite small, if n is large enough.

[14] Xiao Guo-Zhen and Massey, J.L.: *A spectral characterization of correlation-immune combining functions*, IEEE Trans. Info Th. **34**, 1988, pp. 569–571.

The proof of their result follows almost immediately from the following lemma, of which we shall only sketch the proof:

Lemma *The binary variable Z is independent of the m independent binary variables $Y_0, \ldots, Y_{m-1}$ if and only if Z is independent of every nonzero linear combination $\sum_i \lambda_i Y_i$ of the variables Y_i, where $\lambda_i \in \{0, 1\}$.*

Proof The necessity of the stated condition is obvious, as is the sufficiency for the trivial case $m = 1$. We prove the sufficiency only for the case $m = 2$, the general case can be proved by induction from there.[15]

Suppose therefore that Z is independent of Y_0, Y_1 and of $Y_0 + Y_1$. Now consider the following set of equations, where we define p_0 and p_1 as the probabilities that y_0, y_1, respectively, equal 1:

$$\Pr(y_0 = 1|z) = \Pr(y_1 = 0, y_0 = 1|z) + \Pr(y_1 = 1, y_0 = 1|z) = \Pr(y_0 = 1) = p_0,$$

$$\Pr(y_1 = 1|z) = \Pr(y_1 = 1, y_0 = 1|z) + \Pr(y_1 = 1, y_0 = 0|z) = \Pr(y_1 = 1) = p_1,$$

$$\begin{aligned}\Pr(y_0 + y_1 = 1) = \Pr(y_1 + y_0 = 1|z) &= \Pr(y_1 = 1, y_0 = 0|z) + \Pr(y_1 = 0, y_0 = 1|z) \\ &= p_1(1 - p_0) + (1 - p_1)p_0,\end{aligned}$$

$$\begin{aligned}\Pr(y_0 + y_1 = 0) = \Pr(y_1 + y_0 = 0|z) &= \Pr(y_1 = 1, y_0 = 1|z) + \Pr(y_1 = 0, y_0 = 0|z) \\ &= p_1 p_0 + (1 - p_1)(1 - p_0),\end{aligned}$$

where we have used the given statistical independencies.

Solving these four equations for the four unknowns $\Pr(y_1 = a, y_0 = b|z)$ we get

$$\Pr(y_1 = 1, y_0 = 1|z) = p_0 p_1,$$

$$\Pr(y_1 = 1, y_0 = 0|z) = (1 - p_0)p_1,$$

$$\Pr(y_1 = 0, y_0 = 1|z) = p_0(1 - p_1),$$

$$\Pr(y_1 = 0, y_0 = 0|z) = (1 - p_0)(1 - p_1).$$

But this conditional probability distribution is precisely the same as for the pairs $\{y_1, y_0\}$, so this distribution is not influenced by z. Thus the triple $\{Z, Y_1, Y_0\}$ is independent.

[15] We refer the interested reader to the paper by Xiao and Massey referenced in the previous footnote.

Xiao and Massey's theorem below is an immediate consequence of this lemma:

Theorem *The Boolean function* $f(x_{n-1}, \ldots, x_1, x_0)$ *is t-correlation immune if and only if its Walsh transform satisfies* $\hat{F}(\omega) = 0$ *for all* ω *with* $1 \leq W_H(\omega) \leq t$.

A function is called *t-resilient* if it is *t*-correlation immune, as well as balanced. Balance is obviously important (a preponderance of zeros, say, in the output would no doubt help the attacker performing a known plaintext attack), so having a separate definition for this eventuality is undoubtedly justified. Since balancedness is equivalent to $\hat{F}(0) = 0$, we have that f is t-resilient if and only if $\hat{F}(\omega) = 0$ for all ω with $0 \leq W_H(\omega) \leq t$.

We may note here that in the same paper Xiao and Massey give a proof of the result of Siegenthaler[16] using a relationship between the ANF of a function and its Walsh transform. We shall, however, give a version of the original Siegenthaler proof.

For this purpose we introduce the following notations:

$$\mathbf{x} = (x_0, x_1, \ldots, x_{n-1})$$

and therefore

$$f(\mathbf{x}) = f(x_0, x_1, \ldots x_{n-1}).$$

Also

$$S_{01\ldots k} = \{\mathbf{x} | x_{k+1} = \cdots = x_{n-1} = 0\}$$

if $k < n$ and $S_{01\ldots n-1}$ = set of all possible vectors $\mathbf{x}$.

Then the coefficient $a_{01\ldots k}$ in the Algebraic Normal Form is

$$a_{01\ldots k} = \sum_{\mathbf{x} \in S_{01\ldots k}} f(\mathbf{x}).$$

This can be seen as follows: The product $x_{i_0}x_{i_1}\ldots x_{i_j}$, with $0 \leq i_0 < i_1 < \cdots < i_j \leq n-1$, vanishes in this sum if $i_j > k$ and if $i_j \leq k$ then it equals 1 for exactly 2^{k-j} elements of $S_{01\ldots k}$. Hence the product equals 1 if and only if $x_{i_0}x_{i_1}\ldots x_{i_j} = x_0 x_1 \ldots x_k$.

Finally, we define

$$N_{01\ldots k} = \#\{\mathbf{x} \epsilon S_{01\ldots k} | f(\mathbf{x}) = 1\},$$

which means that we can write the expression for $a_{01\ldots k}$ as

$$a_{01\ldots k} = \text{parity}(N_{01\ldots k}),$$

[16]See Sect. 7.10.

where the parity function is defined in the usual way, as being 0 if the input is even and 1 if the input is odd.

Theorem (Siegenthaler) *Let $z = f(\mathbf{x})$. If f is t-correlation-immune then no product of more than $n - t$ variables can appear in the algebraic normal form of f, i.e. the degree of f is at most $n - k$.*

Proof Clearly

$$\Pr(z = 1|x_{k+1} = x_{k+2} = \cdots = x_{n-1} = 0) = \frac{N_{01\ldots k}}{2^{k+1}}$$

for $k = 0, 1, \ldots, n - 2$; and

$$\Pr(z = 1) = \frac{N_{01\ldots n-1}}{2^n}.$$

Now suppose that f is t-correlation immune. Then the probability of z being 1 is independent of the number of x_i being fixed to 0 as long as this number is not greater than t. Therefore:

$$\Pr(z = 1|x_{k+1} = \cdots = x_{n-1} = 0) = \Pr(z = 1)$$

if $n - 1 - k \leq t$, i.e. $n - 1 \geq k \geq n - t$,
and hence

$$\frac{N_{01\ldots n-1}}{2^n} = \cdots = \frac{N_{01\ldots k}}{2^{k+1}} = \frac{N_{01\ldots n-1-t}}{2^{n-t}}.$$

Hence

$$N_{01\ldots k} = 2^{k+1-(n-t)} N_{01\ldots(n-1-t)},$$

which is even if $k + 1 > n - t$, i.e. if $n - t - 1 < k \leq n - 1$.

We therefore have that

$$a_{01\ldots k} = 0 \text{ for } n - t \leq k \leq n - 1.$$

This means that no terms involving products of more than n–t of the variables $x_0, x_1, \ldots, x_k$ can appear in the ANF of f. But the labelling of the variables is purely arbitrary, so what we have in fact proved is that there are no terms of degree greater than $n - t$ in the ANF of f, i.e. that the degree of f is $\leq n - t$. This completes the proof.

We mention as a further observation that if f is balanced, then from the fact that $\Pr(z = 1) = \frac{1}{2}$ we can obtain that the degree of f must be $\leq n - t - 1$. The details of the proof are left to the very diligent reader.

Finally, we note that it follows from the above proof that if $t = n - 1$ then all terms of degree $n - t = 1$ (and no terms of higher degree) must appear in the ANF. There are therefore only two possible functions of n variables which are $(n-1)$-correlation-immune, viz $\sum_{i=0}^{n-1} x_i$ and its complement.

Later Sarkar and Maitra[17] proved the following, which may be regarded as a refinement of the Xiao–Massey result:

If the Boolean function f is t-resilient, then

$$\hat{F}(\omega) \equiv 0 \mod 2^{t+2}$$

for all ω.

8.3.5 Linear Algebraic Gloss

When we consider, instead of a Boolean function f (of n variables, say), its real-valued counterpart $\hat{f}$, and we do the same for the 2^n possible linear functions l_ω (defined, as before, by $l_\omega(x) = \omega \cdot x$) and their counterparts $\widehat{l_\omega}$, we can make an interesting observation.

This observation is based on the fact that if $\omega \neq \nu$, then l_ω and l_ν correspond to distinct rows in the Hadamard matrix used to compute the Walsh transform. This means, as we have noted before, that the (real) vectors $\hat{l}_\omega$ and $\hat{l}_\nu$ are orthogonal to each other in the space $\mathbb{R}^{2^n}$. On that space we can define the usual Euclidean metric, based on the dot product as inner product $< \mathbf{x}, \mathbf{y} > = \sum_{i=0}^{n-1} x_i y_i$, so the norm of a vector is $||\mathbf{x}|| = \sqrt{\sum_{i=0}^{n-1} x_i^2}$.

If we normalise the vectors $\hat{l}_\omega$ we therefore have an orthonormal basis for the space $\mathbb{R}^{2^n}$. Now it is well known that, given an orthonormal basis $\{\mathbf{e}_i | i = 0 \ldots, k\}$ for a vector space of dimension k, then every vector $\mathbf{x}$ in that space can (obviously) be written as a linear combination of those basis vectors, and the coefficients are given by $< \mathbf{x}, \mathbf{e}_i >$. In the case we are considering, therefore, we find that any Boolean function f of n variables corresponds to a real function $\hat{f} : \{0, 1\}^n \longrightarrow \{-1, 1\}$ which can be written as

$$\hat{f} = \sum_{\omega} c_\omega \hat{l}_\omega,$$

where the sum is taken over all $\omega \in \{0, 1\}^n$ and where the $c_\omega = < \hat{f}, \hat{l}_\omega >$ are the normalised values of the Walsh transform of f.

Example The function f

$$[0\ 1\ 1\ 1]$$

[17] Sarkar, P. and Maitra, S.: *Nonlinearity bounds and constructions of resilient Boolean functions*; Proc. Crypto 2000, LNCS 1818, Springer-Verlag, 2000, 516–533.

has Walsh transform

$$[-2\,2\,2\,2]$$

corresponding to the fact that $\hat{f}$ is the linear combination

$$2/4(-[1111] + [1-11-1]+[11-1-1]+[1-1-11])$$

of the basis vectors.

Because of its similarity to the Fourier transform, where a real function is written as an (infinite) linear combination of orthonormal functions (defined in terms of $\sin nx$ and $\cos nx$), the above application of the Walsh–Hadamard transform is sometimes, confusingly, referred to as the finite Fourier transform. Terry Ritter in his Survey of the Walsh–Hadamard Transforms on the Internet (http://www.ciphersbyritter.com/RES/WALHAD.HTM) refers to the version outlined here as "a poor man's version" of the Fast Fourier Transform. To add to the confusion, the Walsh transform is also called the Hadamard transform, the Walsh–Hadamard transform (as we may sometimes call it in this book), the Hadamard–Rademacher–Walsh transform and the Fourier–Walsh transform. (There may still be other combinations of these names.) We shall discuss the Discrete Fourier Transform in the final section of this chapter.

Exercises

1. Find the best linear approximation to the function with truth table

$$0101\ 1011\ 0110\ 1010\ 0111\ 0110\ 0111\ 0001.$$

2. Find all 1-resilient Boolean functions of three variables.
3. Show that the only 2-resilient function of four variables is the function $x_0 \oplus x_1 \oplus x_2 \oplus x_3$.

8.4 Autocorrelation

We can define the autocorrelation $\hat{r}_f$ of a Boolean function f in the same way as we have defined the correlation between f and the linear functions l_ω:

$$\hat{r}_f(s) = \sum_x \hat{f}(x)\hat{f}(x \oplus s),$$

which measures the agreements minus the disagreements between the values of f and the values of f when its inputs are modified by the XOR of s. The vector $[\hat{r}] = [\hat{r}(0)\ \hat{r}(1) \ldots \hat{r}(2^n - 1)]$ is called the *autocorrelation spectrum* of f. Clearly $\hat{r}(0)$ will always have the value 2^n. The *autocorrelation* of f, denoted by Δ_f, is defined as

$$\Delta_f = \max_{s \neq 0} \left| \sum_x \hat{f}(x)\hat{f}(x \oplus s) \right| .$$

This measures the effect of changing the inputs to the function: for obvious reasons one would like this to be close to 0, in that a random change to the input should change, in what appears to be a "random" way, about half the outputs. We shall pursue this line of thought later in this chapter, preparatory to which we note the following theorem, the proof of which is purely computational.

Theorem *Let $\hat{R}(\omega)$ and $\hat{F}(\omega)$ denote the Walsh transform of the functions $\hat{r}_f$ and $\hat{f}$ respectively. Then for any ω*

$$\hat{R}(\omega) = [\hat{F}(\omega)]^2.$$

Proof

$$\begin{aligned}
[\hat{F}(\omega)]^2 &= \sum_x (-1)^{f(x) \oplus \omega.x} \sum_y (-1)^{f(y) \oplus \omega.y} \\
&= \sum_x \sum_y (-1)^{f(x) \oplus f(y) \oplus \omega.(x \oplus y)} \\
&= \sum_s \sum_x (-1)^{f(x) \oplus f(x \oplus s)} (-1)^{\omega.s} \\
&= \sum_s \hat{r}_f(s)(-1)^{\omega.s} \\
&= \hat{R}(\omega).
\end{aligned}$$

This theorem enables us to get $\hat{r}$ easily from the string representing the function. We shall demonstrate by means of an example: Suppose that f is the function

0101 0011 0101 0110.

We construct $\hat{f}$, calculate $\hat{F}$, square this to get $\hat{R} = (\hat{F})^2$, and apply the inverse Walsh transform to get $\hat{r}$.

f	$\hat{f}$	$\hat{F}$	$\hat{R} = \hat{F}^2$	$\hat{r}$
0	1	0	0	16
1	−1	8	64	−8
0	1	4	16	0
1	−1	4	16	−8
0	1	0	0	0
0	1	8	64	0
1	−1	−4	16	0
1	−1	−4	16	0
0	1	0	0	8
1	−1	0	0	−8
0	1	4	16	8
1	−1	−4	16	−8
0	1	0	0	0
1	−1	0	0	0
1	−1	−4	16	0
0	1	4	16	0

We invite the reader to verify that, for example, $f(x)$ agrees with $f(x \oplus 0001)$ 4 times and disagrees with it 12 times.

Note also that

$$\begin{aligned} &\Pr(\hat{f}(x) \neq \hat{f}(x \oplus a)) \\ &= \frac{2^{n-1} - \frac{1}{2}\hat{r}(a)}{2^n} \\ &= \frac{1}{2} - \frac{\hat{r}(a)}{2^{n+1}} \end{aligned}$$

and that

$$\begin{aligned} \sum_a \hat{r}(a) &= \sum_a \sum_x (-1)^{f(x)}(-1)^{f(x \oplus a)} \\ &= \sum_x (-1)^{f(x)} \sum_a (-1)^{f(x \oplus a)} \\ &= \sum_a (-1)^{f(x)} \sum_s (-1)^{f(s)} \\ &= [\hat{F}(0)]^2, \end{aligned}$$

so that if we take the sum of these probabilities, we get

$$\sum_{a \neq 0} \Pr(\hat{f}(x) \neq \hat{f}(x \oplus a)) = 2^{n-1} - \frac{(\hat{F}(0))^2}{2^{n+1}}.$$

If all probabilities are equal to $\frac{1}{2}$, then a simple calculation shows that $|\hat{F}(0)| = 2^{\frac{n}{2}}$, so that f cannot be balanced. On the other hand, if f is balanced, then the average of the probabilities is $\frac{2^{n-1}}{2^n-1}$ which is greater than $\frac{1}{2}$.[18]

Example Consider again the function f of 4 variables with truth table

$$0101\ 0011\ 0101\ 0110.$$

For this function we found the following values for $\hat{r}$

$$16\ -8\ 0\ -8\ 0\ 0\ 0\ 0\ 0\ 8\ -8\ 8\ -8\ 0\ 0\ 0\ 0$$

and the corresponding probabilities of non-equality are

0, 0.75, 0.50, 0.75, 0.50, 0.50, 0.50, 0.50, 0.25, 0.75, 0.25, 0.75, 0.50, 0.50, 0.50, 0.50.

The sum of these values, excluding the case $a = 0$, is 8, with an average value of $\frac{8}{15} =$ 0.53.

8.5 Nonlinearity

We have (repeatedly, by now) seen that $\hat{F}(\omega)$ represents the number of agreements minus the number of disagreements between the (n-variable, say) function f and the linear function l_ω defined by $l_\omega(x) = \omega \cdot x$, and that therefore the distance between these two functions is

$$d(f, l_\omega) = \frac{1}{2}(2^n - \hat{F}(\omega)).$$

If we consider the affine function created by complementing l_ω then the number of agreements minus the number of disagreements will be $-\hat{F}(\omega)$ and the distance between f and this function will be $\frac{1}{2}(2^n + \hat{F}(\omega))$.

When considering the nonlinearity of a function, it seems reasonable to look at the distance between that function and all *affine* functions, rather than just the linear ones.[19] This leads to the following definition:

Definition Let f be a Boolean function of n variables. Its nonlinearity is the minimum distance between f and the set of all linear and affine functions of the same number of variables.[20]

---

[18]These observations appear to be due to B. Preneel, W. van Leeckwijck, L. van Linden, R. Govaerts and J. Vandewalle: *Propagation characteristics of Boolean functions*, Proc. Eurocrypt '90, LNCS 473, Springer-Verlag, 1990.

[19]A linear Boolean function on n variables is one of the form $\sum_{i=0}^{n-1} a_i x_i$, with $a_i \in \{0, 1\}$ for all i. An affine Boolean function is of the form $c + \sum_{i=0}^{n-1} a_i x_i$, with $c, a_i \in \{0, 1\}$.

[20]Sadly, the words "nonlinear" and "nonlinearity" are overworked, and do not always mean the same things in the cryptological literature. Our notion of nonlinearity may be called *functional* nonlinearity, as opposed

It is clear from the above discussion that

$$\text{Nonlinearity of} f = \min_{\omega} \left\{ 2^{n-1} - \frac{1}{2}|\hat{F}(\omega)| \right\}.$$

We shall denote the nonlinearity of a function f by NL(f).

It is not hard to show that if $f(x_{n-1}, \ldots, x_0)$ has nonlinearity N then the function $g(x_n, x_{n-1}, \ldots, x_0) = x_n \oplus f(x_{n-1}, \ldots, x_0)$ has nonlinearity $2N$.

Bent functions are characterised by the fact that $\hat{F}(\omega) = \pm 2^{\frac{n}{2}}$ for all ω. This is, in the light of Parseval's theorem, the smallest value that we can get for $|\hat{F}(\omega)|$ and therefore the bent functions will yield the highest nonlinearity, viz $\frac{1}{2}(2^n - 2^{\frac{n}{2}}) = 2^{n-1} - 2^{\frac{n}{2}-1}$. Unfortunately, it would appear that this only makes sense for even n, and this is correct: bent functions only exist for an even number of variables. Bent functions will be discussed further in Sect. 8.8.

In fact, for odd n the situation is very unsatisfactory: it is not known what the highest achievable nonlinearity is. The following facts are known.[21]

Denote by NL_{max} the maximum nonlinearity.

- For $n = 3, 5, 7$: $NL_{max} = 2^{n-1} - 2^{\frac{n-1}{2}}$.
- For $n = 9, 11, 13$: $2^{n-1} - 2^{\frac{n-1}{2}} \leq NL_{max} \leq 2^{n-1} - 2^{\lfloor \frac{n}{2} \rfloor - 1}$.
- For odd $n \geq 15$: $2^{n-1} - 2^{\frac{n-1}{2}} < NL_{max} \leq 2^{n-1} - 2^{\lfloor \frac{n}{2} \rfloor - 1}$.

8.6 Propagation Criteria

The motivation behind the study of propagation criteria is essentially that of trying to use functions which provide good *diffusion, i.e.* flipping a single bit of the input should have long term effects on the output.[22] This may be likened to the effect of disturbing a small amount of snow on the mountainside and thereby creating a full scale avalanche, and one of the terms that we shall define is correspondingly called the *avalanche criterion.*

---

to *algebraic* nonlinearity, which refers to the degree of the function when expressed in its Algebraic Normal Form. Our definition here gives the only meaning we shall attach to "nonlinearity" as a noun, though when using the adjective we may be a bit more sloppy. So, please be warned.

[21] Our source here is the paper by Filiol and Fontaine: Filiol, E. and Fontaine, C.: *Highly nonlinear balanced functions with a good correlation-immunity*, Proceedings Eurocrypt '98, LNCS 1403, Springer-Verlag. The paper by Sarkar and Maitra, referenced in footnote 17 above, gives some similar bounds for functions which must also satisfy some given correlation immunity constraints. A paper by Fu, S., Sun, B., Li, C. and Qu, L., *Construction of odd-variable resilient Boolean functions with optimal degree* J. Info. Sc. and Eng. **27** (2011) pp. 1931–1942, presents further results considering odd numbers of variables, correlation immunity and the degrees.

[22] Zheng and Zhang (Y. Zheng and X.-M. Zhang: *On relationships among avalanche, nonlinearity and correlation immunity*, Proc. Asiacrypt 2000, LNCS 1976, Springer-Verlag, 2000, pp. 470–482) make the observation, worth quoting, that "high nonlinearity generally has a positive impact on confusion, whereas a high degree of avalanche enhances the effect of diffusion".

One may note here that while creating avalanches in the mountains is, at best, an antisocial thing to do, using the avalanche effect in designing stream ciphers merely represents a sensible approach to information security.

Loosely speaking, a Boolean function f satisfies the avalanche criterion if on flipping a single input bit, the probability of changing the output bit is approximately $\frac{1}{2}$. This may be phrased more precisely as $\Pr(f(\mathbf{x}) = f(\mathbf{x} \oplus \mathbf{a})) \approx \frac{1}{2}$ for all input vectors $\mathbf{a}$ of Hamming weight 1.

Generally, it is preferable to work with the so-called *Strict Avalanche Criterion* (SAC) in which the probability is required to be exactly $\frac{1}{2}$:

Definition A Boolean function f of n variables satisfies the Strict Avalanche Criterion if

$$\#\{\mathbf{x}|f(\mathbf{x}) = f(\mathbf{x} \oplus \mathbf{a})\}$$
$$= 2^{n-1} \quad \forall \mathbf{a} \text{ of Hamming weight 1.}$$

In other words, the function $f(\mathbf{x}) \oplus f(\mathbf{x} \oplus \mathbf{a})$ is balanced whenever $\mathbf{a}$ consists of 0s and a single 1.

There seems to be little reason to restrict oneself to vectors of Hamming weight 1, and a more general notion has therefore been defined:

Definition A Boolean function f of n variables satisfies the *Propagation Criterion of degree* l *(PC(l))* if

$$f(\mathbf{x}) \oplus f(\mathbf{x} \oplus \mathbf{a})$$

is balanced for all $\mathbf{a}$ of Hamming weight $w_H(\mathbf{a})$ with $1 \leq w_H(\mathbf{a}) \leq l$.

SAC is therefore the same thing as $PC(1)$.

In terms of the auto-correlation function defined in Sect. 8.4, we may also define the function f to be $PC(l)$ if

$$\hat{r}_f(\mathbf{a}) = 0 \;\forall \mathbf{a} \text{ with } 1 \leq w_H(\mathbf{a}) \leq l.$$

It seems natural to try and find functions which are $PC(l)$ for all l up to and including n, i.e. functions such that $f(x) \oplus f(x \oplus a)$ is balanced for all nonzero a. Such functions were named *perfect nonlinear* by Meier and Staffelbach.[23] These functions turned out to be precisely the previously mentioned *bent functions*,[24] to which we shall return in Sect. 8.8. They are, as we have noted, characterised by the fact that a function f is bent if and only if $\hat{F}(\omega) = \pm 2^{\frac{n}{2}} \;\forall \omega$. This yields that $\hat{R}(\omega) = 2^n \;\; \forall \omega$, from which an application of the inverse Walsh transform

[23] W. Meier and O. Staffelbach: *Nonlinearity criteria for cryptographic functions*, Proc. Eurocrypt '88, LNCS 330, Springer-Verlag 1989.

[24] Bent functions were discovered by Rothaus (O.S. Rothaus: *On bent functions*, J. Comb. Th. **26**, 1976, pp. 300–305) in a combinatorial context.

yields that $\hat{r}(a) = 0$ for all $a \neq 0$ (and $\hat{r}(0) = 2^{2n}$, of course). They only exist for even values of n, as we have noted before.

For odd values of n the best one can hope to achieve is $PC(n-1)$, but we can in this case achieve balancedness at the same time: In this case $\hat{r}(0) = 2^n$, and since we want $\hat{r}(\mathbf{a}) = 0 \;\; \forall \mathbf{a}$ with $1 \leq w_H(\mathbf{a}) \leq n-1$ and since $\sum_{\mathbf{a}} \hat{r}(\mathbf{a}) = (\hat{F}(0))^2 = 0$ we must have that $\hat{r}$ is of the form

$$\hat{r} = [2^n, 0, 0, 0, \ldots, -2^n].$$

Multiplying by $\mathbf{H}$, as in Sect. 8.3.2, we get that $\hat{R} = (\hat{F})^2$ contains only entries of either 0 or 2^{n+1}, so that $\hat{F}$ contains entries $0, \pm 2^{\frac{n+1}{2}}$. The nonlinearity of such a $PC(n-1)$ function is therefore $\frac{1}{2}(2^n - 2^{\frac{n+1}{2}}) = 2^{n-1} - 2^{\frac{n-1}{2}}$. An example of such a function is the majority function

$$f(x_2, x_1, x_0) = x_0x_1 + x_0x_2 + x_1x_2,$$

i.e. the function 0001 0111, as the reader is invited to verify.

One expects that desirable properties ("desirable" from the cryptographic point of view) of Boolean functions will inevitably lead to trade-offs. Curiously, this is not the case for the properties of (a) good propagation (avalanche effect) and (b) high nonlinearity, as Zheng and Zhang have shown.[25] In fact, they prove that $PC(l)$ for a large value of l implies high nonlinearity, as the following theorem, the proof of which we omit, shows.

Theorem *Let f be a function of n variables, satisfying $PC(l)$. Then*

- $NL(f) \geq 2^{n-1} - 2^{n-1-\frac{1}{2}l}$.
- *Equality holds in this inequality if and only if one of the following two conditions holds:*

  1. *n is odd, $l = n-1$ and*

  $$\begin{aligned} f(x) = \; & g(x_0 \oplus x_{n-1}, \ldots, x_{n-2} \oplus x_{n-1}) \\ & \oplus h(x_0, \ldots x_{n-1}), \end{aligned}$$

  where g is a bent function of $n-1$ variables and h is an affine function of n variables.

  2. *n is even, $l = n$ and f is bent.*

Sadly, when it comes to (a) good propagation and (c) correlation immunity, we find that these two requirements contradict each other so that a trade-off is inevitable. Zheng and Zhang[26] prove the following two results:

Theorem *Let f be a balanced, $PC(l)$, and s-correlation immune function of n variables. Then $l + s \leq n - 2$.*

[25] In their paper referenced in footnote 22.

[26] In the same paper, i.e. the one referenced in footnote 22.

Theorem *Let f be an unbalanced, PC(l), and s-correlation-immune function of n variables. Then*

- $l + s \leq n$;
- *Equality in this inequality holds if and only if n is odd,* $l = n - 1$, $s = 1$ *and*

$$f(x) = g(x_0 \oplus x_{n-1}, \ldots, x_{n-2} \oplus x_{n-1}) \oplus c_0 x_0 \oplus \ldots c_{n-1} x_{n-1} \oplus c,$$

where g is a bent function of $n - 1$ *variables, and the* $c, c_i \in \{0, 1\}$ *satisfy* $\bigoplus_{i=0}^{n-1} c_i = 0$.

8.7 Linear Structures

A vector **a** is called a *linear structure*[27] of a Boolean function f of n variables if

$$\hat{r}_f(\mathbf{a}) = \sum_{\mathbf{x}} \hat{f}(\mathbf{x})\hat{f}(\mathbf{x} \oplus \mathbf{a}) = \pm 2^n,$$

which is equivalent to saying that

$$f(\mathbf{x}) \oplus f(\mathbf{x} \oplus \mathbf{a}) = \text{constant}.$$

8.7.1 Linearity

It is easy to verify that the linear structures of f (together with the zero vector) form a linear subspace of the vector space $(GF(2))^n$. Its dimension is called the *linearity* of f. If f is a linear or affine function of n variables, then this dimension is n, because every nonzero **a** is a linear structure: $f(\mathbf{x} \oplus \mathbf{a}) = f(\mathbf{x}) \oplus f(\mathbf{a})$, after all. This dimension may therefore be taken as a measure of "how linear" f is. From the cryptographic point of view it seems desirable to use functions with zero linearity.

Nevertheless, choosing the term "linearity" for this concept may be considered unfortunate as it would seem to be the negation of the nonlinearity which we defined in Sect. 8.5. In fact, the two concepts are unrelated to some extent. In Exercise 2 you are given two functions f and g with the properties that f is both more linear and more nonlinear than g!

It is not hard to find functions with nonzero linearity without being linear: in fact it does not even have to contain any linear terms in its ANF as the following example shows.

[27]The concept of a linear structure appears to be due to Evertse: Evertse J.-H.: *Linear Structures in Block Ciphers*, Proc. Eurocrypt '87. As an example of such a structure, in vector-valued functions, he mentions the complementation property of the Data Encryption Standard. This is the property that if $c = E_K(m)$ is the DES encryption of plaintext m using key K, then $\overline{c} = E_{\overline{K}}(\overline{m})$, where $\overline{x}$ denotes the bitwise complement of x.

Example Let $f(x_3, x_2, x_1, x_0) = x_1x_0 + x_2x_0 + x_2x_1 + x_2x_1x_0 + x_3x_0 + x_3x_1 + x_3x_1x_0$, with truth table

$$0001\ 0110\ 0110\ 0001.$$

f has linearity 1, since [1,1,0,0] is a linear structure, as is readily verified.

It can be shown that if a function f has k linearly independent structures, then, treating the input $(x_0, \ldots, x_{n-1})$ as a vector $\mathbf{x}$, there exists a non-singular matrix $\mathbf{B}$ such that $f(\mathbf{xB}) = g(\mathbf{y}) + \psi(\mathbf{z})$, where $\mathbf{xB} = (\mathbf{y},\mathbf{z})$, and ψ is an affine function on k variables. The trick[28] lies in changing the basis used for the vector space, from the "obvious" one, viz $\mathcal{B}_0 = \{\mathbf{e}_3 = [1,0,0,0], \mathbf{e}_2 = [0,1,0,0],\ \mathbf{e}_1 = [0,0,1,0], \mathbf{e}_0 = [0,0,0,1]\}$ to another, starting with a basis for the space of linear structures, and extending that to a basis for $\{0,1\}^n$ where $n = 4$ in our example, of course.

Example Continued So we'll take as our new basis the set $\mathcal{B}_1 = \{\mathbf{f}_3 = [1,1,0,0], \mathbf{f}_2 = \mathbf{e}_2, \mathbf{f}_1 = \mathbf{e}_1, \mathbf{f}_0 = \mathbf{e}_0\}$, a choice which should keep the arithmetic simple. In fact, the transformation matrix from $\mathcal{B}_0$ to $\mathcal{B}_1$ is

$$\mathbf{B} = \begin{pmatrix} 1\,1\,0\,0 \\ 0\,1\,0\,0 \\ 0\,0\,1\,0 \\ 0\,0\,0\,1 \end{pmatrix}$$

which, fortuitously, is its own inverse. Thus a vector which has components $[(x_3, x_2, x_1, x_0)]$ with respect to basis $\mathcal{B}_0$ will be $[y_3, y_2, y_1, y_0]$ when expressed in terms of $\mathcal{B}_1$, where

$$x_3 \mapsto y_3 + y_2,$$
$$x_2 \mapsto y_2,$$
$$x_1 \mapsto y_1,$$
$$x_0 \mapsto Y_0.$$

Making this substitution in the ANF of f leads to $g(\mathbf{y}) = y_1y_0 + y_3y_0 + y_3y_1 + y_3y_1y_0 + \mathbf{O}(y_2)$, where $\mathbf{O}$ is the (linear!) function $\mathbf{O}(y_2) \equiv 0$. This last function arises from the fact that $f(\mathbf{x} + \mathbf{f}_3) + f(\mathbf{x}) = 0$ for all $\mathbf{x}$.

From the comments of the previous section we recall that if the number of variables n is odd, and the function f is $PC(n-1)$, then $\hat{r} = [2^n, 0, 0, \ldots, 0, -2^n]$, so in this case the all-1 vector is a linear structure (and the only non-trivial one). On the other hand, if n is even, then the bent functions turn out to be precisely those without any (non-trivial) linear structures.

[28] A proof of a more general result (dealing with functions on $(GF(p))^n$ rather than just $(GF(2))^n$, may be found in Lai, X.: *Additive and linear structures of cryptographic functions*, Proc. Fast Software Encryption 1994, LNCS 1008, Springer-Verlag, 75–85.

8.7.2 Another Measure of Nonlinearity

Non-linearity of a function f was defined as the distance between f and the set of all linear and affine functions (of the same number of Boolean variables, of course). Acting on the principle that a function with linear structure is at least "partially" linear, and should therefore be avoided, it seems reasonable to consider, instead, the distance between the function f, and the set of all functions which have a linear structure. This was proposed many years ago in a paper by Meier and Staffelbach[29] but this approach does not seem to have found many followers. In the absence of an accepted name, we shall denote this by AFNL(f) (for "Another Form of Non-Linearity"). Thus:

$$\mathrm{AFNL}(f) = \min_{g \in \Lambda} d(f, g),$$

where Λ is the set of all Boolean functions with a linear structure.

Determining AFNM(f) where f is a function of n variables is a slightly tedious affair. For any nonzero **a**, consider all 2^{n-1} unordered pairs $(\mathbf{x}, \mathbf{x} \oplus \mathbf{a})$. Let $n_0 = \#\{\mathbf{x} | f(\mathbf{x}) = f(\mathbf{x} \oplus \mathbf{a}\}$ and $n_1 = \#\{\mathbf{x} | f(\mathbf{x}) \neq f(\mathbf{x} \oplus \mathbf{a}\}$. Now by altering n_0 or n_1 of the entries in the truth table of f, f is changed into a function which has **a** as a linear structure. Thus the distance between f and the nearest function which has **a** as a linear structure is

$$n_f(\mathbf{a}) = \min(n_0, n_1).$$

It follows that

$$\mathrm{AFNL}(f) = \min_{\mathbf{a} \neq 0} n_f(\mathbf{a}).$$

Example Consider the function $f = x_1 + x_0x_1 + x_2 + x_0x_1x_3 + x_1x_2x_3$, which has truth table

0010 1101 0011 1001.

If one flips the values for $f(1)$ and $f(5)$ one gets the function f':

0110 1001 0011 1001

which has the property that $f'(\mathbf{x} \oplus 0010) \oplus f'(\mathbf{x}) = 1$ for all $\mathbf{x}$, i.e. it has 0010 as a linear structure. Two flips is, in fact, the minimum number of flips required to obtain a function with a linear structure, so we have that $\mathrm{AFNL}(f) = 2$.

[29] Meier, W. and Staffelbach, O.: *Non-linearity criteria for Boolean functions*, Proc. Eurocrypt '89, LNCS 434, 549–562.

Exercises

1. Find the linear structures of the function

$$f(\mathbf{x}) = x_0x_2 + x_1x_2 + x_0x_3 + x_1x_3 + x_0x_4 \\ +x_1x_4 + x_2x_4 + x_3x_4$$

   and write this f in the form above, viz as the sum of an affine function and a function without any non-trivial linear structures.
2. Let $f(x) = x_0 + x_1x_2 + x_1x_3 + x_2x_3$ and let $g(x) = f(x) + x_1x_2x_3$. Show that f has both a higher linearity and a higher nonlinearity than g.
3. Let $\mathbf{A}$ be an $n \times n$ matrix with all entries from $GF(2)$ and define the Boolean (quadratic) function f by $f(\mathbf{x}) = \mathbf{x}(\mathbf{A} + \mathbf{A}^{\mathbf{t}})\mathbf{x}^{\mathbf{t}}$. Show[30] that the linearity of f is $n - \text{rank}(\mathbf{A} + \mathbf{A}^t)$.

8.8 Bent Functions

The concept of the bent function keeps cropping up in our discussions. These functions were initially introduced by Rothaus in connection with certain combinatorial problems.[31] There are a number of properties of bent functions, any one of which might be taken as a definition. Since we have to start somewhere, we shall take the following as our definition:

Definition A Boolean function f of n variables is called *bent* if

$$\hat{F}(\omega) = \pm 2^{\frac{n}{2}} \ \forall \omega \in (GF(2))^n,$$

i.e.

$$\sum_{x \in (GF(2))^n} (-1)^{f(x) \oplus \omega . x} = \pm 2^{\frac{n}{2}} \ \forall \omega \in (GF(2))^n.$$

We have the following theorem, the proof of which is left as an exercise to the diligent reader:

Theorem *Let f be a function of n variables. The following are equivalent:*

1. *f is bent;*
2. *$\hat{f} \cdot \hat{l}_\omega = \pm 2^{\frac{n}{2}} \ \forall \omega \in (GF(2))^n$;*
3. *$f(x) \oplus f(x \oplus a)$ is balanced for any nonzero $a \in (GF(2))^n$;*
4. *$f(x) \oplus l_\omega(x)$ assumes the value 1 $2^{n-1} \pm 2^{\frac{n}{2}-1}$ times for any $\omega \in (GF(2))^n$;*
5. *$\hat{r}_f(a) = 0 \ \forall a$ with $1 \le W_H(a) \le n$.*

[30] Nyberg, K.: *On the construction of highly nonlinear permutations*, Proc. Eurocrypt '92.

[31] The most complete analysis of bent functions still appears to be that of Dillon, which first appeared in the in-house journal of the National Security Agency in 1972, (Dillon, J.F.: *A Survey of Bent Functions*, NSA Techn. J., 1972) and which has been updated periodically since then.

We re-emphasize that bent functions only exist for an even number of variables and that, as the above theorem makes clear, bent functions cannot be balanced: for a balanced function f, we have $\hat{F}(0) = 0$, whereas for a bent function $\hat{F}(\omega) \neq 0$ for all ω.

Various methods of constructing bent functions are known, of which we shall only give the two most important ones. After that, we shall investigate how to use these to obtain Boolean functions with various desirable properties.[32]

For future use we define the matrix $[\hat{f}]$ of a function f of n variables as follows:

$$[\hat{f}]_{i,j} = \hat{f}(i \oplus j) \quad 0 \leq i, j \leq 2^n - 1.$$

It is evident that such a matrix will be symmetrical.

We shall shortly need the following observation, which we demonstrate by means of a trivial example.

Example Let $f(x_{n-1}, \ldots, x_0)$ be a Boolean function of n variables, and consider the function g of $n+1$ variables, defined by $g(x_n, x_{n-1}, \ldots, x_0) = x_n \oplus f(x_{n-1}, \ldots, x_0)$. Use the notations $0||x$ and $1||x$ for x with either a 0 or a 1 prepended. Then we see that

$$g(0||x) = f(x) \text{ i.e. } \hat{g}(0||x) = \hat{f}(x),$$
$$g(1||x) = 1 \oplus f(x) \text{ i.e. } \hat{g}(1||x) = -\hat{f}(x)$$

and hence that for the matrix $[\hat{g}]$ we get

$$[\hat{g}]_{0||i,0||j} = [\hat{f}]_{i,j},$$
$$[\hat{g}]_{0||i,1||j} = -[\hat{f}]_{i,j},$$
$$[\hat{g}]_{1||i,0||j} = -[\hat{f}]_{i,j},$$
$$[\hat{g}]_{1||i,1||j} = [\hat{f}]_{i,j},$$

so that, in fact, $[\hat{g}]$ is the Kronecker product of the matrix

$$\begin{pmatrix} 1 & -1 \\ -1 & 1 \end{pmatrix},$$

which is the matrix $[\hat{i}]$ of the function $i(x_n) = x_n$, and the matrix $[\hat{f}]$:

$$[\hat{g}] = [\hat{i}] \otimes [\hat{f}].$$

[32] I should stop using this phrase: it makes me sound too much like an estate agent.

Much more generally (but it is rather messy to write out a decent proof, even though it is conceptually quite simple) we have the following fact:

Let $f(x_{n-1},\ldots,x_0)$ and $g(y_{m-1},\ldots,y_0)$ be two Boolean functions, and define $h(y_{m-1},\ldots,y_0,x_{n-1},\ldots,x_0) = g(y_{m-1},\ldots,y_0) \oplus f(x_{n-1},\ldots,x_0)$. Then $[\hat{h}] = [\hat{g}] \otimes [\hat{f}]$; in other words, the matrix representing the sum of two functions defined on disjoint sets of variables is the Kronecker product of the matrices representing these two functions.

Now suppose f is bent, then the (i,j)th entry of the product $[\hat{f}][\hat{f}]$ is

$$
\begin{aligned}
([\hat{f}][\hat{f}])_{i,j} &= \sum_k \hat{f}(i \oplus k)\hat{f}(k \oplus j) \\
&= \sum_k \hat{f}(k)\hat{f}(i \oplus j \oplus k) \\
&= \hat{r}_f(i \oplus j) \\
&= 2^n \delta_{ij},
\end{aligned}
$$

where δ_{ij} is the Kronecker delta, using the 5th of the characterizations of bent functions in the preceding theorem. Thus: if f is a bent function, then its matrix $\hat{f}$ is a Hadamard matrix, and conversely, as is easily seen. (So we could have listed this as number 6 in the theorem.)

8.8.1 The Simplest Bent function

The simplest bent function is one of two variables: $f(x_1, x_0) = x_1x_0$. This is represented by the sequence f = [0 0 0 1] or by $\hat{f}$ = [1 1 1 -1]. One can easily, almost in one's head, verify that f is indeed bent.

8.8.2 The "Dot-Product" Bent Functions

We can generalise the simple case as follows: The function

$$f(\mathbf{x}, \mathbf{y}) = \mathbf{x} \cdot \mathbf{y} = x_0y_0 + x_1y_1 + \cdots + x_{n-1}y_{n-1}$$

is bent on $(GF(2))^{2n}$.

Proof This follows from the fact that the sum of two bent functions, defined on disjoint sets of variables, is itself bent, which in turn follows from our observation that if $f(x_{n-1},\ldots,x_0)$ and $g(y_{m-1},\ldots,y_0)$ are bent, then the matrix $[\hat{h}]$ of the function $h = g \oplus f$ is the Kronecker product of $[\hat{g}]$ and $[\hat{f}]$. But the Kronecker product of two Hadamard matrices is itself a Hadamard matrix.

It can, in fact, be proved that the "dot-product" bent functions are the only quadratic bent functions. Rothaus proved that for *any* function $g(x)$ the function

$$f(\mathbf{x},\mathbf{y}) = \mathbf{x} \cdot \mathbf{y} \oplus g(\mathbf{x})$$

with $\mathbf{x},\mathbf{y} \in \{0,1\}^n$, is a bent function on $\{0,1\}^{2n}$. The proof is again left to the diligent reader, if she is still with us. It follows immediately that there are bent functions f of $2n$ variables of every degree $2 \leq \text{degree}(f) \leq n$.

8.8.3 The Maiorama Construction

The family of bent functions discovered by Maiorama[33] forms a natural extension of the above family. He proved:

Theorem

$$f(x,y) = \pi(x) \cdot y \oplus g(y) \quad \text{is a bent function on } (GF(2))^{2n},$$

where $g(x)$ is an arbitrary function, and where π is an arbitrary permutation of the set $(GF(2))^n$.

The proof is similar to (but even more messy than) the proof of the dot-product case. This construction is also known as the Maiorama–McFarland construction.[34]

Maiorama also proved the following result, relevant to this construction:

Lemma *The function*

$$\pi : (GF(2))^n \longrightarrow (GF(2))^n : x \longmapsto (P_{n-1}(x), \ldots, P_0(x))$$

is a permutation of $(GF(2))^n$ if and only if every nonzero linear combination of the P_i is a balanced function.

8.8.4 Other Constructions

Not only because of their importance in cryptology, but also *inter alia* because of their interesting combinatorial properties, the existence of bent functions has been widely studied, and various other methods of constructing them have been found. Some of these are closely

[33] Maiorama, J.A.: *A class of bent functions*, R41 Technical Paper, 1970.

[34] McFarland, R.L.: *A family of difference sets in non-cyclic groups*, J. Combinatorial Th., Series A **15**, 1973.

related to the study of finite fields, and in particular make extensive use of the trace function. We shall describe one of these constructions.

Let n be an even integer, and i an integer satisfying $1 \leq i \leq \frac{n}{2}$, and such that $\frac{n}{\gcd(n,i)}$ is even. Then for some values of $a \in GF(2^n)$, the function

$$f : GF(2^n) \longrightarrow GF(2) : \ x \mapsto \mathrm{Tr}(ax^s)$$

where $s = 2^i + 1$, is bent. Here the identification of the input vector $(x_{n-1}, \ldots, x_1, x_0)$ with the field element $x_{n-1}\alpha^{n-1} + \cdots + x_1\alpha + x_0$ is taken as implied.

Example Taking the simplest possible example, let $n = 4$, which forces $i = 1$, so that $s = 3$. We shall work with the field $GF(2^4)$ as defined in Sect. 5.6. Considering the functions $f(x) = \mathrm{Tr}(ax^3)$, we find that these are bent for all $a \in GF(2^4)\backslash\{0, 1, \alpha^3, \alpha^3 + \alpha, \alpha^3 + \alpha^2, \alpha^3 + \alpha^2 + \alpha + 1\}$. Taking $a = \alpha$, for example, we obtain the function

$$0\,0\,0\,1\,1\,0\,0\,0\,0\,1\,1\,1\,0\,0\,0\,1.$$

The algebraic normal form of this function is $f(x_4, x_3, x_2, x_1, x_0) = x_0x_1 + x_2 + x_0x_2 + x_1x_2 + x_0x_3 + x_1x_3 + x_2x_3$. We leave it to the reader to verify that this is not a Maiorama-type function.

The functions of this example are known as *monomial* bent functions, for a reason which becomes clear when told that $\mathrm{Tr}(\alpha \cdot x^{2^k+1} + x^{3 \cdot 2^{k-1}-1})$ (where $n = 2k$ and $\alpha + \alpha^{2^k} = 1$) is a binomial bent function. There are several other constructions. All binomial bent functions of 20 or fewer variables are known.[35]

Exercise With n even, and i relatively prime to n, it has been shown that there exist $a \in GF(2^n)$ such that f defined on the set $GF(2^n)$ by

$$f(x) = \mathrm{Tr}(a \cdot x^{2^{2i}-2^i+1})$$

is bent. Use this fact to find another bent function of four variables. (If you are enthusiastic enough, you are, of course, welcome to work with 6 or 8 variables.)

8.8.5 Extensions of Bent Functions

One of the most useful properties of bent functions is their propagation behaviour: a single bit flip in the input is guaranteed to cause flips of half the output bits. However, the unpleasant facts remain that they are only defined for even n (= number of input variables) and that they are unbalanced: one of the most important properties of a stream cipher generator is surely that in encryption flipping a plaintext bit must be just as likely as leaving it unchanged. We

[35] Kocak, O., Kurt, O., Öztop, N. and Saygi, Z.: *Notes on bent functions in polynomial form*, Int. J. Info. Sec. Sc. **1** (2012), 43–48.

shall in this subsection discuss some ways in which bent functions may be used as building blocks for other Boolean functions.

The first construction uses two bent functions f_1 and f_2, both of $2k$ variables, to construct, by means of concatenation, a function g of $2k+1$ variables:

$$g(x_{2k+1}, x_{2k}, \dots, x_0) = (1 \oplus x_{2k+1}) f_1(x_{2k}, \dots, x_0) \\ \oplus x_{2k+1} f_2(x_{2k}, \dots, x_0).$$

It is not hard to see that g has nonlinearity $\mathrm{NL}(g) \geq 2^{2k} - 2^k$: Think of the last step in the computation of $\hat{G}$. In the previous step every entry was $\pm 2^k$, so that in the last step every entry will be either the sum or the difference of two of these. Hence $|\hat{G}(\omega)| \leq 2^{k+1}$, and $\mathrm{NL}(g) \geq \frac{1}{2}(2^{2k+1} - 2^{k+1}) = 2^{2k} - 2^k$.

By choosing f_1 and f_2 suitably, one can even ensure that g is balanced. Take, for example $f_1(x_1, x_0) = x_0 x_1$ and $f_2(x_1, x_0) = 1 + x_0 + x_0 x_1$. Then g becomes

$$g(x_2, x_1, x_0) = (1 + x_2)(x_0 x_1) + x_2(1 + x_0 + x_0 x_2) \\ = x_2 + x_0 x_1 + x_0 x_2,$$

which is balanced.

An even simpler way of obtaining a nice balanced function with "reasonable" propagation characteristics is the following: Let f be a bent function of $2k$ variables, and define g by

$$g(x_{2k}, x_{2k-1}, \dots, x_0) = x_{2k} \oplus f(x_{2k-1}, \dots, x_0).$$

Then g is balanced, as can easily be seen, and, moreover, g is pretty close to being $PC(2k+1)$: in fact $\hat{r}(\gamma) = 0$ for all nonzero γ except $\gamma = \alpha = (1, 0, 0, \dots, 0)$. Thus g has linearity 1, the vector $\alpha = (1, 0, 0, \dots, 0)$ being a linear structure. The proof of this statement is simple: just verify that as long as $\gamma \neq \alpha$, $g(x) \oplus g(x \oplus \gamma)$ is balanced, which follows from the fact that f is bent. It also follows, as before, that $\mathrm{NL}(g) \geq 2^{2k} - 2^k$.

In exactly the same way one can use a bent function of $2k$ variables to obtain a balanced function of $2k+2$ variables: Let f be as before and define g by

$$g(x_{2k+1}, x_{2k}, x_{2k-1}, \dots, x_0) \\ = x_{2k+1} \oplus x_{2k} \oplus f(x_{2k-1}, \dots, x_0).$$

Again, g is balanced, but its linearity is now 2, with both $\alpha = (1, 0, 0, \dots, 0)$ and $\beta = (0, 1, 0, \dots, 0)$ as well as their sum $(1, 1, 0, \dots, 0)$ being linear structures. But g satisfies the propagation criterion for all other nonzero vectors, i.e. $\hat{r}(\gamma) = 0$ for all $\gamma \notin \{0, \alpha, \beta, \alpha \oplus \beta\}$. The nonlinearity of g is at least $2^{2k-1} - 2^k$.

The above constructions are well known, and rather obvious, really, but first seem to have been considered by Seberry et al.[36] They also pointed out that by applying an affine

[36] Seberry, J., Zhang, X.-M. and Zheng, Y.: *Nonlinearly balanced Boolean functions and their propagation characteristics*, Proc. Crypto '93, Springer-Verlag.

transformation to the input vector x, i.e. by considering $g(x\mathbf{A} \oplus b)$ instead of $g(x)$, where $\mathbf{A}$ is a non-singular (binary) matrix and b is any vector of appropriate length, one can move the linear structures to wherever one wants, without destroying the propagation and nonlinearity properties.

A more complicated construction appears in a paper by Kurosawa and Satoh.[37] There has, in fact, been lots of research in the area of finding Boolean functions which in some way or other find optimal trade-offs between the requirements that cryptography imposes on such functions. This does not seem to be the right place to pursue this matter further, except for some more or less general comments in the penultimate section of this chapter.

8.9 Algebraic Immunity

So far we have concentrated on using Boolean functions which do not correlate well with linear functions in the (perhaps unstated) belief that while systems of linear equations are easy to solve, life becomes much more difficult once one is faced with having to solve systems of nonlinear Boolean equations. This is, to some—limited—extent, a vain hope, since there are techniques available for solving such systems, either through the use of so-called Gröbner basis techniques or, more easily understandably, through a process called *linearisation*. If the system contains terms (monomials) of degree higher than 1, linearisation involves replacing each of these monomials by a new unknown, and then solving, or attempting to solve, the resulting system of linear equations. Of course, this will imply that the system of equations needs to be *overdetermined* in the sense that we will need more equations than there are unknowns, because of all the unknowns that are introduced to cope with the nonlinear terms. If, for example, the system contains n unknowns, but also terms of degree up to the dth, and there might be as many as $\binom{n}{a}$ for each a up to and including d, we might need $\sum_{a=1}^{d} \binom{n}{a}$ equations to find the solution (essentially because we ignore such facts as that if $x_0 = 1$ and $x_1 = 1$ then x_0x_1 also has to equal 1, or that $x_0x_1 = 0$ implies that at least one of x_0, x_1 must equal 0).

For example, the system of equations

$$\begin{aligned} x_0 + x_1 + x_0x_1 + x_0x_2 + x_1x_2 &= 0, \\ x_1 + x_0x_1 &= 1, \\ x_1 \qquad + x_0x_2 &= 1 \end{aligned}$$

has a unique solution, which we can find by doing an exhaustive search (or something pretty much like one) on x_0, x_1 and x_2, but it is much easier to find that solution if we have two more

[37] Kurosawa, K. and Satoh, T.: *Design of SAC/PC(l) of order k Boolean functions and three other cryptographic criteria*, Proc. Eurocrypt '97, Springer-Verlag.

equations

$$
\begin{aligned}
x_1 \qquad\qquad + x_1x_2 &= 0, \\
x_0x_1 + x_0x_2 + x_1x_2 &= 1
\end{aligned}
$$

and solving the system as if they were linear equations, obtaining $x_0 = 0, x_1 = 1, x_0x_1 = 0, x_0x_2 = 0, x_1x_2 = 1$ (from which, to state the obvious, we get $x_2 = 1$).

Now consider an LFSR with connection polynomial (state update function) L and let f be a filter function on it. Then, in its essential form, the problem faced by the cryptanalyst is the following: Let $(k_0, k_1, \ldots, k_{n-1})$ be the initial state. Then the output bits are given by

$$
\begin{aligned}
b_0 &= f(k_0, k_1, \ldots, k_{n-1}), \\
b_1 &= f(L(k_0, k_1, \ldots, k_{n-1})), \\
b_2 &= f(L^2(k_0, k_1, \ldots, k_{n-1})), \\
&\ldots
\end{aligned}
$$

If, in the process of cryptanalyzing a stream cipher, the cryptanalyst can obtain enough equations, he or she can in principle (but only in principle, because of the difficulty in general of solving systems of nonlinear equations) break the system. Now efficient techniques for solving such large systems of equations (some or all of which may only hold with a probability close to 1) exist, the currently most promising one being known as XL,[38] with some variations on its theme also being available. These techniques employ linearisation, and therefore run the risk of explosive growth in the number of unknowns. It will be clear that, in order to keep the number of unknowns in the linearisation within reasonable bounds, it would be useful (for the cryptanalyst, anyway) if the degrees of the monomials in the equations could be kept low.

In their paper *Algebraic attacks on stream ciphers with linear feedback*[39] Courtois and Meier discuss a method of achieving this. We quote from their paper:

> At the time t, the current keystream bit gives an equation $f(s) = b_t$ with s being the current state. The main new idea consists of multiplying $f(s)$, that is usually of high degree, by a well chosen polynomial $g(s)$, such that fg is of substantially lower degree[...] Then, for example, if $b_t = 0$, we get an equation of low degree $f(s)g(s) = 0$. This, in turn, gives a multivariate equation of low degree on the initial state bits k_i. If we get one such equation for each of sufficiently many keystream bits, we obtain a very overdefined system of multivariate equations that can be solved efficiently.

Courtois and Meier applied these ideas successfully to the stream cipher LILI-128, which had been submitted to the Japanese government's Cryptrec initiative. They show, moreover, "that all very traditional stream ciphers, with linear feedback, and a highly nonlinear (stateless) filtering function are insecure, for example when they use only a small subset of state bits".

---

[38]Shamir, A., Patarin, J., Courtois, N. and Klimov, A.: *Efficient algorithms for solving overdefined systems of multivariate polynomial equations*, Proc. Eurocrypt 2000, LNCS 1807, Springer-Verlag.

[39]Courtois, N. and Meier, W.: *Algebraic attacks on stream ciphers with linear feedback*, Proc. Eurocrypt 2003, LNCS 2656, Springer-Verlag.

The following fundamental theorem appears in their paper:

Theorem *Let $f : (GF(2))^k \longrightarrow GF(2)$ be a (nonzero) Boolean function. Then there exists a Boolean function $g \neq 0$ of degree at most $\lceil \frac{k}{2} \rceil$ such that $f(x)g(x)$ is of degree at most $\lceil \frac{k+1}{2} \rceil$.*

Proof Let A be the set of all monomials of degree $\leq \lceil \frac{k}{2} \rceil$. Then

$$\#(A) = \sum_{i=0}^{\lceil k/2 \rceil} \binom{k}{i} \geq \frac{1}{2} 2^k.$$

Let B be the set of all polynomials formed by multiplying f by all the monomials of degree $\leq \lceil \frac{k+1}{2} \rceil$. Then

$$\#(B) = \sum_{i=0}^{\lceil (k+1)/2 \rceil} \binom{k}{i} > \frac{1}{2} 2^k.$$

If $C = A \cup B$, then $\#(C) > 2^k$, so that the elements of C (considered as binary strings of length 2^k, if you like) must be linearly dependent. But the elements of A are linearly independent, so there exists a g such that $f(x)g(x)$ is nonzero and has degree $\leq \lceil \frac{k+1}{2} \rceil$.

The consequence of this theorem is that for any stream cipher with linear feedback for which the filter function uses k variables, it is possible to generate at least one equation of degree $\approx \frac{k}{2}$ in the keystream bits. The resulting equations can then be solved by linearisation, more easily than the original equations.

We also note another observation: Suppose we have found a function g of degree d such that the product $h = gf$ is also of degree d or less. Then we have, because for a Boolean function $f^2(x) = f(x) \cdot f(x) = f(x)$, that $hf = gf^2 = gf = h$, so $(g + h)f = 0$, so that $g' = g + h$ is an annihilator of f again of degree at most d. Our search for low degree multiples is therefore also one for low degree annihilators.

One way of finding such low degree annihilators is to look for common factors in the terms of high degree. We demonstrate this by means of a toy example: Let

$$f(x) = x_0 + x_4 + x_0x_1x_2 + x_0x_1x_3.$$

Then using the common factor x_0 of the two monomials of degree 3:

$$(1 + x_0)f(x) = x_4 + x_0x_4.$$

Repeating this trick with the common factor x_4, we find

$$(1 + x_4)(1 + x_0)f(x) = 0,$$

so that we have a polynomial of degree 2 which annihilates the function f.

A more systematic way of finding annihilators is by considering the truth tables of the functions: if for any particular value x of the input into function f one has that $f(x) = 1$, then any annihilator g of f must satisfy $g(x) = 0$. But if $f(x) = 0$, $g(x)$ can be either 0 or 1. So we can run through all possible functions g which satisfy the first condition and choose the one of lowest degree. If f is a function of n variables and of weight k (i.e. $f(x) = 1$ for k values of x), then this process would involve finding the Algebraic Normal Form of each of 2^{n-k} possible choices for g. This method is not efficient, but it is the only straightforward technique currently available for finding an optimal annihilator.

The theorem we have proved can be sharpened slightly: For any Boolean function of n variables, there exists a function g of degree at most $\lceil \frac{n}{2} \rceil$ which annihilates it.

From the point of view of cryptographic strength, it is obviously desirable to require a high degree for any annihilator. For the purpose of making all this precise, we make the following definitions[40]:

Definitions For a Boolean function of n variables, a function g of the same variables is called an *annihilator* of f if $f(x)g(x) \equiv 0$. The set of annihilators of f will be denoted by AN(f). The *algebraic immunity*[41] of f is

$$\min\{\deg(g)|g \in \mathrm{AN}(f)\}.$$

Thus the algebraic immunity of a Boolean function of n variables is at most $\lceil \frac{n}{2} \rceil$. Dalai et al. give a construction of a function achieving this upper bound (but are at some pains to point out that this does not imply that their function is suitable for cryptographic purposes, at least not without some modification such as composition with other functions with other cryptographic properties). Their construction is very simple:

If n is odd, define f by

$$f(x) = 0 \text{ if } W_H(x) \leq \left\lfloor \frac{n}{2} \right\rfloor,$$

$$f(x) = 1 \text{ if } W_H(x) \geq \left\lceil \frac{n}{2} \right\rceil.$$

If n is even, define f by

$$f(x) = 0 \text{ if } W_H(x) < \frac{n}{2},$$

$$f(x) = 1 \text{ if } W_H(x) > \frac{n}{2},$$

$$f(x) = b \in \{0, 1\} \text{ if } W_H(x) = \frac{n}{2}.$$

[40]These definitions are taken from the paper by Dalai, D.K., Maitra, S. and Sarkar, S.: *Basic theory in construction of Boolean functions with maximum annihilator immunity*, Designs, Codes and Cryptogrpahy **40** (2006), pp. 41–58.

[41]Dalai et al. use the term *annihilator immunity* for reasons explained in their paper.

Note that (at least for odd n) this function is *symmetrical* in the sense that any permutation of the input bits will leave the value of the function unchanged. A detailed study of the algebraic immunity properties of symmetric functions has been carried out by Braeken and Preneel.[42]

To counter algebraic attacks Courtois and Meier[43] make the following recommendations in regard to filtering and combining functions which are to be used in LFSR-based stream ciphers. These recommendations must, of course, be seen as additional to the other requirements that we have discussed.

- The filtering function should use many state bits.
- It should not be sparse; in particular it should have many terms of very high degree.
- The part of high degree should not have a low degree factor, and should itself use many state bits.
- "Ideally, to resist algebraic attacks, given the function f, for every function g of 'reasonable' size, the degree of fg should always be equal to $\deg(f) + \deg(g)$."

Exercise Find an annihilator of lowest degree of the function $1 + x_2 + x_3 + x_1x_3 + x_0x_2x_3$.

8.10 Completeness

In this chapter we have so far only discussed functions with $GF(2)$ or $\mathbb{Z}_2$ as their codomain. In the next we shall deal with block ciphers, and in those confusion, and to a more limited extent diffusion, are obtained through the use of substitution boxes (S-boxes), which are essentially functions which map bitstrings of length n onto bitstrings of length m, or, in other words, functions from $GF(2)^n$ to $GF(2)^m$. Frequently $m = n$ and in such a case it may be more convenient to describe the S-box as an operation on the set $GF(2^n)$. This is, for example, the case with the encryption function used in the Advanced Encryption Standard, where it is a permutation in fact.

We have listed a number of properties that we expect from a Boolean function which is to have a cryptological application. As a function from $\mathbb{Z}^n$ to $\mathbb{Z}^m$ can obviously be described in terms of its component functions, it seems obvious that these component functions should have, as far as possible, the properties that we have considered. One more factor should, however, be considered in this more general case.

This is the property of *completeness*: Every output bit should be dependent on every input bit, i.e. every component function should be a function of *all* the input variables. This can be phrased as follows: For every i $(0 \leq i \leq n-1)$ and every j $(0 \leq j \leq m-1)$ there should exist $x, x' \in \{0,1\}^n$ which differ only in their ith component such that $f(x)$ and $f(x')$ differ in (at least) their jth component.

---

[42] Braeken, A. and Preneel, B.: *On the algebraic immunity of symmetric Boolean functions*, Proc. Indocrypt 2005, LNCS 3797, Springer-Verlag, pp. 35–48.

[43] *Op. cit.*

This property, first enunciated by Kam and Davida,[44] has received rather less attention than its importance would seem to warrant, perhaps because it seems so obvious, and, in addition, is quite easily satisfied.

8.11 The Discrete Fourier Transform

Having started this chapter with the Walsh–Hadamard transform, we feel that we cannot end it without mentioning another (and generally better known) transform, viz the Discrete Fourier Transform or DFT.

8.11.1 Introduction

Let k be a positive integer and let $\mathbb{F}$ be a field in which 1 has a kth root; let $\zeta \in \mathbb{F}$ be a primitive root of 1. Now define the matrix $\mathbf{Z}$ as

$$\mathbf{Z} = \begin{pmatrix} 1 & 1 & 1 & \dots & 1 \\ 1 & \zeta & \zeta^2 & \dots & \zeta^{k-1} \\ 1 & \zeta^2 & \zeta^4 & \dots & \zeta^{2(k-1)} \\ \dots & \dots & \dots & \dots & \dots \\ 1 & \zeta^{k-1} & \zeta^{2(k-1)} & \dots & \zeta^{(k-1)^2} \end{pmatrix}.$$

We consider the ring of polynomials over $\mathbb{F}$, with multiplication and addition modulo the ideal generated by $x^k - 1$, i.e. consider the ring

$$R = \mathbb{F}[x]/ < x^k - 1 > .$$

For any element $a_{k-1}x^{k-1} + \dots + a_1 x + a_0$ in R we now define the Fourier vector as

$$\mathbf{A} = (a_0\ a_1 \dots a_{k-1})\mathbf{Z}$$

or, in other words, the vector $(A_0, A_1, \dots, A_{k-1})$ with

$$A_j = \sum_{i=0}^{k-1} a_i \zeta^{ij} \text{ for } j = 0, \dots, k-1.$$

This transformation from k-dimensional vectors to k-dimensional vectors is called the discrete Fourier transform (DFT), and the coefficients A_j are called the *Fourier coefficients.*

---

[44]Kam, J.B. and Davida, G.I: Structured design of substitution-permutation encryption networks, in IEEE Trans. Computers **28** (1979).

If the vectors are considered as the coefficients of polynomials (as we did when we started this discussion) the DFT of a polynomial (modulo $x^k - 1$, to be precise) is referred to as, somewhat confusingly, a Mattson–Solomon polynomial in the Coding Theory literature.

The DFT is then a transformation from the ring $\mathbb{F}[x]/ < x^k - 1 >$ to the ring $\mathbb{F}^k$, where the operations in the latter are to be taken componentwise, obviously. This transformation is in fact an isomorphism: that it maps sums onto sums is obvious, so we only need to check that it maps products onto products and that it is one-to-one and onto.

Regarding the first of these requirements, we note that A_j is really just the evaluation of the polynomial $a(x)$ at the value ζ^j: $A_j = a(\zeta^j)$. Therefore, if $c(x) = a(x)b(x)$ (modulo $x^k - 1$), then $C_j = c(\zeta^j) = a(\zeta^j)b(\zeta^j) = A_jB_j$. As far as the second requirement is concerned, it is sufficient to prove that the DFT is invertible. A slightly laborious, but easy, computation shows, using the fact that $\zeta^k = 1$, that the inverse of the DFT is given by the equation

$$a_j = \frac{1}{k}\sum_{i=0}^{k-1} A_i\zeta^{-ij} \text{ for } j = 0, \ldots, k-1.$$

In line with our usual (cryptological) preferences, we are mainly interested in the case of finite fields of characteristic 2, although many, if not most, applications of the DFT elsewhere are concerned with primitive kth roots of 1 in the field $\mathbb{C}$ of complex numbers.

8.11.2 Linear Complexity Revisited

We shall restrict ourselves for the time being to just one application of the DFT:

Theorem *Let* $\mathbf{s} = s_0s_1s_2\ldots$ *be a binary sequence of period* ν, *where* ν *is odd, and let* ζ *be a primitive* νth *root of 1 in some field of characteristic 2. Then the linear complexity of* $\mathbf{s}$ *equals the number of nonzero coefficients* S_j *of the DFT of the polynomial* $s(x) = s_0 + s_1x + \cdots + s_{\nu-1}x^{\nu-1}$.

Proof We know from Sect. 7.5 that the linear complexity of **s** is the rank of the matrix

$$\begin{pmatrix} s_0 & s_1 & \ldots & s_{\nu-1} \\ s_1 & s_2 & \ldots & s_\nu \\ \ldots & \ldots & \ldots & \ldots \\ s_{\nu-1} & s_\nu & \ldots & s_{2\nu-2} \end{pmatrix} = \begin{pmatrix} s_0 & s_1 & \ldots & s_{\nu-1} \\ s_1 & s_2 & \ldots & s_0 \\ \ldots & \ldots & \ldots & \ldots \\ s_{\nu-1} & s_0 & \ldots & s_{\nu-2} \end{pmatrix},$$

where we have used the fact that the period of the sequence is ν, so that $s_{i+\nu} = s_i$. This is the same as the rank of the matrix

$$M = \begin{pmatrix} s_0 & s_1 & \ldots & s_{\nu-1} \\ s_{\nu-1} & s_0 & \ldots & s_0 \\ \ldots & \ldots & \ldots & \ldots \\ s_1 & s_2 & \ldots & s_0 \end{pmatrix},$$

which differs only in the order of the rows.

Considering the entries m_{ij} of this matrix, we have the further property that

$$m_{i,j} = m_{i+k,j-k} \ \forall k = 0, \ldots, \nu - 1,$$

where the indices are taken modulo ν.

Now consider the column vectors

$$\mathbf{e}_j = \begin{pmatrix} 1 \\ \zeta^j \\ \zeta^{2j} \\ \ldots \\ \zeta^{(\nu-1)j} \end{pmatrix}$$

of the DFT matrix $\mathbf{Z}$.

Computing the ith entry of $M\mathbf{e}_j$ we get

$$\begin{aligned} \sum_{h=0}^{\nu-1} m_{ih}\zeta^{hj} &= \textstyle\sum_{h=0}^{\nu-1} m_{0,h-i}\zeta^{hj} \\ &= \sum_{h=0}^{\nu-1} m_{0,h-i}\zeta^{(h-i)j}\zeta^{ij} \\ &= \left(\sum_{h=0}^{\nu-1} s_h(\zeta^{hj})\right)\zeta^{ij} = s(\zeta^j)\zeta^{ij}. \end{aligned}$$

Thus the vector $\mathbf{e}_j$ is an eigenvector of the matrix M, with eigenvalue $s(\zeta^j)$ = jth Fourier coefficient of $s(x)$. It is clear from their definition that these eigenvectors $\mathbf{e}_j$ are linearly independent, and the matrix M is therefore similar to the diagonal matrix with the Fourier coefficients along the main diagonal. Hence the rank of M is equal to the number of nonzero coefficients.

Example Consider the sequence 1 0 0 1 0 1 0 0 1 0 ... of period 5. We seek a primitive 5th root of 1 in some binary field: the element $\beta = \alpha^3$ in the field $GF(2^4)$ will do (where α is, as in our discussion of finite fields, a long time ago,[45] a root of the irreducible polynomial $x^4 + x + 1$).

Then we get the Fourier coefficients as

$$\begin{aligned} &(1\,0\,0\,1\,0)\begin{pmatrix} 1 & 1 & 1 & 1 & 1 \\ 1 & \beta & \beta^2 & \beta^4 & \beta^3 \\ 1 & \beta^2 & \beta^4 & \beta^3 & \beta \\ 1 & \beta^4 & \beta^3 & \beta & \beta^2 \\ 1 & \beta^3 & \beta & \beta^2 & \beta^4 \end{pmatrix} \\ &= (0,\ 1+\beta^4,\ 1+\beta^3,\ 1+\beta,\ 1+\beta^2) \end{aligned}$$

[45] Section 5.6, to be precise.

from which we conclude that the linear complexity of the sequence is 4. (In fact, the connection polynomial of the shortest LFSR which can generate this sequence is $x^4 + x^3 + x^2 + x + 1$, of degree 4, as expected.)

8.11.3 DFT over a Finite Field

The following exposition is due to my friend and colleague Prof. G.J. Kühn and his contribution is gratefully acknowledged.

If we consider the domain of the function to be a finite field $GF(q)$ where $q = p^n$ for some prime p and some integer $n \geq 1$, we can define a similar, but not identical, version of a Fourier transform. Let $f : GF(q) \longrightarrow GF(q)$ be a function. We now define the Fourier transform as the transform that maps f onto the coefficients of the power series expansion:

$$f(x) = \sum_{i=0}^{q-1} F_i x^i$$

where the coefficients $F_0, F_1, \ldots, F_{q-1}$ lie in $GF(q)$. We shall concern ourselves only with the cases where $p = 2$. Then the difference between this Fourier transform and the one previously discussed amounts to the difference between having the domain of the function as the field $GF(2^n)$ or the vector space $(GF(2))^n$.

We show how this Fourier transform can be computed, or, in other words, how the coefficients F_i are determined.

We need in our derivation the fact that if $n = 2^k - 1$ for some k, then all binomial coefficients $\binom{n}{l}$, $(0 \leq l \leq n)$, are odd. We prove this in the appendix to this section, but the reader is welcome to accept this on faith or after computing the coefficients $\binom{7}{l}$ and $\binom{15}{l}$ for some values of l.

Thus, in a field $GF(q)$ of characteristic 2, one has that

$$(x - a)^{q-1} = \sum_{l=0}^{q-1} x^l a^{q-1-l}.$$

Now we start with the observation that

$$f(x) = \sum_{x_j \in GF(q)} \left(1 - (x - x_j)^{q-1}\right) f(x_j).$$

Using the above binomial expansion for $(x - x_j)^{q-1}$ and collecting terms of the same degree, we get

$$F_0 = f(0),$$

$$F_i = \sum_{x_j \in GF(q)\setminus\{0\}} x_j^{-i} f(x_j) \text{ for } 1 \leq i \leq q-2,$$

$$F_{q-1} = \sum_{x_j \in GF(q)\setminus\{0\}} f(x_j).$$

If the function f is balanced, the last equation shows that F_{q-1} will then be 0.

Finally, we note that if $f(0) = 0$, the Fourier transform discussed here reduces to the form discussed earlier. For in this case we can choose a primitive element α (which has order $q-1$, of course) and in the summations run through the elements of $GF(q)$ in the order in which they are generated if we use α as generator; in other words, we take $x_0 = 1, x_1 = \alpha, \ldots, x_{q-2} = \alpha^{q-2}$. Then the coefficient of x_i is given by

$$F_i = \sum_{j=0}^{q-2} \alpha^{-ij} f(\alpha^j) \text{ for } 0 \leq i \leq q-2,$$

as we had before. But keep your wits about you: the order of the elements considered here is the order of the powers of α and *not* their numerical (or lexicographical) order! For example, while the list of values starts with the value at 0 (which must be zero), the next entry is not the value at 1, but the value at α. The value at 1 comes at the very end of the list, namely as the value at α^{q-1}. In the tables you might want to use, such as those for the substitution boxes of block ciphers, the order will almost certainly be numerical.

Appendix

To prove our assertion that all the binomial coefficients $\binom{2^n-1}{k}$ are odd, we need the following

Lemma *Define $\psi(n)$ as the greatest integer k such that $2^k | n!$. Then*

1. *If n is odd, then $\psi(n) = \psi(n-1)$;*
2. *If $n = 2^{l_1} + 2^{l_2} + \ldots, 2^{l_r}$, where $l_1 > l_2 > \cdots > l_r > 0$, then $\psi(n) = n - \#\{l_1, l_2, \ldots, , l_r\}$. Thus $\psi(n) = n - w_H(n)$.*

Proof Part 1 of the conclusion is obvious. We only prove part 2.

Consider first of all the case where n is a power of 2, say $n = 2^k$. Of the set $\{1, 2, \ldots, 2^k\}$ exactly half, i.e. 2^{k-1}, are divisible by 2. Terms divisible by 4 each contribute further factors of 2, viz $\psi(2^{k-1})$ of them. We therefore have that

$$\psi(2^k) = 2^{k-1} + \psi(2^{k-1}).$$

Since obviously $\psi(2) = 1$, it is easily verified that $\psi(2^n) = 2^n - 1$ for every $n \geq 1$.

Now consider $2^n + a$, where $a < 2^n$. If we compare $(2^n + a)!$ and $(2^n)!$ we note that extra factors of 2 are introduced by some of the terms $2^n + a, 2^n + a - 1 \ldots, 2^n + 1$, but that there will be exactly $\psi(a)$ of them doing so. Hence

$$\begin{aligned}\text{For } a < 2^n, \quad \psi(2^n + a) &= \psi(2^n) + \psi(a) \\ &= 2^n - 1 + \psi(a).\end{aligned}$$

The result now follows: whenever a 1 appears in the binary representation of the number n we add the corresponding power of 2 to the answer and subtract a 1, leading to the required result. This completes the proof.

Our conclusion regarding the parity of $\binom{2^n-1}{k}$ now follows from the fact that $(2^n - 1) - k = (2^n - 1) \oplus k$. (The fact that this only works for $2^n - 1$ explains why the world is full of even binomial coefficients.) We therefore have that $n = w_H(2^n - 1) = w_H(2^n - 1 - k) + w_H(k)$. Now the highest power of 2 that divides into $\binom{2^n-1}{k} = \frac{(2^N-1)!}{(2^n-1-k)!k!}$ will be $n - (\psi(2^n - 1 - k) + \psi(k)) = n - (n - w_H(k) + w_H(k)) = 0$, so $\binom{2^n-1}{k}$ is odd.

9

Applications to Block Ciphers

Finite fields have for a long time been important in Applied Algebra, in particular in the theory of error correcting codes. In more recent times, they have assumed an equally important role in Cryptography, initially mainly in the generation of pseudorandom sequences and the design of stream ciphers, as we have seen. But more recently, especially since the selection of *Rijndael* in 2000/2001 as the *Advanced Encryption Standard* (AES), they have assumed a vitally important role in the design of block ciphers as well. In this chapter we discuss some aspects of these further applications.

9.1 Block Ciphers

Symmetric ciphers are the work horses of applied cryptography. We noted in Chap. 1 the seminal work of Claude Elwood Shannon published immediately after World War 2, in which the principles of *confusion* and *diffusion* were for the first time explicitly stated. The first of these is required to disguise the statistical properties of the plaintext, and to make the relationship between plaintext, key and ciphertext as complex as possible; the second serves to "diffuse" differences in the input plaintexts and the effect of the key bits as widely as possible.

Diffusion is usually achieved through the use of linear operations such as exclusive-or[1] ($\oplus$), bit shifts and bit rotations, and multiplication by constants or matrices. Confusion is established through the use of permutations or through substitution boxes ("S-boxes") which

[1]This is as good a place as any to remind the reader and to re-emphasise the rather obvious fact that the logical/Boolean exclusive-or operation is really nothing else than addition modulo 2. Addition modulo 2 is, in turn, nothing but addition in $GF(2)$ or in any field of characteristic 2. This explains why such fields are so popular in cryptology.

We shall continue to abbreviate 'exclusive-or' to 'XOR', and use the symbol '$\oplus$' to denote this operation.

A.R. Meijer, *Algebra for Cryptologists*, Springer Undergraduate Texts in Mathematics and Technology,
DOI 10.1007/978-3-319-30396-3_9

frequently are permutations, but need not be. For example, in the Data Encryption Standard (DES) the S-boxes were 6-bit input, 4-bit output functions.[2]

Block ciphers are normally *iterative*, in that a *round function*, normally key-dependent, is applied repeatedly. The input into the first round is the plaintext, its output forms the input into the second round, and so on, until the output of the last round constitutes the ciphertext. Naturally, since decryption is to be possible, the round function has to be a one-to-one function. The round function may, cryptographically speaking, be quite weak, in the sense that an attacker would probably have no difficulty at all in breaking it. However, the expectation is that by having enough rounds, and mixing in unknown key bits in each round, the cipher can be made strong enough. (There is, of course, the counter-argument that having a lot of rounds may make the cipher very slow in its execution. A compromise will therefore need to be found. In addition, it is not quite true that lots of rounds will make the cipher secure eventually; the so-called slide attacks[3] for example are independent of the number of rounds, but depend on all the rounds being the same.)

Two possible structures for the round function have taken pre-eminence: the Feistel structure (of which the Data Encryption Standard is the supreme example) and its generalisations, and the so-called *Substitution-Permutation Networks* (SPNs), as exemplified by the Advanced Encryption Standard (AES), also known by its original name of *Rijndael*. Following the AES "competition", an initiative of the U.S. National Institute of Standards and Technology (NIST) to find a replacement for the ageing and increasingly less secure DES, the current trend appears to be in favour of SPNs, which may, in part at any rate, be due to the fact that the diffusion in SPNs is more efficient than in Feistel-type ciphers.

A third structure, which may be called the Lai–Massey structure, is less common: it was used by the two designers in their famous *IDEA* cipher, which was at one time the default in Pretty Good Privacy (PGP), as well as in their SAFER and SAFER+. The family of ciphers FOX, later renamed IDEA NXT by Junod and Vaudenay, are based on the IDEA structure. An interesting feature of these ciphers was their use of incompatible (i.e. algebraically meaningless) operations on bitstrings. This property is completely absent in *Rijndael*, as we shall see.

In the next section we shall discuss the algebraic aspects of *Rijndael* in some detail, and see that the theory of finite fields plays a fundamental role in that design.

9.2 Finite Fields in *Rijndael*

We shall use the names "Rijndael" and "AES" interchangeably, since the difference between these is minimal, being restricted to the permissible block lengths and key lengths.

---

[2]At first sight, it seems odd that non-invertible functions can be used in an encryption function. After all, decryption must be possible, so the encryption function must be invertible. But that does not mean that every component needs to be! The Data Encryption Standard's *Feistel structure* allowed for any function whatsoever to be used in its round function, as the nonlinear function only operates on half the round input.

[3]Biryukov, A. and Wagner, D.: *Slide Attacks*, Proc. 6th International Workshop on Fast Software Encryption (FSE '99), LNCS 1636, Springer-Verlag, 245–259.

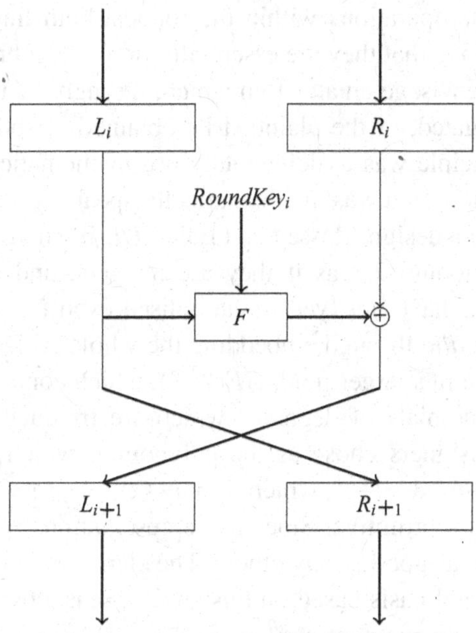

Structure of a round in a Feistel structure

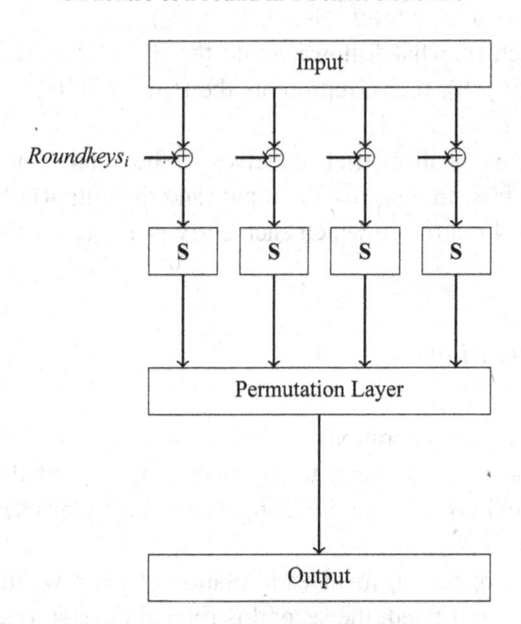

Structure of a round in a Substitution-Permutation Network

In *Rijndael* both the linear operations and the S-box (there is only one, which is used 16 times in parallel in each round of the cipher) are based on algebraic operations in the field $GF(2^8)$. In this field, as we have seen, elements can be represented by "octets" of 0s and 1s, in other words by *bytes*. Bytes are fundamental objects in any computer system, which accounts for the choice of $n = 8$ in $GF(2^n)$.

The fact that all the operations within the rounds, both linear and nonlinear, take place within the same field, i.e. that they are essentially proper algebraic (field) operations, may be a weakness: one can envisage an attack in which, through the use of some clever arithmetic, the key may be calculated, or the plaintext be obtained in some other way. (In many other ciphers, a design principle was to deliberately use mathematically incompatible operations, so that the effect of the round was, mathematically speaking, meaningless. We already noted as an example of such a design Massey and Lai's *IDEA*, where within a round, 16-bit strings are XORed, added modulo 2^{16} as if they are integers, and multiplied modulo $2^{16} + 1$.) Such algebraic attacks have not (yet?) materialised even though attempts have been made in this direction, *inter alia* through embedding the whole AES in a larger cipher in which all calculations take place in a larger field, $GF(2^{256})$, which contains $GF(2^8)$.

There are 30 polynomials of degree 8 which are irreducible over $GF(2)$, of which 16 are primitive. The designers chose as the polynomial which defines the field $GF(2^8)$ the polynomial $x^8 + x^4 + x^3 + x + 1$, which, it may be necessary to note, is irreducible (it jolly well has to be!) but not primitive. Since all representations of $GF(2^8)$ are isomorphic, this choice is, presumably, as good as any other.[4] The elements of $GF(2^8)$ will be represented by means of the polynomial basis based on this choice; thus a byte $(a_7a_6a_5a_4a_3a_2a_1a_0)$ (where $a_i \in \{0, 1\}$) represents the element $a_7\alpha^7 + a_6\alpha^6 + a_5\alpha^5 + a_4\alpha^4 + a_3\alpha^3 + a_2\alpha^2 + a_1\alpha + a_0$, with α being a root of $x^8 + x^4 + x^3 + x + 1$.

We shall, in much of what follows, write the bytes in hexadecimal notation. Thus, for example, the byte $5b_{16}$ = `0x5b` represents the byte 01011011 which in turn represents the element $\alpha^6 + \alpha^4 + \alpha^3 + \alpha + 1$.

In what follows, we shall restrict ourselves to the version of *AES/Rijndael* in which the block length is 128 bits. In this case the input (and the output) of each round may be written in the form of a 4×4 matrix, in which each entry is a byte (or an element of $GF(2^8)$).

9.2.1 Linear Operations

There are three linear operations in each round, viz `AddKey`, `ShiftRows` and `MixColumns`. There is not much to say about the first of these, which consists simply of adding (in the field $GF(2^8)$, i.e. XORing, $\oplus$ing, bit by bit) key bytes to the entries in the matrix.

The `ShiftRows` operation involves a rotation of the rows through differing numbers of bytes: the first row is left fixed, the second is rotated through one byte (to the left), the third

[4]The question needed to be asked why the designers chose this particular polynomial. The answer is reputedly that it was the first entry in a list of irreducible polynomials of degree 8 that the designers had available.

through two bytes, and the last through three bytes. Thus the matrix

$$\begin{pmatrix} a_{00} & a_{01} & a_{02} & a_{03} \\ a_{10} & a_{11} & a_{12} & a_{13} \\ a_{20} & a_{21} & a_{22} & a_{23} \\ a_{30} & a_{31} & a_{32} & a_{33} \end{pmatrix},$$

where the a_{ij} are bytes, is mapped by `ShiftRows` onto the matrix

$$\begin{pmatrix} a_{00} & a_{01} & a_{02} & a_{03} \\ a_{11} & a_{12} & a_{13} & a_{10} \\ a_{22} & a_{23} & a_{20} & a_{21} \\ a_{33} & a_{30} & a_{31} & a_{32} \end{pmatrix}.$$

More interesting than either of these is the `MixColumns` operation which takes each column, interprets it as a polynomial in $GF(2^8)[x]$ (*e.g.* the column $(a_{00}, a_{10}, a_{20}, a_{30})^T$ is interpreted as the polynomial $a_{00} + a_{10}x + a_{20}x^2 + a_{30}x^3$), multiplies it by the polynomial $c(x) = 03_{16}x^3 + 01_{16}x^2 + 01_{16}x + 02_{16}$, and reduces the product (which conceivably might have degree 6) modulo $x^4 + 1$. Since $x^4 + 1$ and $c(x)$ are relatively prime, this whole operation is invertible (which is, as we have already observed, essential if one wants to be able to decrypt).

More simply, the `MixColumns` operation can equivalently be described as multiplying the 4×4 input matrix **S** on the left by the matrix (in hexadecimal)

$$\mathbf{A} = \begin{pmatrix} 02 & 03 & 01 & 01 \\ 01 & 02 & 03 & 01 \\ 01 & 01 & 02 & 03 \\ 03 & 01 & 01 & 02 \end{pmatrix},$$

where the entries are interpreted as described earlier. Thus the result of the `MixColumn` operation applied to **S** is the matrix

$$\mathbf{S'} = \mathbf{AS}.$$

It will be noted that the entries in this matrix all have low Hamming weight: this will lead to efficient performance of the `MixColumns` operation. It is not quite possible for both this matrix and its inverse to have this desirable property, so that decryption in *Rijndael* is slightly slower than encryption, though that is mainly as a result of the key expansion routines[5] actually. The inverse of the `MixColumns` matrix, as required in decryption, is,

[5]The key expansion is the process of obtaining the round keys, each of 128 bits, from the cipher key itself.

in hexadecimal,

$$\begin{pmatrix} 0e & 0b & 0d & 09 \\ 09 & 0e & 0b & 0d \\ 0d & 09 & 0e & 0b \\ 0b & 0d & 09 & 0e \end{pmatrix}.$$

The question needs to be asked: how good is this function at achieving diffusion? The answer is: very good. A change in 1 input byte will always lead to changes in all 4 output bytes, 2 changes in the input bytes will lead to changes in at least 3 of the output bytes, and so on. We call the sum of these numbers of changes the *branch number* of the linear transformation. Thus the `MixColumns` matrix of *Rijndael* has branch number equal to 5; it is easy to see that this is in fact the maximum possible for a (4×4) matrix multiplication.

9.2.2 *The* Rijndael *Substitution Box*

The (nonlinear) confusion in *Rijndael* is in essence achieved through a remarkably simple operation: the S-box maps every nonzero element of $GF(2^8)$ onto its inverse, and 0 onto itself:

$$S(x) = \begin{cases} x^{-1} & \text{if } x \neq 0, \\ 0 & \text{if } x = 0. \end{cases}$$

This can also be read as $S(x) = x^{254}$ for all x, since $x^{255} = 1$ for all nonzero $x \in GF(2^8)$.

(We have actually simplified matters a little here: in Rijndael this inversion function is combined with a subsequent permutation performed on the bits, and a nonzero constant byte **a** is XORed with the result. These operations do not affect the properties with which we shall concern ourselves.)

9.2.3 *Properties of the* Rijndael *S-box*

9.2.3.1 Nonlinearity

Since linear equations are the easiest (by far!) type of equation to solve, it is important that every round function used in a block cipher contains some nonlinear function. In fact more than that: it should contain a function which cannot even be approximated, to any reasonable level of approximation, by a linear function; such an approximation would open the possibility of using statistical attacks. So how "nonlinear" is the *Rijndael* S-box?

We recall that the *distance* between two Boolean functions f and g, both on n variables, say, is defined to be the number of inputs x for which the outputs $f(x)$ and $g(x)$ differ, i.e.

$$d(f,g) = \#\{x \in (GF(2))^n | f(x) \neq g(x)\}.$$

The nonlinearity of a Boolean function f of n variables is defined to be

$$NL(f) = \min_A d(f,A),$$

where the minimum is taken over the set A of all affine[6] functions.

An S-box is, however, not a Boolean function: the output of the S-box in *Rijndael* is not a single bit, but a byte. In order to make an assessment of the nonlinearity of an S-box, we consider all affine functions of the *output bits*, and we define

$$NL(S) = \min_{\omega \neq 0} \min_A \{d(\mathrm{Tr}(\omega S(x)), A(x))\},$$

the inner minimum being taken over all nonzero affine functions, as before. We have used here the observation made in Sect. 6.3 that all linear functions on $GF(2^n)$ are of the form $\mathrm{Tr}(\omega x)$ for some ω.

In the case of the function $S(x) = x^{-1}\ (x \neq 0),\ \ S(0) = 0$ on $GF(2^n)$, Nyberg[7] states that

$$NL(S) \geq 2^{n-1} - 2^{\frac{n}{2}},$$

which is pretty good, since it can be proved that the nonlinearity of any Boolean function is at most $2^{n-1} - 2^{\frac{n}{2}-1}$.

The following non-proof of Nyberg's statement uses the Walsh–Hadamard transform which we dealt with in Chap. 7.

Theorem *Let* $S : GF(2^n) \longrightarrow GF(2^n)$ *be defined by* $S(x) = \begin{cases} x^{-1} & x \neq 0, \\ 0 & x = 0. \end{cases}$ *Then* S *has nonlinearity* $NL(S) \geq 2^{n-1} - 2^{\frac{n}{2}}$.

[6]At the risk of offending the reader, through implying that he or she does not remember anything from the chapter on Boolean functions: An affine Boolean function is a function of the form $\sum_{i=0}^{n-1} a_i x_i \oplus c$, where $a_i, c \in GF(2)$. Thus a linear function is an affine function for which the constant $c = 0$.

[7]Nyberg, K.: *Differentially uniform mappings for cryptography*, Proc. Eurocrypt '93, Lecture Notes in Computer Science 765, Springer-Verlag, Berlin, Heidelberg, New York, 1994. The selection of the *Rijndael* S-box was, I suspect, based on the results obtained by Nyberg and published in this paper. The current discussion is lifted from Nyberg's results. Nyberg does not provide a proof of the nonlinearity property, but refers to a paper by Carlitz and Uchiyama, which appeared in the Duke Mathematical Journal **24**, 1957.

Proof Let

$$f_t(x) = \begin{cases} \mathrm{Tr}(tx^{-1}) & \text{if } x \neq 0, \\ 0 & \text{if } x = 0, \end{cases}$$

and let $\widehat{F}_t(w) = \sum_x (-1)^{f_t(x)+w\cdot x}$ denote its Walsh–Hadamard transform. It is sufficient to prove that

$$|\hat{F}_t(w)| \leq 2^{\frac{n}{2}+1}$$

for all w.

Now

$$\begin{aligned}
|\hat{F}_t(w)|^2 &= \sum_x (-1)^{f_t(x)+w\cdot x} \sum_y (-1)^{f_t(y)+w\cdot y} \\
&= \sum_x \sum_y (-1)^{f_t(x)+f_t(y)+w\cdot(x+y)} \\
&= \sum_x \sum_s (-1)^{w\cdot s+f_t(x)+f_t(x+s)} \\
&= \sum_s (-1)^{w\cdot s} \sum_x (-1)^{f_t(x)+f_t(x+s)}.
\end{aligned}$$

We now attempt to find the inner sum (keeping s constant), i.e. we need to estimate the lack of balance in the function $\mathrm{Tr}(tx^{-1} + t(x+s)^{-1}) = \mathrm{Tr}\,(t(x^{-1} + (x+s)^{-1}))$, for the cases where $x \neq 0, s$. It is sufficient to consider only the case where $t = 1$.

We do this by considering the equation

$$x^{-1} + (x+s)^{-1} = \gamma$$

which, if $x \neq 0$ and $x \neq s$, is equivalent to the equation

$$\gamma x^2 + \gamma s x + s = 0$$

or

$$x^2 + sx + s\gamma^{-1} = 0$$

which, through the substitution $z = sx$ becomes

$$z^2 + z + (s\gamma)^{-1} = 0.$$

From our discussion on solving quadratic equations, we know that this equation will have two solutions if $\mathrm{Tr}((s\gamma)^{-1}) = 0$, and no solutions otherwise. Now if γ runs through all its possible nonzero values, then so do $s\gamma$ and $(s\gamma)^{-1}$, and since the trace function is balanced this will yield $2^{n-1} - 1$ values of γ for which a solution exists, or $2^n - 2$ distinct values of x. For $2^{n-1} - 1$ of these solutions γ will have trace 0 (or 1)[8] so the contribution to the inner sum has absolute value at most 2.

Added to this must be $2 \times \mathrm{Tr}(s^{-1})$, which represents the values in the two cases $x = 0$ and $x = s$. Hence

$$\left|\sum_x (-1)^{f_t(x)+f_t(x+s)}\right| \leq 4,$$

whence

$$\begin{aligned}|\hat{F}_t(w)|^2 &= \textstyle\sum_x (-1)^{f_t(x)+w\cdot x} \sum_y (-1)^{f_t(y)+w\cdot y}| \\ &\leq 4 \cdot 2^n = 2^{n+2},\end{aligned}$$

as required.

9.2.3.2 Differential 4-Uniformity

In differential cryptanalysis, the cryptanalyst exploits the fact that some differences in input may lead to some differences in output being more probable than others, and will choose input differences which with "high" probability will lead to some *known* output differences. If we consider a single round of a block cipher, the linear components cannot contribute to making such attacks difficult, since a difference δa in the input will always, with probability 1, lead to a difference $L(\delta a)$ in the output:

$$L(x + \delta a) = L(x) \oplus L(\delta a)$$

for all inputs x if L is a linear function. We must therefore look for the nonlinear component, *i.e.* the S-box or S-boxes, to make such attacks more difficult. Thus we must make, for any δa and δb, the probability that $S(x \oplus \delta a) = S(x) \oplus \delta b$ as small as possible. This amounts to making all the sets

$$\{x | S(x \oplus \delta a) \oplus S(x) = \delta b\}$$

as small as possible. Note that such a set (if it isn't empty) always contains at least two elements, for if x belongs to the set, then so does $x \oplus \delta a$.

[8] We are cheating here: we claim that the fact that we restrict ourselves to the subset $\mathrm{Tr}((s\gamma)^{-1}) = 0$ does not affect the distribution of $\mathrm{Tr}(\gamma)$. This seems to be true.

We call a function $\chi : GF(2^n) \longrightarrow GF(2^n)$ *differentially δ-uniform* if

$$\#\{x \in GF(2^n) | \chi(x \oplus \delta a) \oplus \chi(x) = \delta b\} \leq \delta$$

for all nonzero δa, and all δb.

Nyberg proved the following

Theorem *The function S of Rijndael is differentially 4-uniform.*

Proof We shall actually prove a little more than is stated.

Let $\delta a, \delta b \in GF(2^n)$, with $\delta a \neq 0$. Consider the equation

$$(x \oplus \delta a)^{-1} \oplus x^{-1} = \delta b.$$

If $x \neq \delta a$ and $x \neq 0$, then we can rewrite this equation as

$$\delta b \cdot x^2 \oplus \delta a \cdot \delta b x \oplus \delta a = 0.$$

This is a quadratic equation, which has at most two solutions.

However, if either $x = 0$ or $x = \delta a$ is a solution to the original equation $(x \oplus \delta a)^{-1} \oplus x^{-1} = \delta b$, then both of them are, so we get two further solutions. In this case we must have $\delta b = \delta a^{-1}$ and our original equation becomes

$$(x \oplus \delta a)^{-1} \oplus x^{-1} = \delta a^{-1}$$

or

$$x^2 \oplus \delta a \cdot x \oplus \delta a^2 = 0.$$

Putting $z = \delta a \cdot x$, we get the equation $z^2 + z + 1 = 0$, which, as we know, only has a solution if $\mathrm{Tr}(1) = 0$. This is the case if and only if n is even.

Hence we have:

- If n is odd, then the function $S(x) = x^{-1}$ for nonzero x, $S(0) = 0$ is differentially 2-uniform.
- If n is even, then this function is differentially 4-uniform.

By way of example, we consider the inverse function defined on the smaller field $GF(2^4)$, and we construct the table of solutions of $x^{-1} \oplus (x \oplus a)^{-1} = b$, for all possible pairs (a, b). We define the field by means of the primitive polynomial $x^4 + x + 1$. According to the theory, all the entries in this table should be even and at most equal to 4. We find (and the reader is welcome to verify this) the following table, in which the row represents a and the column b

(or the other way round—see below):

$$
\begin{array}{cccccccccccccccc}
16 & 0 & 0 & 0 & 0 & 0 & 0 & 0 & 0 & 0 & 0 & 0 & 0 & 0 & 0 & 0 \\
0 & 4 & 0 & 0 & 0 & 0 & 2 & 2 & 0 & 2 & 0 & 2 & 0 & 2 & 2 & 0 \\
0 & 0 & 0 & 2 & 0 & 0 & 0 & 2 & 0 & 4 & 2 & 0 & 2 & 2 & 0 & 2 \\
0 & 0 & 2 & 0 & 0 & 0 & 0 & 2 & 2 & 0 & 2 & 2 & 2 & 0 & 4 & 0 \\
0 & 0 & 0 & 0 & 0 & 2 & 2 & 0 & 2 & 0 & 2 & 0 & 0 & 4 & 2 & 2 \\
0 & 0 & 0 & 0 & 2 & 0 & 2 & 0 & 2 & 2 & 0 & 4 & 2 & 0 & 0 & 2 \\
0 & 2 & 0 & 0 & 2 & 2 & 2 & 4 & 0 & 0 & 2 & 0 & 2 & 0 & 0 & 0 \\
0 & 2 & 2 & 2 & 0 & 0 & 4 & 2 & 2 & 0 & 0 & 0 & 0 & 0 & 0 & 2 \\
0 & 0 & 0 & 2 & 2 & 2 & 0 & 2 & 0 & 0 & 0 & 2 & 0 & 0 & 2 & 4 \\
0 & 2 & 4 & 0 & 0 & 2 & 0 & 0 & 0 & 2 & 0 & 0 & 2 & 0 & 2 & 2 \\
0 & 0 & 2 & 2 & 2 & 0 & 2 & 0 & 0 & 0 & 0 & 0 & 4 & 2 & 2 & 0 \\
0 & 2 & 0 & 2 & 0 & 4 & 0 & 0 & 2 & 0 & 0 & 2 & 2 & 2 & 0 & 0 \\
0 & 0 & 2 & 2 & 0 & 2 & 2 & 0 & 0 & 2 & 4 & 2 & 0 & 0 & 0 & 0 \\
0 & 2 & 2 & 0 & 4 & 0 & 0 & 0 & 0 & 0 & 2 & 2 & 0 & 2 & 0 & 2 \\
0 & 2 & 0 & 4 & 2 & 0 & 0 & 0 & 2 & 2 & 2 & 0 & 0 & 0 & 2 & 0 \\
0 & 0 & 2 & 0 & 2 & 2 & 0 & 2 & 4 & 2 & 0 & 0 & 0 & 2 & 0 & 0
\end{array}
$$

As an aside, note that the table is symmetrical, indicating that the pair (a, b) yields as many solutions as the pair (b, a). This follows quite easily from the fact that if x satisfies the equation $x^{-1} \oplus (x \oplus a)^{-1} = b$, then $y = x^{-1}$ is a solution of the equation $y^{-1} \oplus (y \oplus b)^{-1} = a$.

It is apparent from the stated result that, from the point of view of strength against differential cryptanalysis, it would have been better if the computations in *Rijndael* had been performed in (say) $GF(2^7)$ or $GF(2^9)$. However, working with strings of odd length would have been inconvenient for implementation. Computer scientists prefer working with words of even length (in fact, they prefer working with powers of 2). In addition, values like $n = 11$ or 13 or even $n = 9$ would lead to larger look-up tables[9] which might make the cipher less suitable for implementation under circumstances where there are space constraints.

9.2.4 *Properties of the* Rijndael *Round Function*

The round function on n bits is an invertible mapping, i.e. a permutation of the set of all bitstrings of length n. Thus a round function, with a fixed round key, is an element of the group S_{2^n}. Wernsdorf[10] has proved the interesting fact that the round functions of *Rijndael* generate the entire alternating[11] group A_N where $N = 2^{128}$, in the sense that any even

[9] We have already noted that, in order to achieve adequate speed, the S-box of a block cipher is frequently, or even usually, implemented as a simple table, rather than in the mathematical form, if any, by which it is defined.

[10] Wernsdorf, R.: *The round functions of RIJNDAEL generate the alternating group*, Proc. Fast Software Encryption, 2002, LNCS 2365, Springer-Verlag, 2002, pp. 143–148.

[11] See Sect. 3.2 for the definitions of S_N and A_N, in case you have forgotten what these are.

permutation of the set of 128-bit strings can be written as a finite product of such round functions.

Wernsdorf comments "This result implies that from the algebraic point of view some thinkable weaknesses of RIJNDAEL can be excluded (if the generated group were smaller, then this would point to regularities in the algorithm)". In fact, it would appear to imply that the set of permutations represented by instantiations of *Rijndael* is about as large as it could possibly be. Note, however, that Wernsdorf did *not* prove that the permutations induced by the different keys of *Rijndael* form a group, i.e. it is probably not true that if K_1 and K_2 are keys, then there exists a key K such that $E_{K_2}(E_{K_1}(P)) = E_K(P)$ for all plaintext blocks P. This is of more than just theoretical importance, since if the permutations induced by the keys form a group, then double encryption, or even triple or higher multiple encryption would offer no extra protection against exhaustive search. This was for some time an important area of research for DES—the Data Encryption Standard—but it was eventually proved that the DES permutations do not form a group.

9.3 Properties of the Function $F(x) = x^{2^k+1}$

Having considered inversion in a finite field as a method of constructing cryptographically strong substitution boxes for use in block ciphers, we shall in this section discuss the cryptologically significant properties, i.e. differential uniformity and nonlinearity, of another class of functions, viz those defined on $GF(2^n)$ by $F(x) = x^{2^k+1}$.

We shall several times have need of the following facts, the proof of the first of which we left as an exercise a long time ago (Exercise 7 of Sect. 2.6):

Proposition 1 *The congruence* $kx \equiv a \mod n$ *has either no solutions or* s *solutions modulo* n, *where* $s = gcd(k, n)$.

Proposition 2 *An integer (or a polynomial)* $x^k - 1$ *is a divisor of the integer (or polynomial, respectively)* $x^n - 1$ *if and only if* $k|n$. *The greatest common divisor of the integers (or polynomials)* $x^k - 1$ *and* $x^n - 1$ *is* $x^{\gcd(k,n)} - 1$.

Proof Dividing n by k and thus $x^n - 1$ by $x^k - 1$, one gets

$$x^n - 1 = (x^k - 1)(x^{n-k} + x^{n-2k} + x^{n-3k} + \cdots + x^{n-qk}) + (x^{n-qk} - 1),$$

where we have put $q = \lfloor \frac{n}{k} \rfloor$.

The first part of the statement follows immediately.

Looking at the exponents involved (n, qk and $r = n - qk$), one sees that these are just the terms one would have in an application of the classical Euclidean algorithm, and one can continue the analogue of that algorithm until one has a zero remainder, by which time the divisor is just $x^{\gcd(k,n)} - 1$.

Of particular importance in what follows is the next corollary, which follows immediately from the previous two propositions:

Corollary *In the finite field $GF(2^n)$ the equation $x^{2^k-1} = 1$ has either no solutions or $2^{\gcd(k,n)} - 1$ solutions.*

Proof This follows from the fact that in $GF(2^n)$ every nonzero element satisfies the equation $x^{2^n-1} - 1 = 0$, and that $GF(2^n)$ contains a subfield $GF(2^k)$ if and only if $k|n$. Then $GF(2^k)$ consists of 0 and precisely all the $(2^k - 1)$th roots of 1.

Examples Thus $GF(2^{16})$ contains the subfields $GF(2), GF(2^2), GF(2^4)$ and $GF(2^8)$, but $GF(2^3)$ and $GF(2^{127})$ do not contain any subfield other than $GF(2)$, because 3 and 127 are primes.

We can now discuss the cryptologically important features of the power functions in the same way as we did for the inversion function of *Rijndael*'s S-box.

9.3.1 Differential Uniformity

Nyberg's result here is the following:

Theorem *Define on $GF(2^n)$ the power polynomial $F(x) = x^{2^k+1}$. F is differentially 2^s-uniform, where $s = gcd(n, k)$.*

Proof With $a, b \in GF(2^n)$ the equation

$$(x \oplus a)^{2^k+1} \oplus x^{2^k+1} = b$$

has either no solutions, or at least two. (If x is a solution, then so is $x \oplus a$.) Now suppose that x_1 and x_2 are two distinct solutions. Then rewriting, using the Freshman's Theorem,

$$(x_1 \oplus a)^{2^k+1} \oplus x_1^{2^k+1} = b$$

as

$$(x_1 \oplus a)(x_1^{2^k} \oplus a^{2^k}) \oplus x_1^{2^k+1} = b$$

or

$$x_1 a^{2^k} \oplus a^{2^k+1} \oplus x_1^{2^k} a = b$$

and similarly

$$x_2 a^{2^k} \oplus a^{2^k+1} \oplus x_2^{2^k} a = b$$

and adding the last two equations, we get

$$(x_1 \oplus x_2)a^{2^k} \oplus (x_1 \oplus x_2)^{2^k} a = 0,$$

so that

$$(x_1 \oplus x_2)^{2^k - 1} = a^{2^k - 1}.$$

But then it follows from the corollary above that $x_1 \oplus x_2 = ay^r$, with $r = 0, 1, 2, \ldots, 2^s - 1$ where y is a primitive sth root of 1 in $GF(2^n)$.

9.3.2 Nonlinearity

With F defined on $GF(2^n)$ as before by $F(x) = x^{2^k+1}$, we define, for any $\omega \in GF(2^n)$, a function

$$f_\omega(x) = L_\omega(F(x)) = \omega \cdot F(x)$$

(where we consider $GF(2^n)$ merely as an n-dimensional vector space over $GF(2)$, and where the "·" now represents the dot product (scalar product)). Thus, we consider 2^n linear Boolean functions. As we did in the case of the *Rijndael* S-box, we wish to see how close these linear combinations of the output bits of the S-box can come to linear functions of the input bits, or equivalently we consider the correlations between the f_ω and linear functions of the input bits. This, by the way, is precisely what is done in linear cryptanalysis, where such correlations (picking the "best" linear combinations in both cases, i.e. those with the highest correlations) are exploited.

As before, we use the Walsh–Hadamard transform: For each ω we consider the Walsh–Hadamard transform of f_ω:

$$\hat{F}_\omega(t) = \sum_x (-1)^{t \cdot x \oplus f_\omega(x)}$$

and we assert that

$$\max_t |\hat{F}_\omega(t)| = 2^{\frac{n+s}{2}}$$

where $s = \gcd(k, n)$.

This implies the following

Theorem *For each nonzero $\omega \in (GF(2))^n$, the nonlinearity of f_ω is $2^{n-1} - 2^{\frac{n+s}{2}-1}$, or, in terms of correlations: For all $\omega, t \in GF(2^n)$*

$$c(f_\omega, L_t) = \frac{\hat{F}_\omega(t)}{2^n} \leq \frac{2^{(n+s)/2}}{2^n} = 2^{\frac{-n+s}{2}}.$$

We now prove our assertion:

$$\hat{F}_\omega(t) = \sum_x (-1)^{f_\omega(x) \oplus t \cdot x}$$

so, on squaring, we obtain

$$\begin{aligned}(\hat{F}_\omega(t))^2 &= \left(\sum_x (-1)^{f_\omega(x)\oplus t\cdot x}\right)\left(\sum_y (-1)^{f_\omega(y)\oplus t\cdot y}\right)\\ &= \sum_x (-1)^{f_\omega(x)\oplus t\cdot x} \sum_y (-1)^{f_\omega(x+y)\oplus t\cdot(x+y)}\\ &= \sum_y (-1)^{t\cdot y} \sum_x (-1)^{f_\omega(x+y)\oplus f_\omega(x)}. \end{aligned} \tag{*}$$

For $y \neq 0$, define the mapping

$$x \mapsto F(x+y) + F(x) + F(y)$$

i.e. as in the previous proof

$$x \mapsto x^{2^k} y + x y^{2^k}.$$

This is a linear mapping, and as before one shows that its kernel is the set $\{0\} \cup \{yz | z^{2^s-1} = 1\}$, which clearly has dimension s, since the set $\{z | z^{2^s-1} = 1\}$ is the subfield $GF(2^s)$ of $GF(2^n)$. Hence, using the notation E_y for the image of this mapping, we have that E_y is a subspace of $GF(2^n)$ of dimension $n - s$.

Now note that, for any $y \neq 0$, either

(i) $\omega \cdot z = 0$ for all $z \in E_y$ or
(ii) $\omega \cdot z$ is balanced on E_y, in which case $\sum_{z \in E_y} (-1)^{\omega \cdot z} = 0$.

As far as (i) is concerned, consider the set

$$\begin{aligned} Y &= \{y | \omega \cdot z = 0 \ \forall\, z \in E_y\}\\ &= \{y | \omega \cdot (F(x+y) + F(x) + F(y) = 0 \ \forall x\}\\ &= \{y | \omega \cdot (x^{2^k} y + x y^{2^k}) = 0 \ \forall x\}. \end{aligned}$$

We may also write

$$Y = \{y | f_\omega(x) \oplus f_\omega(y) \oplus f_\omega(x+y) = 0 \ \forall x\}$$

i.e. f_ω is linear on Y.

Returning to equation (*), we have

$$(\hat{F}_\omega(t))^2 = \sum_y (-1)^{t\cdot y} \sum_x (-1)^{f_\omega(x\oplus y)\oplus f_\omega(x)}$$

$$= \sum_y (-1)^{t\cdot y\oplus f_\omega(y)} \sum_x (-1)^{f_\omega(x\oplus y)\oplus f_\omega(x)\oplus f_\omega(y)}$$

$$= 2^n + \sum_{y\neq 0} (-1)^{t\cdot y\oplus f_\omega(y)} \sum_x (-1)^{f_\omega(x\oplus y)\oplus f_\omega(x)\oplus f_\omega(y)}.$$

But if $y \notin Y$, then the inner sum is 0, since $f_\omega(x+y) \oplus f_\omega(x) \oplus f_\omega(y)$ is a balanced function of x. Hence

$$(\hat{F}_\omega(t))^2 = 2^n + \sum_{y\in Y\setminus\{0\}} (-1)^{t\cdot y+f_\omega(y)} \times 2^n,$$

each term in the inner sum being $(-1)^0$.

But f_ω is linear on Y, and therefore so is $t \cdot y \oplus f_\omega(y)$, and therefore they are both balanced on Y. Hence

$$\sum_{y\in Y} (-1)^{t\cdot y+f_\omega(y)} = 0,$$

whence

$$\sum_{y\in Y\setminus\{0\}} (-1)^{t\cdot y\oplus f_\omega(y)} = (-1)^{t\cdot 0\oplus f_\omega(0)} = 1.$$

Thus we have that

$$(\hat{F}_\omega(t))^2 = 2^n + 2^n \times \#(Y - \{0\}) = 2^n \times \#Y.$$

It remains to show that Y contains 2^s elements.

For convenience we now shift to using the $\mathrm{Tr}(\nu x)$ formulation of looking at linear Boolean functions: write $\omega \cdot x$ as $\mathrm{Tr}(\nu x)$. Let $y \in Y$. Then we get from $\omega \cdot (xy^{2^k} + yx^{2^k}) = 0$ that $\omega \cdot (yx^{2^k}) = \omega \cdot (xy^{2^k})$ for all $x \in GF(2^n)$, or

$$\begin{aligned}\mathrm{Tr}(\nu y x^{2^k}) &= \mathrm{Tr}(\nu y^{2^k} x)\\ &= \mathrm{Tr}(\nu^{2^k} y^{2^{2k}} x^{2^k}) \;\; \forall\, x \in GF(2^n),\end{aligned}$$

where the second equality follows from the properties of the Trace function.

Hence

$$\mathrm{Tr}((\nu y - \nu^{2^k} y^{2^{2k}})x^{2^k}) = 0 \;\; \forall x \in GF(2^n).$$

This implies (see Exercise 4 of Sect. 6.3) that $\nu y = \nu^{2^k} y^{2^{2k}}$ or $\nu^{2^k-1} y^{2^{2k}-1} = 1$. Consequently

$$(\nu F(y))^{2^k-1} = (\nu y^{2^k+1})^{2^k-1} = 1.$$

As before, and using the fact that F is a permutation, we have an equation of the form $\zeta^{2^k-1} = 1$ in a multiplicative group of order $2^n - 1$, which, as we have seen, has $2^s - 1$ nonzero solutions where $s = \gcd(k, n)$. Thus $\#(Y) = 2^s$, as required. This completes the proof.

9.3.3 Degree

In order to avoid algebraic attacks through linearisation, for example, Boolean functions used in ciphers should preferably have high degree; in fact, the algebraic normal form (ANF) should have many nonzero terms, including many of high degree. In this respect, the function F defined on $GF(2^n)$ falls short. Nyberg quotes without proof the following theorem:

Theorem *If the mapping $x \mapsto x^e$ is a permutation of $GF(2^n)$ and if $\nu \in GF(2^n)$, then the function $Tr(\nu x^e)$ has algebraic degree equal to the Hamming weight of the exponent e.*

Thus the functions $f_\omega = L_\omega \circ F$ are only quadratic. However, the inverse of F is much better in this respect:

Theorem *The inverse of the function $F(x) = x^{2^k+1}$ on $GF(2^n)$, n odd, if it exists, is $F^{-1}(x) = x^l$, where $l = \sum_{i=0}^{(n-1)/2} 2^{2ik}$.*

Corollary *The component functions of F^{-1} (and all other Boolean functions of the form $f_\omega = L_\omega \circ F^{-1}$ for some $\omega \in GF(2^n)$, n odd) have degree $\frac{n+1}{2}$.*

Proof of Theorem

$$\begin{aligned}(2^k+1)l &= \sum_{i=0}^{(n-1)/2} 2^{(2i+1)k} + \sum_{i=0}^{(n-1)/2} 2^{2ik} \\ &= 2^k + 2^{3k} + 2^{5k} + \cdots + 2^{nk} + 1 + 2^{2k} + 2^{4k} \\ &\quad + \cdots + 2^{(n-1)k} \\ &= \sum_{i=0}^{n} 2^{ik} \\ &= \sum_{i=0}^{n} 2^i\end{aligned}$$

since the mapping $i \mapsto ki$ is a permutation of $\{0, 1, 2, \ldots, n-1\}$ if k and n are relatively prime.

Hence, $(2^k + 1)l = 2^{n+1} - 1 \equiv 2(2^n - 1) + 1 \equiv 1 \mod (2^n - 1)$.

Thus returning to our function F, we have that

$$(x^{2^k+1})^l = x \ \forall x \in GF(2^n)$$

or $F^{-1}(x) = x^l$.

9.4 Perfect Nonlinear S-boxes

As noted when we discussed the *Rijndael* S-box, differential cryptanalysis is based on a certain kind of asymmetry in the nonlinear component of the round function of a block cipher: certain differences in the input of the S-box lead to some differences in the output being more probable than others. In the case of an n-input, m-output S-box this amounts to saying that for some Δ the difference

$$S(x \oplus \Delta) \oplus S(x)$$

assumes some values more often than others.

In the case of a Boolean function $f : (GF(2))^{2m} \longrightarrow GF(2)$ this amounts to saying that $f(x \oplus \Delta) \oplus f(x)$ is unbalanced. We have seen (Sect. 8.8) that the only Boolean functions f for which the difference $f(x \oplus \Delta) \oplus f(x)$ is balanced (for all nonzero Δ) are the bent functions; recall that these may be constructed by means of the Maiorama–McFarland construction:

$$f(x_1, x_2) = \pi(x_1) \cdot x_2 + g(x_2),$$

where $x_1, x_2 \in (GF(2))^m$, π is any permutation of $(GF(2))^m$, g is any Boolean function defined on $(GF(2))^m$ and "$\cdot$" denotes the dot (or scalar) product of two vectors. We have noted that bent functions only exist for even numbers of variables and that, sadly, they are unbalanced: in the case we are considering here, the function will have output 0 $2^{2m-1} + 2^{m-1}$ times and output 1 $2^{2m-1} - 2^{m-1}$ times (or the other way round, depending on g).

The concept of a *perfect nonlinear transformation*, as defined by Nyberg,[12] is an extension of this idea to functions from $(GF(2))^n$ to $(GF(2))^m$:

Definition A function $f : (GF(2))^n \longrightarrow (GF(2))^m$ is called *perfect nonlinear* if for every fixed $\Delta \in (GF(2))^n$ the difference

$$f(x \oplus \Delta) \oplus f(x)$$

assumes every value $y \in (GF(2))^m$ exactly 2^{n-m} times.

[12] Yes, her again. This time in a paper entitled *Perfect Nonlinear S-boxes*, presented at Eurocrypt '91; LNCS 547, Springer-Verlag.

Nyberg shows that perfect nonlinear functions can only exist if $n \geq 2m$, and offers the following construction, which may be considered a generalisation of the Maiorama–McFarland construction. We denote by f_i the coordinate functions of f, and construct $f : (GF(2))^{2m} \longrightarrow (GF(2))^m$ in such a way that each of these is a bent function:

$$f_i(x) = f_i(x_1, x_2) = \pi_i(x_1) \cdot x_2 \oplus g_i(x_1),$$

where each of the π_i is a permutation of $(GF(2))^m$ and the g_i are Boolean functions defined on $(GF(2))^m$. For efficiency, the choice $g_i(x) = 0$ for all i and all x is recommended.

Nyberg then suggests that the permutations π_i be selected as multiplications by α, a primitive element of the field $GF(2^m)$, or, more precisely, that we define π_i by $\pi_i(x) = \alpha^i x$. Thus, the least significant bit $f_0(x)$ of the output is $x_1 \cdot x_2$, the next bit of the output is obtained by moving x_1 through one bit of the Galois field counter representing multiplying by α and computing $(\alpha x_1) \cdot x_2$, and so on. Since linear shift registers/Galois field counters are easily and efficiently implemented (as described in Sect. 7.5), the resulting function is moderately efficient.

A fairly serious problem with this construction remains, however, in the fact that the output of the function is not balanced. For the sake of convenience, let us again assume that $g_i(x_2)$ is the zero function. Then (for nonzero Δ) $f(x \oplus \Delta) \oplus f(x)$ takes on all values equal numbers of times, but the function f itself takes on the value $(0, 0, 0, , \ldots, 0)$ $2^{m+1} - 1$ times, and all other values $2^m - 1$ times.

One should also note that the fact that the inputs and the outputs of such a perfect nonlinear S-box are of different lengths makes them rather awkward for use in Substitution-Permutation Networks (SPNs). On the other hand, they may well be useful as components in the round function of block ciphers with a (possibly modified) Feistel structure.

9.5 Diffusion

We have so far in this chapter concentrated on the confusion aspect of block cipher design, i.e. on the nonlinear components of the round function. In the remainder we shall discuss the usage of finite fields in the diffusion layer of a round.

We need to do quite a bit of preparation, as will become clear.

9.5.1 Introduction

MDS matrices originate in Coding Theory, or to be precise, in the study of error correcting linear codes, about which more shortly. They have assumed an important role in Cryptography through their use in the design of block ciphers.

In block ciphers the usual, if not the universal, method of obtaining diffusion is by including a linear transformation as one of the components of the round function. In ciphers with a Feistel structure this is a simple XOR, in Substitution-Permutation Networks, the entire permutation layer is probably linear. (Confusion, on the other hand, is obtained through the use of a nonlinear function, frequently implemented as a simple look-up table.)

Suppose that the input into the linear layer is an n-dimensional vector $\mathbf{x}$, and its output an m-dimensional vector $\mathbf{y}$. In most cases $m = n$, and we shall in what follows assume that this is indeed the case. If the diffusion layer is indeed linear then it follows that there exists a matrix A such that

$$\mathbf{y} = \mathbf{x}A.$$

Note that we write the state vectors (input and output) as row vectors. This is different from our notation when we discussed *Rijndael*/AES, but more in accordance with the practice in Coding Theory. It will be clear, however, that the construction described here corresponds precisely with what happens in *Rijndael*. When discussing the round structure of *Rijndael*, we noted that the branch number of the 4×4 matrix used there equals 5, and that this is the maximum possible, where the branch number of a matrix $\mathbf{A}$ is defined as

$$\mathcal{B}(\mathbf{A}) = \min_{\mathbf{a},\mathbf{a}\neq\mathbf{0}} [\text{weight}(\mathbf{a}) + \text{weight}(\mathbf{a}A)],$$

where the weight of a vector is defined to be the number of its nonzero components, which may be bits, but probably are bytes or even longer words.

For a square $n \times n$ matrix it is clear that the branch number cannot be greater than $n + 1$, since a nonzero input can have weight 1, and the maximum weight of the output is n. On the other hand, if the branch number is $n + 1$, then for any input (of weight at least 1, but at most n obviously), the output cannot be the zero vector. Thus a matrix with branch number $n + 1$ must be non-singular.

But a matrix with branch number $n + 1$ has a further property: Every square submatrix of an MDS matrix is non-singular. For if an input vector has weight $n - j$, $1 \leq j \leq n - 1$, then the output vector must have weight at least $j + 1$. This implies that the sum of any $n - j$ rows has weight at least $j + 1$, so even if we only consider $n - j$ of the columns, the weight of the remaining entries is still at least 1. In other words, no restriction of $n - j$ rows of $n - j$ columns can give a matrix which can map a nonzero vector of length $n - j$ onto the zero vector of that length: every $(n - j) \times (n - j)$ submatrix of an MDS matrix is non-singular.

In particular, an MDS matrix cannot contain any zero entries.

Conversely, it is easy to see that an $n \times n$ matrix with the property that every square submatrix is non-singular must have branch number $n + 1$.

Such matrices are called *Maximum Distance Separable* or MDS for short.

The reason for this terminology comes from Coding Theory, as we shall now outline.

9.5.2 Linear Codes

Let $\mathbb{F}$ be a field. An $[n, k, d]$ code C is a k-dimensional subspace of the vector space $\mathbb{F}^n$, such that every nonzero element of C has weight at least d. As mentioned above, we shall denote the elements of C as row vectors, whose components are elements of $\mathbb{F}$. In what follows we shall always assume that $\mathbb{F}$ is a field of characteristic 2, so that we can consider addition of elements to be the componentwise exclusive or operation $\oplus$.

n is called the *length* of the code. The elements of C are called *codewords*. Thus, if C is an $[n, k, d]$-code then it contains 2^k codewords.

Let us consider a basis for the vector space C. Without any real loss of generality, it is convenient to write the elements of this basis as the rows of a $k \times n$ matrix G of the following form:

$$G = \begin{pmatrix} 1\ 0\ 0 \ldots 0 & a_{0,k} & a_{0,k+1} & \ldots & a_{0,n-1} \\ 0\ 1\ 0 \ldots 0 & a_{1,k} & a_{1,k+1} & \ldots & a_{1,n-1} \\ 0\ 0\ 1 \ldots 0 & a_{2,k} & a_{2,k+1} & \ldots & a_{2,n-1} \\ \ldots & & & & \\ 0\ 0\ 0 \ldots 1 & a_{k-1,k} & a_{k-1,k+1} & \ldots & a_{k-1,n-1} \end{pmatrix}$$

G is then called the generator matrix of the code.

The intention is that a message word $\mathbf{m}$ is, prior to transmission, converted into a codeword $\mathbf{c}$ by multiplying it by G:

$$\mathbf{c} = \mathbf{m}G,$$

where clearly the first k symbols of $\mathbf{c}$ carry the same information as $\mathbf{m}$, and the remaining ones play the role of parity check bits. This can be seen as follows:

Write G in block matrix form, denoting by I_r the $r \times r$ identity matrix,

$$G = (I_k : A),$$

where A is a $k \times (n - k)$ matrix; put

$$H = (A^t : I_{n-k}).$$

Then $GH^t = 0$, the $k \times (n - k)$ zero matrix.[13] H is called the parity check matrix; an n-character string $\mathbf{x}$ is a codeword if and only if $\mathbf{x}H^t = 0$. The value $\mathbf{x}H^t$ is called the *syndrome* of $\mathbf{x}$. Put another way: if the syndrome $\mathbf{x}H^t$ of $\mathbf{x}$ is not equal to 0, then we know that an error in transmission occurred. Now what can we do about that?

If we define the *distance* $d(\mathbf{a},\mathbf{b})$ between two codewords $\mathbf{a}$ and $\mathbf{b}$ to be the number of components in which they differ, then it follows that

$$d(\mathbf{a},\mathbf{b}) = \text{weight}(\mathbf{a} \oplus \mathbf{b})$$

and since $\mathbf{a} \oplus \mathbf{b} \in C$, this implies that the distance between any two distinct codewords is at least d, i.e. d can also be interpreted as the minimum distance between any two words in the code.

[13]The superscript ^{t} denotes, as usual, the transpose.

9.5.3 Error Detection and Correction

If, during transmission, d errors occur, i.e. d components of the transmitted vectors are corrupted, then it is possible that the transmitted codeword gets changed into another codeword, so multiplying by the parity check gives the "correct" answer 0. In that case the fact that errors occurred will not be detected. But if fewer than d errors occurred, this will be detected, because $\mathbf{x}$ fails the parity check test. Thus an $[n, k, d]$-code is $(d-1)$-*error detecting*.

But we can actually do better than that: If e errors occur, where $e < \frac{d}{2}$, then the errors can be corrected! For suppose that $\mathbf{x}$ is a vector in which e components have gone wrong. Then there can only be at most one codeword which is within distance e from $\mathbf{x}$. Under the not unreasonable condition that the communication channel is such that correct transmission of a component is more probable than its corruption, that codeword is the most likely, and the error can therefore be corrected. Thus an $[n, k, d]$-code is $\lfloor \frac{d}{2} \rfloor$-*error correcting*.

It is therefore evident that the minimum distance of a code is important, and we would like to make optimal use of it. Note that the parity check symbols, and there are $n - k$ of them, carry no information, but are there purely in order to detect and possibly correct errors. What we want is a code which gives us the maximum distance between codewords for given values of n and k. Such codes are called *Maximum Distance Separable* or *MDS* codes. It is remarkable that MDS codes are seldom used in practice; very few are actually known.[14] But their generating matrices are of great interest in cryptography, as we shall see.

Their importance lies in the relationship between MDS codes and MDS matrices. Again, not surprisingly, this in turn lies in the parity bits of the generator matrix of the code.

We still need a bit more theory.

9.5.4 Some Properties of MDS Matrices

We follow the arguments of MacWilliams and Sloane in the very first chapter of their book.[15]

Theorem 1 *If H is the parity check matrix of a code of length n, then the dimension of the code is $n - r$ if and only if some r columns of H are linearly independent and no $r + 1$ columns are.*

Proof This is clear from the relationship between the parity check matrix H and the generator matrix G.

Theorem 2 *If H is the parity check matrix of a code of length n, then the code has minimum distance d if and only if every $d - 1$ columns of H are linearly independent and some d columns are linearly dependent.*

---

[14] Junod, P. and Vaudenay, S.: Perfect diffusion matrices for block ciphers; Proc. Selected Areas in Cryptography, SAC 2004, LNCS 3357, Springer-Verlag 2005.

[15] MacWilliams, F.J. and Sloane, N.J.A.: *The Theory of Error-Correcting Codes*, North-Holland Publishing Co., Amsterdam, 1977.

Proof The minimum distance d is the minimum weight of the nonzero codewords. Now a codeword $\mathbf{x}$ of weight d exists if and only if $\mathbf{x}H^t = 0$ for some $\mathbf{x}$ of weight d, which means d rows of H^t are linearly dependent. By the same argument no $d - 1$ rows of H^t can be linearly dependent.

Theorem 3 (The Singleton Bound) *If there exists a binary linear* $[n, k, d]$*-code then* $k \leq n - d + 1$.

Proof Consider a table formed from the encodings of all the 2^k words as rows. Since the distance between any two of them is at least d, we can delete the first $d - 1$ columns of this table, and still have a one-to-one relationship between the input strings of length k and the rows. Since there are at most $2^{n-(d-1)}$ strings in our table, we must have that $2^k \leq 2^{n-d+1}$, and the result follows.

Note that if, instead of binary codes, we consider a code in which the codewords are strings formed from the elements of some field $GF(q)$, exactly the same argument shows that

$$q^k \leq q^{n-d+1}$$

or

$$k \leq n - d + 1.$$

Definition Any $[n, k, d]$-code over $GF(q)$ such that $k = n - d + 1$ (i.e. one in which the Singleton bound is met) is called *Maximum Distance Separable* or MDS, indicating that it offers the maximum number of n-character code words for the given distance.

A warning: If $GF(q) \neq GF(2)$ the distance between two strings (e.g. codewords) $\mathbf{a_0}$ and $\mathbf{a_1}$, with $\mathbf{a_i} = (a_{i0}, a_{i1}, \ldots, a_{i,n-1})$, where $a_{ij} \in GF(q)$, is defined to be $\#\{j | a_{0j} \neq a_{1j}\}$. This is different from the Hamming weight as we have used it up to now, even in the case where $q = 2^n$ for some $n > 1$.

Theorem 4 *Consider an* $[n, k, d]$*-code over* $GF(q)$ *with parity check matrix* H^t*. This code is MDS if and only if every set of* $n - k$ *columns of* H *(i.e. rows of* H^t*) is linearly independent.*

Proof Suppose t rows of H^t are linearly dependent. This means that there is a string $\mathbf{x}$ of length n and weight t such that $\mathbf{x}H^t = 0$, i.e. $\mathbf{x}$ is a codeword. Since the minimum weight in the code is $n - k + 1$, we must have that $t \geq d > n - k$.

Conversely, if every set of $n - k$ columns of H is independent, then no nonzero string of weight less than $n - k + 1$ can be a codeword, so the minimum distance is at least $n - k + 1$. But the Singleton bound tells us that the minimum distance cannot be greater than that.

If the matrix $G = (I|A)$ generates a code C, then the code generated by the matrix $H = (A^t|I)$ is called the *dual code* $C^\perp$ of C. If C is an $[n, k, n - k + 1]$-code, then $C^\perp$ is an $[n, n - k, k + 1]$-code.

Example Define $GF(2^2) = \{0, 1, \alpha, \beta\}$, where $\beta = \alpha^2$ and $\alpha + \beta = 1$, and consider the code with generator matrix $G = \begin{pmatrix} 1 & 0 & 1 & \alpha & \beta \\ 0 & 1 & 1 & \beta & \alpha \end{pmatrix}$. There are 16 codewords in this code, of length 5, and the dimension of the code is 2, and it can be verified that the minimum distance between codewords (= minimum weight of the codewords) is 4. Thus we have a [5,2,4]-code over $GF(2^2)$, so it is an MDS-code.

The code dual to one generated by the matrix $(I|A)$ is the code with generator matrix $(A^t|I)$, which in this case is $H = \begin{pmatrix} 1 & 1 & 1 & 0 & 0 \\ \alpha & \beta & 0 & 1 & 0 \\ \beta & \alpha & 0 & 0 & 1 \end{pmatrix}$.

Elementary row operations can reduce this to what is known as *standard form*[16]: The dual code is generated by $H' = \begin{pmatrix} 1 & 0 & 0 & \alpha & \beta \\ 0 & 1 & 0 & \beta & \alpha \\ 0 & 0 & 1 & 1 & 1 \end{pmatrix}$. This is a [5, 3, 3]-code, so it too is MDS.

Note that in G any two columns are linearly independent, whereas in H (and therefore in H') any three columns are independent.

We need one more observation about MDS codes. If we write the generator matrix in standard form as $(I|A)$, then we have proved that A has maximal rank. But more is actually true: every square submatrix of A is also non-singular.[17] This property provides a possible test for verifying whether or not a given square matrix is MDS.

To simplify the proof of this assertion, we restrict ourselves to the case where the singular square submatrix is in the upper right-hand corner of A; the general case can be reduced to that by a process easier to understand than writing it out in full. In a further simplification of the proof we shall assume that the singular submatrix is 3×3. Denote this submatrix by N.

Let the dimension of the code be k as before. Consider the submatrix of the generator matrix formed from the three columns which contain N as well as the $k - 3$ columns to the immediate left of that. This submatrix therefore has the form

$$\begin{pmatrix} \underset{3\times(k-3)}{0} & \underset{3\times 3}{N} \\ \\ \underset{(k-3)\times(k-3)}{I} & \underset{3\times(k-3)}{N'} \end{pmatrix},$$

where N' represents the remainder of the columns of which N forms the top part. The determinant of this matrix, which equals the determinant of N, is nonzero (every k columns of the generator matrix being linearly independent), so N is non-singular.

[16] If the standard form is used in a code, the code is called *systematic*. When considering the code purely in its sense of a linear subspace, the form of the matrix does not matter, of course; the vital point is what subspace the row vectors actually span.

[17] In particular every 1×1 submatrix is non-singular, so an MDS matrix cannot contain any 0 entries. This explains why in consideration of MDS matrices we have to use fields other than $GF(2)$, even if we restrict ourselves to fields of characteristic 2. The only binary MDS codes, i.e. codes over the alphabet $\{0, 1\}$, are the repetition codes with generator matrices of the form

$$(111\ldots1)$$

and their duals, the even weight codes. We leave the proof of this as an easy exercise. But see Sect. 9.5.7.

It is clear that this proof can be generalised to show that every square submatrix is indeed non-singular. Writing out the proof of the generalisation in detail will be messy and contribute nothing to our understanding.

It follows from this that, as in the example we have looked at, the dual of an MDS code is itself an MDS code. We show that any $k = n - (n - k)$ columns of the parity check matrix $G = (I|A)$ of $G^{\perp}$ are linearly independent: taking any k columns of G, we have a matrix $\mathbf{X} = (\mathbf{e}_1 \ldots \mathbf{e}_i\ \mathbf{a}_{i+1} \ldots \mathbf{a}_k)$ of i columns of the identity matrix and $k - i$ columns of A. Expanding the determinant of this along those columns of I in the first steps we find that the value of the determinant $|\mathbf{X}|$ equals the determinant of a $(k - i) \times (k - i)$ submatrix of $\mathbf{A}$, which we know to be nonzero.

As a final observation, we prove that if A is a square MDS matrix, then so is A^{-1}. After all, if the rows of $G = \begin{pmatrix} \underset{k \times k}{I} & \underset{k \times k}{A} \end{pmatrix}$ span the code (remember, a linear code is a vector space!), then so do the rows of $A^{-1}G = (A^{-1}\ I)$, and re-arranging the components of the codewords (so that the generator matrix becomes $(I\ A^{-1})$) will not change the minimum distance of the code, which remains $k + 1$.

The reader may have thought to himself or herself that determining whether or not all the square submatrices of a given matrix are non-singular is computationally rather demanding. It is therefore good to know that a "smallish" randomly selected matrix without any zero entries is likely to be MDS, although this probability drops with the size of the matrix. It is not hard to verify that the probability of a randomly selected 2×2 matrix with entries from $GF(2^8)$ being MDS is 0.98. Rough Monte Carlo tests of 1000000 randomly selected 3×3 matrices over the same field shows that the probability of the MDS property drops to about 0.938, and this drops further to about 0.814 for 4×4 matrices. The explanation lies in the fact that there are more square submatrices to consider, all of which have to be non-singular; in other words, the larger the matrix, the harsher are the conditions it must satisfy.

On the other hand, working over a larger field raises the probability of a randomly selected matrix being MDS: In fact, the probability that a randomly selected 4×4 matrix with entries from $GF(2^{16})$ is MDS is 0.995.

In the case of 4×4 matrices with entries from $GF(2^n)$, proving the MDS property is simplified by the fact that the following theorem holds[18]:

Theorem *A nonsingular* 4×4 *matrix A with entries from* $GF(2^n)$ *is MDS if and only if all entries are nonzero and all* 2×2 *submatrices are nonsingular.*

9.5.5 MDS Matrices in Block Ciphers

After that lengthy introduction, the use of MDS matrices in block cipher design, such as that of *Rijndael*, is now easily explained. If we consider an MDS code with its generator matrix G

[18]Gupta, K.C. and Ray, I.G.: On constructions of MDS matrices from companion matrices for lightweight cryptography; IACR ePrint Archive 2013/056.

in standard form

$$G = \begin{pmatrix} I_{k\times k} & A_{k\times(n-k)} \end{pmatrix}$$

then every codeword will have weight at least $k+1$. Consider a vector $\mathbf{x}$ of length k and weight w. The product $\mathbf{x}G = (\mathbf{x}||\mathbf{x}A)$, so $\mathbf{x}A$ must have weight at least $k+1-w$.

It follows therefore that the matrix A has branch number $k+1$. (Conversely, if a $k\times(n-k)$ matrix has branch number $k+1$, then $G = \begin{pmatrix} I_{k\times k} & A_{k\times(n-k)} \end{pmatrix}$ generates an MDS code, clearly.)

Since $k+1$ is the maximum branch number for $k\times k$ matrices, the usefulness of MDS matrices in achieving good diffusion in block ciphers is clear. This usage is not confined to substitution-permutation networks; the AES finalist block cipher *Twofish* uses an MDS matrix in the round function of its Feistel structure.

9.5.6 Examples of MDS Matrices

We shall discuss a few classes of MDS matrices which are in use or have been suggested for use in block ciphers.[19]

9.5.6.1 Circulant MDS Matrices

Recall that the `MixColumns` operation in *Rijndael* consists of multiplication of the state matrix by the MDS matrix

$$\begin{pmatrix} \alpha & 1+\alpha & 1 & 1 \\ 1 & \alpha & 1+\alpha & 1 \\ 1 & 1 & \alpha & 1+\alpha \\ 1+\alpha & 1 & 1 & \alpha \end{pmatrix},$$

where α is a root of the polynomial $x^8+x^4+x^3+x+1$. This is a *circulant* matrix, in that every row is obtained from the previous one by a rotation to the right. Clearly a circulant matrix is completely defined by its size and by the entries in its top (or any other) row.

Circulant MDS matrices are used in quite a few block ciphers other than *Rijndael* as well as in some hash functions, including *Whirlpool.*

When attempting to find an MDS matrix, restricting the search to circulant matrices has the advantage that the number of submatrices that have to be tested for non-singularity is considerably smaller, because of the pattern in the rows and columns. For example: in the

[19]See Xiao, L. and Heys, H.M.: Hardware design and analysis of block cipher components. Proc. 5th International Conference on Information Security and Cryptology—ICISC 2002, Seoul, LNCS 2587, Springer-Verlag.

case of 4×4 matrices, there are 70 submatrices to be checked, but if the matrix is circulant, this drops to 17. In the case of 8×8 matrices the drop is from 12869 to only 935.[20]

9.5.6.2 Hadamard Matrices

Given a set $\{\alpha_0, \alpha_1, \ldots, \alpha_{k-1}\}$ of k elements of the field $GF(2^n)$, a matrix M is formed by setting $M_{i,j} = \alpha_{i \oplus j}$. This matrix has the property that $M^2 = c \cdot I$, where $c = \sum_{i=0}^{k-1} \alpha_i^2$. In the particular case where $c = 1$, this means that M is an involution ($M^2 = I$), which may be useful in applications where the saving in storage space through not having to implement M and its inverse separately is desirable. This property is also required if the block cipher as a whole is designed to be involutory.

Again, having selected a matrix of this form, it is necessary to check for the MDS property, which for $k > 4$ implies looking at all submatrices. Hadamard matrices are used in the block ciphers *Khazad* and *Anubis*, among others.

9.5.6.3 Cauchy Matrices

As opposed to the previous two constructions, Cauchy matrices are guaranteed to be MDS. It is rather unfortunate that it is very hard to construct one which has any specified properties, such as low weight of the entries or the self-inverse (involutory) property.

A Cauchy matrix is constructed by choosing $2k$ elements $\{\alpha_0, \alpha_1, \ldots, \alpha_{k-1}, \beta_0, \ldots, \beta_{k-1}\}$ from $GF(2^n)$. All elements of this set must be distinct. Now define M as the $k \times k$ matrix with

$$m_{i,j} = (\alpha_i \oplus \beta_j)^{-1}.$$

9.5.6.4 VanderMonde Matrices

The VanderMonde matrices are a well known family with the property that all submatrices are non-singular. A VanderMonde matrix is a matrix of the following form:

$$\begin{pmatrix} 1 & 1 & \ldots & 1 \\ \alpha_0 & \alpha_0^2 & \ldots & \alpha_0^{n-1} \\ \alpha_1 & \alpha_1^2 & \ldots & \alpha_1^{n-1} \\ \ldots & \ldots & \ldots & \ldots \\ \alpha_{m-1} & \alpha_{m-1}^2 & \ldots & \alpha_{m-1}^{n-1} \end{pmatrix},$$

where the α_i are distinct elements from some finite field. It can be shown that every square submatrix of such a matrix is non-singular.

VanderMonde matrices are seldom, if ever, used in the design of block ciphers.

9.5.6.5 MDS Matrices from Companion Matrices

A currently very active area of research in cryptology is that of "lightweight cryptography", as a result of the spreading need for implementations in the expanding "Internet of Things". The

[20]These figures are from Daemen, J., Knudsen, L. and Rijmen, V.: *The block cipher SQUARE*, Proc. Fast Software Encryption (FSE) 1997, LNCS 1267, Springer-Verlag. pp. 149–165. (SQUARE was an ancestor of *Rijndael*.)

need is for algorithms and/or adaptations of implementations of existing algorithms which can run on low cost, high volume devices with limited computing power, such as secure RFID tags, sensor nodes, and Internet-enabled devices.[21]

In the required block cipher designs it has been proposed that good diffusion can be obtained through the iteration of simple operations using the companion matrices of polynomials with low weight coefficients. Thus the complexity of the linear layer can be reduced to the iteration of a simple linear operation.

Recall that if $f(x) = a_0 + a_1x + \cdots + a_{k-1}x^{k-1} + x^k \in GF(2^n)[x]$, then the *companion matrix* of this polynomial is

$$C_f = \begin{pmatrix} 0 & 0 & \dots & 0 & a_0 \\ 1 & 0 & \dots & 0 & a_1 \\ 0 & 1 & \dots & 0 & a_2 \\ \dots & \dots & \dots & \dots & \dots \\ 0 & 0 & \dots & 1 & a_{k-1} \end{pmatrix}.$$

Note that the effect

$$\begin{aligned} &(x_0, x_1, x_2, \dots, x_{k-1}) \\ &\mapsto (x_0, x_1, \dots, x_{k-1})C_f \\ &= (x_1, x_2, \dots, x_{k-1}, a_0x_0 + a_1x_1 + \cdots + a_{k-1}x_{k-1}) \end{aligned}$$

of C_f on the input vector can in hardware be easily implemented as an LFSR over $GF(2^n)$.

The inverse of C_f is the matrix

$$C_f^{-1} = \begin{pmatrix} a_1a_0^{-1} & 1 & 0 & \dots & 0 \\ a_2a_0^{-1} & 0 & 1 & \dots & 0 \\ \dots & \dots & \dots & \dots & \dots \\ a_{k-1}a_0^{-1} & 0 & 0 & \dots & 1 \\ a_0^{-1} & 0 & 0 & \dots & 0 \end{pmatrix}.$$

With a suitably chosen polynomial $f(x)$ it is possible to obtain for C_f^k a $k \times k$ MDS matrix, so we have reduced one computation by a matrix with entries of high Hamming weight to a k-fold application of a lightweight matrix.

[21] See Gligor, V.D.: *Light-Weight Cryptography—How Light is Light?*; Keynote presentation at the Information Security Summer School, Florida State University. Available for download at http://www.sait.fsu.edu /conferences/2005/is3/resources/slides/gligorv-cryptolite. ppt, May 2005.

An investigation of how suitable currently standardised algorithms are for these purposes has been carried out by a team from the University of Luxembourg: Dinu, D., Le Corre, Y., Khovratovich, D., Perrin, L., Johann Großsädl, J. and Biryukov, A.: *Triathlon of lightweight block ciphers for the Internet of Things*, IACR ePrint Archive, 2015/209.

Example If $f(x) = 1 + \alpha^3 x + \alpha x^2 + \alpha^3 x^3 + x^4 \in GF(2^8)[x]$, where α is a primitive element of $GF(2^8)$, defined by $\alpha^8 = \alpha^4 + \alpha^3 + \alpha + 1$, then (in hexadecimal)

$$C_f = \begin{pmatrix} 0\,0\,0\,1 \\ 1\,0\,0\,8 \\ 0\,1\,0\,2 \\ 0\,0\,1\,8 \end{pmatrix},$$

which has only entries of very low Hamming weight, and C_f^4 is the MDS matrix

$$\begin{pmatrix} 1 & 8 & 42 & 3e \\ 8 & 41 & 2e & a9 \\ 2 & 18 & c5 & 52 \\ 8 & 42 & 3e & 2e \end{pmatrix}.$$

Moreover

$$C_f^{-1} = \begin{pmatrix} 8\,1\,0\,0 \\ 2\,0\,1\,0 \\ 8\,0\,0\,1 \\ 1\,0\,0\,0 \end{pmatrix}$$

again contains only entries of low weight.

MDS matrices arising from companion matrices are a fairly recent development. They are used in the *PHOTON* family of cryptographic primitives.

9.5.7 Binary Codes

The reader will have noted that in reducing the matter of good diffusion/high branch number to coding theory, and in particular to an investigation of MDS matrices, we changed from considering purely binary codes to codes in which the alphabet used was generalised from $\{0, 1\}$ to the elements of $GF(2^n)$. This leaves open the question: how good is the diffusion we can get if the linear layer allows only 0s and 1s as entries in the matrix defining that layer?

Or, phrasing the question slightly differently: How good is the diffusion we can obtain if we operate on inputs (bytes or words of greater length) if we consider only operations at the bit level?

We again consider this question from the view of coding theory, considering only square matrices: If we have a purely binary code, what is the maximum distance d for which a

$[2n, n, d]$-code exists? Interestingly, the answer to this question is unknown for n in general. The data in the table are obtained from a paper by Kwon et al.[22]:

n	d_{max}
1	2
4	4
8	5
16	8
24	12
32	Unknown, $12 \leq d_{max} \leq 16$

9.5.8 Boolean Functions and Codes

We note here, without proof, an interesting relationship between binary linear codes and Boolean functions. We recall from Sect. 8.3 that a Boolean function f is *t-resilient* if it is balanced and

$$\Pr(f(x) = 1) = \Pr(f(x) = 1|x_{i_0}, \ldots, x_{i_{t-1}}).$$

Slightly more generally we can define a function $f : \mathbb{Z}_2^n \longrightarrow \mathbb{Z}_2^m$ to be t-resilient if for any $y \in \mathbb{Z}^m$

$$\Pr(f(x) = y) = \Pr(f(x) = y|x_{i_0}, \ldots, x_{i_{t-1}}).$$

We shall call such a function an (n, m, t)-function.

Then it can be proved that an $[n, k, d]$ code exists if and only if there exists an $(n, n-k, d-1)$-function.

For example, the code generated by the matrix

$$\mathbf{G} = \begin{pmatrix} 1\,0\,0\,0\,0\,1\,1 \\ 0\,1\,0\,0\,1\,0\,1 \\ 0\,0\,1\,0\,1\,1\,0 \\ 0\,0\,0\,1\,1\,1\,1 \end{pmatrix}$$

[22]Kwon, D., Sung, S.H., Song, J.H. and Park, S.: *Design of block ciphers and coding theory*, Trends in Mathematics **8** (2005), 13–20.

is a [7,4,3] code[23] and has the matrix

$$\mathbf{H} = \begin{pmatrix} 0\,1\,1\,1\,1\,0\,0 \\ 1\,0\,1\,1\,0\,1\,0 \\ 1\,1\,0\,1\,0\,0\,1 \end{pmatrix}$$

as its parity check matrix.

It is easy to verify that the function $f : \mathbb{Z}_2^7 \longrightarrow \mathbb{Z}_2^3$ defined by $f(\mathbf{x}) = \mathbf{x}H^t$ is a 2-resilient linear function.

[23]The well-known [7,4,3] Hamming code, in fact.

10
Number Theory in Public Key Cryptography

Our main emphasis in the last few chapters has been on *symmetric*, i.e. secret-key, cryptography. We shall, however, in this chapter return to the Number Theory we touched upon in Chap. 2.

We have already dealt with two of the most important applications of number theory, namely the difficulty of factoring as used in the RSA public key system, and the difficulty of the discrete logarithm problem in Diffie–Hellman key establishment and ElGamal encryption. It is worthwhile recalling that until as recently as 1976, when Diffie and Hellman's famous paper appeared, it was generally thought to be impossible to encrypt a message from Alice to Bob, if Alice and Bob had not previously obtained a secret key. Until then number theory was seen as the "Queen of Mathematics", which is how Gauss described it, and as one would like a queen to be, this could (somewhat rudely) be construed as meaning "beautiful, but of no practical value".

In fact, G.H. Hardy in his book *A Mathematician's Apology* said

> Science works for evil as well as for good (and particularly, of course, in time of war); and both Gauss and lesser mathematicians may be justified in rejoicing that there is one science at any rate, and that their own, whose very remoteness from ordinary human activities should keep it gentle and clean.

In this chapter we show how Number Theory, contrary to what Hardy would have expected, can provide the equipment to perform some other activities, which might at one time not so long ago have seemed impossible. Favourite among these, as far as my personal taste is concerned, is a protocol which enables Alice and Bob to perform a virtual "coin flip". This depends on the theory of quadratic residues, which just happened to be one of Gauss's favourite topics.

Coin flipping by telephone is an early instance of a more general cryptological concept, namely that of commitment, which in turn is a useful concept in, for example, devising practical protocols for bidding for contracts. By which time we have descended into commerce, a very ordinary human activity, not particularly gentle or clean.

In the first two sections of this chapter we deal with the *sharing* of secrets; this is different from what we did in Chap. 4 where we dealt with the *establishment* of jointly held secrets. The exposition in Sects. 10.3–10.6 is very heavily based on that of Vinogradov in his book

A.R. Meijer, *Algebra for Cryptologists*, Springer Undergraduate Texts in Mathematics and Technology,
DOI 10.1007/978-3-319-30396-3_10

Elements of Number Theory, the English translation of which was published in 1954 by Dover Publications Inc. This may seem like a very old source, but let us remember—and here Gauss and Hardy would undoubtedly agree—mathematical truths are forever. And Vinogradov never knew what we would be able to do with his theorems. Nor did Gauss or Hardy, for that matter.

10.1 Secret Sharing: Shamir's Mechanism

There are many situations in which it is desirable that a group of people be able to do something, but where the task is too important to allow a single individual (who may go off, either on holiday or his head) to perform it. Examples that spring to mind are opening a bank vault or releasing a nuclear armed intercontinental missile. In such cases one needs a form of shared control: the action can only be instituted if both (or all, if there are more than two) the authorised parties agree.

One way of achieving this might be to give parties A, B and C, say, secrets s_A, s_B, s_C, and having $s_A \oplus s_B \oplus s_C$ as the master secret.

At the same time, it may be desirable that stand-ins can act on behalf of these authorised parties in the event that one or more of them are unable to perform their task or tasks. It would be inconvenient for the customers if the bank is unable to serve them because the branch manager is recovering in bed from the effect of eating some superannuated calamari. If each of A, B, C above has a stand-in A', B', C', who holds exactly the same share of the secret as the prime representative, one has a solution of sorts. But this solution implies that if A and A' enjoyed the same calamari meal, the vault still cannot be opened.

A more elegant solution would be to distribute the secret in such a way that any subset of, say, m members of an n-member set can reconstruct it. We shall follow Shamir,[1] in showing how this can be done, restricting ourselves to the case $m = 2$, although it will be obvious how the method can be generalised.

The fundamental idea is that in order to completely determine a linear function $f(x) = mx + c$, one needs at least two points on its graph. If we take the secret to be c, the point of intersection of the graph with the y-axis, then any two individuals who know distinct points on the graph can determine it. On the other hand, a single individual, knowing only one point on the straight line, has no information whatsoever about where the line crosses the y-axis.

As noted, the generalisation is obvious: if one requires that the secret can only be computed by m individuals, one uses a polynomial function of degree $m-1$ instead of a linear function. These schemes are known as *threshold* schemes, for the simple reason that the secret value cannot be found unless the number of individuals is above a certain threshold.

Note that there is no reason to consider real polynomials of a real variable: the same process works over any field—and, in the case of computer applications, probably better over a (large) field of characteristic 2. In case you wondered: the techniques for polynomial interpolation—in this case the technique for finding the equation of a curve of degree $m - 1$ when

[1] Shamir, A.: *How to share a secret* Comm. ACM **22** (1979), pp. 612–613.

given m points on the curve—are the same, regardless of the field over which these curves are defined.

But threshold schemes do not cover (quite realistic) situations where mere numbers are not enough, but where ranks are important (e.g. two generals, or one general and two colonels) or other conditions must hold (e.g. two Americans and two Bulgarians). Such more complicated scenarios will be discussed in the next section.

10.2 Access Structures and the Chinese Remainder Theorem

10.2.1 Access Structures

Suppose we have a set P of participants. There are some groups of members (subsets of P) who are granted access to the secret (whatever it may be: the numerical code to open the vault, or the private telephone number of the editor of the *Daily Blaster*), and all other groupings of people are excluded. We assume that the scheme is *monotonic*, in the sense that if a certain set of participants can derive the secret, then any set which properly contains it will also be able to do so. (This may not always be a good idea: it would be nice to arrange that if a spy joins an authorised group, the ability to access the secret falls away, but that is *much* harder to arrange.)

So what we have is a subset $\mathcal{A}$ of the set of subsets of P with the following property:

$$\text{If } M \in \mathcal{A} \text{ and } M \subseteq N, \text{ then } N \in \mathcal{A}.$$

Such a family of subsets is called an *access structure*. It is frequently convenient, as well as quite natural, to specify $\mathcal{A}$ by listing its minimal elements, i.e. elements $B \in \mathcal{A}$ which also satisfy

$$B \in \mathcal{A} \text{ and if } X \subset B \text{ then } X \notin \mathcal{A}.$$

10.2.2 Disjunctive and Conjunctive Normal Forms

We shall explain how to share a secret, given an access structure, only by means of an example. But to do so, it will be convenient if we temporarily use the following notation from symbolic logic (but note the similarity with the corresponding notation for sets): $\wedge$ and $\vee$ are placed between propositions with

$\wedge$ denoting the logical AND (logical *conjunction*),

$\vee$ denoting the logical OR (logical *disjunction*).

The distributive laws and De Morgan's laws hold (with negation denoted by $\neg$, and logical equivalence by $\equiv$):

$$a \wedge (b \vee c) \equiv (a \wedge b) \vee (a \wedge c),$$
$$a \vee (b \wedge c) \equiv (a \vee b) \wedge (a \vee c),$$
$$\neg(a \wedge b) \equiv \neg a \vee \neg b,$$
$$\neg(a \vee b) \equiv \neg a \wedge \neg b.$$

Also note the "absorbtion laws":

$$a \wedge (a \vee b) \equiv a,$$
$$a \vee (a \wedge b) \equiv a.$$

Using these identities, a proposition consisting of the disjunction of a number of expressions each of which is a conjunction, can be transformed into an equivalent proposition formed as a conjunction of a number of expressions each of which is a disjunction. Let us clear this up by giving the promised

Example

$$\begin{aligned} &(a \wedge b) \vee (b \wedge c \wedge d) \\ &\equiv (a \vee (b \wedge c \wedge d)) \wedge (b \vee (b \wedge c \wedge d)) \\ &\equiv (a \vee b) \wedge (a \vee c) \wedge (a \vee d) \wedge b \\ &\equiv b \wedge (a \vee c) \wedge (a \vee d), \end{aligned}$$

where the first equivalence follows from the distributive law and the second and third from absorbtion. The first line gives us the so-called *disjunctive normal form*, the final result is in *conjunctive normal form.*

Let us now show how this ties in with the secret sharing problem.

Example At the Last International Bank, the vault has to be opened every morning. This must be done by at least two persons in each other's company: either by the local Manager, Mr Abercrombie-Fitch, together with his chief accountant Ms Booker, or, if Mr Abercrombie is out playing golf with (or without) one of his important customers, by Ms Booker, together with two of the tellers, Mr Cash and Mrs De Bitte.

Thus the authorised sets are $\{A, B\}$ and $\{B, C, D\}$ and the vault can be opened if one OR the other set is present. But this is exactly the proposition of the previous example, and as we saw there, it is equivalent to three conditions, ALL of which must be satisfied, viz

1. The presence of A or C;
2. The presence of A or D;
3. The presence of B.

So the solution to the problem lies in splitting the secret (the numerical value, say, of the code to the vault) into three parts, all of which must be known in order to reconstruct it. To do so, we allocate part 1 to A and C, part 2 to A and D, and part 3 to B.

10.2.3 *Yet Another Appearance of the CRT*

One way of doing this splitting up process of the secret is by using the Chinese remainder theorem. Let X be the secret number and choose four integers p_1, p_2, p_3 which are relatively prime, and such that $X < p_1p_2p_3$. Now we define part i of the secret as the value $share_i = X \bmod p_i$. Thus

1. A receives the values $share_1$ and $share_2$, which can be combined into a single number $X \bmod p_1p_2$;
2. B receives the value $share_3$;
3. C receives $share_1$;
4. D receives $share_2$.

10.2.4 *The Conjunctive Normal Form*

The tedious, and error prone, part of the procedure we have described is the rewriting of a logical expression given in disjunctive normal form into one in conjunctive normal form: we want to express a set of expressions, only one of which needs to be satisfied, into a set of expressions all of which must be satisfied. For this purpose we describe a simple algorithm, which is essentially based on De Morgan's laws. The basic trick is the following:

$$(a \wedge b) \vee c \equiv \neg\neg[(a \wedge b) \vee c] \equiv \neg[(\neg(a \wedge b) \wedge \neg c]$$
$$\equiv \neg[(\neg a \vee \neg b) \wedge \neg c].$$

We may therefore write

$$\neg[(a \wedge b) \vee c] \equiv (\neg a \vee \neg b) \wedge \neg c.$$

Thus the negation of a disjunctive form becomes the conjunctive normal form of negations. We explain the use of this tautology further by means of an

Example To write $P = (a \wedge b) \vee (b \wedge c) \vee (a \wedge c \wedge d)$ in conjunctive normal form we construct the following table, running through all possible truth values of a, b, c and d and noting the truth value of P. (The first column is merely for reference purposes.)

	a	b	c	d	P
0	0	0	0	0	0
1	0	0	0	1	0
2	0	0	1	0	0
3	0	0	1	1	0
4	0	1	0	0	0
5	0	1	0	1	0
6	0	1	1	0	1
7	0	1	1	1	1
8	1	0	0	0	0
9	1	0	0	1	0
10	1	0	1	0	0
11	1	0	1	1	1
12	1	1	0	0	1
13	1	1	0	1	1
14	1	1	1	0	1
15	1	1	1	1	1

In accordance with our observation, we now take the disjunctions of those *false* values of the variables which give a *false* result for P, and take their conjunction. This gives us the conjunction of the expressions in lines 0,1,2,3,4,5,8,9 and 10:

$$\begin{gathered}(a \vee b \vee c \vee d) \wedge (a \vee b \vee c) \wedge (a \vee b \vee d) \\ \wedge (a \vee b) \wedge (a \vee c \vee d) \wedge (a \vee c) \wedge (b \vee c \vee d) \\ \wedge (b \vee c) \wedge (b \vee d).\end{gathered}$$

Using the absorbtion laws, we can simplify this to $(a \vee b) \wedge (a \vee c) \wedge (b \vee c) \wedge (b \vee d)$.

Exercises

1. The reader may have noticed in the example above that Ms Booker is absolutely indispensable: if she is absent, the vault cannot be opened. Change the requirements so that Abercrombie-Fitch and the two tellers are also authorised.
2. In the case of the original example, assume that all the primes concerned are approximately 80 bits long. Approximately how many bits of information does each of the participants receive when the shares are distributed? And in the distribution of the previous exercise?
3. Write in conjunctive normal form:

$$(a \wedge b) \vee (b \wedge c) \vee (a \wedge c).$$

4. Find a CRT method of dividing a secret value X among four people so that any three of them can determine it.

10.3 Quadratic Residues

If a congruence of the form $x^m \equiv a \bmod n$ has a solution, then a is called an mth power residue modulo n. In particular, if $m = 2$, a is called a *quadratic residue* modulo n.

If the modulus is a prime p ($p > 2$), and if a is a quadratic residue, then the congruence $x^2 \equiv a \bmod p$ will have two solutions, for if x_1 is a solution, then so is $x_2 = p - x_1$. On the other hand, there cannot be more than two solutions: for if $x_1^2 \equiv x_2^2 \bmod p$, then $p|(x_1 - x_2)(x_1 + x_2)$, so that either $x_1 \equiv x_2 \bmod p$ or $x_1 \equiv -x_2 \bmod p$.

It follows easily that, denoting the group of nonzero elements of $\mathbb{Z}_p$ by $\mathbb{Z}_p^*$, exactly half the elements of $\mathbb{Z}_p^*$ are quadratic residues. The other elements of $\mathbb{Z}_p^*$ are called quadratic non-residues.

It is also trivially easy to prove that the quadratic residues form a subgroup of the group $\mathbb{Z}_p^*$. The order of this subgroup is $\frac{p-1}{2}$.

Example If $p = 11$, then $\mathbb{Z}_p^* = \{1, 2, 3, \ldots, 10\}$, and the quadratic residues are 1, 4, 9, 5 and 3. The multiplication table of the subgroup is

	1	4	9	5	3
1	1	4	9	5	3
4	4	5	3	9	1
9	9	3	4	1	5
5	5	9	1	3	4
3	3	1	5	4	9

The following lemma gives an easy, and useful, characterization of quadratic residues:

Lemma *Let $p > 2$ be a prime. Let a be an integer such that $p \nmid a$. Then a is a quadratic residue modulo p if and only if $a^{\frac{p-1}{2}} \equiv 1 \bmod p$, and a is a quadratic non-residue if and only if $a^{\frac{p-1}{2}} \equiv -1 \bmod p$.*

Proof First note that if a is a quadratic residue, i.e. if $a \equiv b^2 \bmod p$ for some b, then $a^{\frac{p-1}{2}} \equiv b^{p-1} \equiv 1 \bmod p$, by Fermat's Little Theorem. Thus every quadratic residue modulo p is a solution of the congruence $x^{\frac{p-1}{2}} \equiv 1 \bmod p$. But such a congruence has at most $\frac{p-1}{2}$ solutions, so these solutions must be exactly the quadratic residues.

On the other hand if a is not a quadratic residue, then, because

$$(a^{\frac{p-1}{2}} - 1)(a^{\frac{p-1}{2}} + 1) = a^{p-1} - 1 \equiv 0 \bmod p,$$

which is true for all a, and

$$a^{\frac{p-1}{2}} - 1 \not\equiv 0 \bmod p,$$

which is true by the above, we must have that

$$a^{\frac{p-1}{2}} + 1 \equiv 0 \bmod p.$$

10.4 The Legendre Symbol

Definition If $p > 2$ is a prime and $a \in \mathbb{Z}$ is not a multiple of p, then the Legendre symbol $\left(\frac{a}{p}\right)$ is defined by

$$\left(\frac{a}{p}\right) = \begin{cases} 1 & \text{if } a \text{ is a quadratic residue mod } p, \\ -1 & \text{otherwise.} \end{cases}$$

We should perhaps emphasize here that we are talking about residue classes, so that $(\frac{a}{p})$ is the same thing as $(\frac{a+kp}{p})$. So we are really considering a property which half of the elements of the field $GF(p)$ possesses, and the other half doesn't.

We can consider the subgroup QR_p of $\mathbb{Z}_p^*$ consisting of the quadratic residues modulo p in another way. Consider the group homomorphism[2]

$$\phi : \mathbb{Z}_p^* \longrightarrow \mathbb{Z}_p^* : a \mapsto a^2.$$

Then $\text{Ker}(\phi)$ consists of those integers (modulo p) which are mapped onto $1 \in \mathbb{Z}_p^*$, which is the set $\{p-1, 1\}$, and it follows that the index of QR_p in $\mathbb{Z}_p^*$ is two. More than that: we get that $\mathbb{Z}_p^*/QR_p$ is isomorphic to the multiplicative group of order 2. But that implies immediately that for any two elements $x, y \in \mathbb{Z}_p^*$

$$\left(\frac{x}{p}\right)\left(\frac{y}{p}\right) = \left(\frac{xy}{p}\right).$$

But a simpler way of deriving this fact, without using any group-theoretical notions, is from the earlier observation that $a \in \mathbb{Z}_p^*$ is a quadratic residue if and only if $a^{\frac{p-1}{2}} \equiv 1 \bmod p$, and $a^{\frac{p-1}{2}} \equiv -1 \bmod p$ otherwise. Thus

$$\left(\frac{a}{p}\right) = a^{\frac{p-1}{2}}.$$

The following are immediate easy consequences of this:

Property 1 $\left(\frac{1}{p}\right) = 1.$

Property 2 $\left(\frac{-1}{p}\right) = (-1)^{\frac{p-1}{2}} = \begin{cases} 1 & \text{if } p \equiv 1 \bmod 4, \\ -1 & \text{if } p \equiv 3 \bmod 4. \end{cases}$

---

[2]Note that we are only dealing with a *group* homomorphism. We are not concerned with ring properties here.

Property 3 $\left(\frac{a_1 a_2 \ldots a_n}{p}\right) = \left(\frac{a_1}{p}\right)\left(\frac{a_2}{p}\right)\ldots\left(\frac{a_n}{p}\right)$.

Property 4 In particular $\left(\frac{ab^2}{p}\right) = \left(\frac{a}{p}\right)$.

In order to derive some further useful properties we need the following

Theorem *Let $p > 2$ be a prime, and $p = 2p_1 + 1$. Let a be any integer not divisible by p. Then*

$$\left(\frac{a}{p}\right) = (-1)^S,$$

where $S = \sum_{i=1}^{p_1} \lfloor \frac{2a \cdot i}{p} \rfloor$.

Proof Consider the congruences (modulo p):

$$a \cdot 1 \equiv \epsilon_1 r_1,$$
$$a \cdot 2 \equiv \epsilon_2 r_2,$$
$$\ldots$$
$$a \cdot p_1 \equiv \epsilon_{p_1} r_{p_i},$$

where the r_i are the smallest *absolute* remainders when $a \cdot i$ is divided by p; in other words, $1 \le r_i \le p_1$, and each of the ϵ_i is either $+1$ or -1. As i runs through all values from 1 to p_1, r_i runs through p_1 different values, since $a \cdot i \equiv x$ and $a \cdot j \equiv -x$ would imply that $a \cdot (i + j) = 0$, so that $i = p - j$ or $j = p - i$, which is impossible, since both i and j are strictly less than $p/2$. But if all the r_i are different, we must have that they assume all possible values from 1 to p_1.

Now, on multiplying all these congruences, and "dividing" by the product $1 \cdot 2 \cdot 3 \cdot \cdots \cdot p_1$ (i.e. multiplying modulo p by the inverse, modulo p, of $1 \cdot 2 \cdot 3 \cdot \cdots \cdot p_1$), we get

$$a^{\frac{p-1}{2}} = a^{p_1} = \epsilon_1 \epsilon_2 \ldots \epsilon_{p_1},$$
$$\text{i.e} \left(\frac{a}{p}\right) = \epsilon_1 \epsilon_2 \ldots \epsilon_{p_1}.$$

We shall denote the fractional part of a number x by $\{x\}$, so $x = \lfloor x \rfloor + \{x\}$. Then we can write

$$\left\lfloor \frac{2ai}{p} \right\rfloor = \left\lfloor \left\lfloor 2\frac{ai}{p} \right\rfloor + 2\left\{\frac{ai}{p}\right\} \right\rfloor$$
$$= 2\left\lfloor \frac{ai}{p} \right\rfloor + \left\lfloor 2\left\{\frac{ai}{p}\right\} \right\rfloor,$$

which is *even* precisely when $\left\{\frac{ai}{p}\right\} < \frac{1}{2}$, or, in other words, when the least absolute residue of ai modulo p is less than $\frac{1}{2}p$, that is, when $\epsilon_i = +1$. And it is *odd* precisely when $\left\{\frac{ai}{p}\right\} > \frac{1}{2}$, that is, when the least absolute residue of ai modulo p is negative (because the least positive residue is greater than p_1), i.e. when $\epsilon_i = -1$.

We therefore have that $\epsilon_i = (-1)^{\lfloor \frac{2ai}{p} \rfloor}$ and the result follows from our statement that $\left(\frac{a}{p}\right) = \epsilon_1\epsilon_2 \ldots \epsilon_{p_1}$, completing the proof.

As a first consequence of this theorem we have the following

Lemma *If $p > 2$ is a prime, and a is an odd integer ($p \not| a$), then*

$$\left(\frac{2}{p}\right)\left(\frac{a}{p}\right) = (-1)^{\sum_{i=1}^{p_1} \lfloor \frac{ai}{p} \rfloor + \frac{p^2-1}{8}},$$

where, as before, $p_1 = \frac{p-1}{2}$.

Proof $\left(\frac{2}{p}\right)\left(\frac{a}{p}\right) = \left(\frac{2a}{p}\right) = \left(\frac{2a+2p}{p}\right) = \left(\frac{4\times\frac{a+p}{2}}{p}\right)$ because $a + p$ is even, which yields $\left(\frac{4}{p}\right)\left(\frac{\frac{a+p}{2}}{p}\right) = \left(\frac{\frac{a+p}{2}}{p}\right) = (-1)^{\sum_{i=1}^{p_1} \lfloor \frac{ai}{p} \rfloor + i}$. Since $\sum_{i=1}^{p_1} i = \frac{p_1(p_1+1)}{2} = \frac{(p-1)(p+1)}{8} = \frac{p^2-1}{8}$, the result follows.

Putting $a = 1$ in this result yields the following useful observation:

Property 5

$$\left(\frac{2}{p}\right) = (-1)^{\frac{p^2-1}{8}} = \begin{cases} +1 & \text{if } p \equiv \pm 1 \bmod 8, \\ -1 & \text{if } p \equiv 3 \text{ or } p \equiv 5 \bmod 8. \end{cases}$$

For if $p = 8k \pm 1$, then $\frac{p^2-1}{8} = 8k^2 \pm 2k$, which is even, and if $p \equiv 3$ or $p \equiv 5$ modulo 8, then $\frac{p^2-1}{8}$ is odd.

For example, since $41 \equiv 1 \bmod 8$, $(\frac{2}{41}) = 1$, and if one takes the time to check, it turns out that $17^2 \equiv 2 \bmod 41$. On the other hand, since $43 \equiv 3 \bmod 8$, the congruence $x^2 \equiv 2 \bmod 43$ does not have a solution.

We finally note the following result, which we obtain by substituting the above result into the lemma:

Corollary *If $p > 2$ is a prime, with $p_1 = \frac{p-1}{2}$, and if a is any odd integer not divisible by p, then*

$$\left(\frac{a}{p}\right) = (-1)^{\sum_{i=1}^{p_1} \lfloor \frac{ai}{p} \rfloor}.$$

At this stage you may be wondering whether we are going anywhere, but try to keep your spirits up.

Exercises

1. For $p = 71, 73, 101, 103$, and $a = -1, 2$ determine which of the following congruences have solutions:

$$x^2 \equiv a \bmod p.$$

2. Show that the congruence $x^8 \equiv 16 \bmod p$ has at least one solution for every prime p.
3. If p is a prime and $p|x^4 + 1$ for some integer x, prove that $p \equiv 1 \bmod 8$.
4. If $p = 2q + 1$ is a prime, where q is itself a prime, and $q \equiv 1 \bmod 4$, show that $2^q \equiv -1 \bmod p$.

10.5 The Quadratic Reciprocity Theorem

The quadratic reciprocity theorem was first proved by Gauss, and is reputed to be the theorem of which he was most proud: in the course of his career, which covered many aspects of both pure and applied mathematics, he came back to it on many occasions, and provided no fewer than eight different proofs for it.

Here it is:

Theorem *If p and q are distinct odd primes, then*

$$\left(\frac{p}{q}\right)\left(\frac{q}{p}\right) = (-1)^{\frac{p-1}{2} \times \frac{q-1}{2}}.$$

Before proving this result, we shall consider another way of stating it:

If one or both p and q are of the form $4k + 1$, then the exponent on the right-hand side of this equation is even, which implies that

$$\left(\frac{p}{q}\right)\left(\frac{q}{p}\right) = 1$$

or, equivalently

$$\left(\frac{p}{q}\right) = \left(\frac{q}{p}\right),$$

so p is a quadratic residue modulo q if and only if q is a quadratic residue modulo p.

On the other hand, if both p and q are of the form $4k + 3$, then $\frac{p-1}{2}$, $\frac{q-1}{2}$ and their product are all odd, so

$$\left(\frac{p}{q}\right)\left(\frac{q}{p}\right) = -1,$$

so that

$$\left(\frac{p}{q}\right) = -\left(\frac{q}{p}\right),$$

or in other words p is a quadratic residue modulo p if and only if q is not a quadratic residue modulo p.

For example: $(\frac{17}{13}) = 1$ and $(\frac{13}{17}) = 1$, whereas $(\frac{19}{3}) = 1$ and $(\frac{3}{19}) = -1$.

Proof of the Theorem Put $p_1 = \frac{p-1}{2}$ and $q_1 = \frac{q-1}{2}$. We consider all $p_1 \times q_1$ pairs of the form (qx, py), where $x \in \{1, 2, \ldots, p_1\}$ and similarly $y \in \{1, 2, \ldots, q_1\}$. Now we define S_1 as the number of such pairs in which $qx < py$, and S_2 as the number of pairs with $qx > py$. Note that $qx = py$ is impossible, so that the total number of pairs is $S_1 + S_2 = p_1 q_1 = \frac{p-1}{2} \times \frac{q-1}{2}$.

For any given y, the number of pairs (qx, py) that belong to S_1 is the number of values of x that satisfy the inequality $x < \frac{p}{q}y$, which is the same as (because x is an integer) the number of values of x satisfying $x < \left\lfloor \frac{p}{q}y \right\rfloor$.

Hence

$$S_1 = \sum_{y=1}^{q_1} \left\lfloor \frac{p}{q}y \right\rfloor.$$

Similarly

$$S_2 = \sum_{x=1}^{p_1} \left\lfloor \frac{q}{p}x \right\rfloor.$$

At this stage we recall the lemma with which the previous section ended[3] and recognise that what we have is that

$$\left(\frac{p}{q}\right) = (-1)^{S_1},$$

$$\left(\frac{q}{p}\right) = (-1)^{S_2},$$

so that

$$\left(\frac{p}{q}\right)\left(\frac{q}{p}\right) = (-1)^{S_1+S_2} = (-1)^{\frac{p-1}{2} \times \frac{q-1}{2}}.$$

[3] You thought we weren't getting anywhere, didn't you?!

Exercises

1. Find all primes p such that $(\frac{5}{p}) = 1$.
2. Find all primes p such that $(\frac{10}{p}) = 1$.
3. Prove that if p is a prime of the form $28k + 1$, then the congruence $x^2 \equiv 7 \bmod p$ has a solution.
4. Let $q = 4^n + 1$. Show that q is prime if and only if $3^{\frac{q-1}{2}} \equiv -1 \bmod q$.

10.6 The Jacobi Symbol

The Jacobi symbol provides what appears to be an easy generalisation of the Legendre symbol, and it will turn out to be, among other things, useful in the computation of Legendre symbols, something which we can as yet not do terribly well. It will also turn out to have properties which follow easily from and are astonishingly like the properties we have derived about the Legendre symbol.

Definition Let $P = p_1 p_2 \dots p_n$ be any odd natural number, where the p_i are primes, not necessarily distinct. If a is any integer, relatively prime to P, then its Jacobi symbol $(\frac{a}{P})$ is defined as

$$\left(\frac{a}{P}\right) = \prod_{i=1}^{n} \left(\frac{a}{p_i}\right).$$

It is clear that if $x^2 \equiv a \bmod P$ has a solution, then a must be a quadratic residue mod p_i for every i, $i = 1, 2, \dots, n$. But the converse is easily seen not to be true: for example $(\frac{2}{15}) = (\frac{2}{3})(\frac{2}{5}) = (-1)(-1) = 1$, but 2 is a quadratic residue neither modulo 3 nor modulo 5, so it cannot be a quadratic residue modulo 15.

To cover such cases, we shall call every a with $(\frac{a}{P}) = 1$ which is not a square modulo P a *pseudosquare modulo P*.

We list the most important properties of the Jacobi symbol. Using these, one can, as we shall show in the examples following, evaluate $(\frac{a}{P})$ for any a relatively prime to P. It will also be clear from the examples that the complexity of this technique is the same as that of the Euclidean algorithm; in other words, it is highly efficient.

In the following list P is an odd natural number, and a, b, a_i, etc. are integers relatively prime to P. The first four properties follow immediately from the corresponding properties of the Legendre symbol.

Property 1 If $a_1 \equiv a_2 \bmod P$, then $(\frac{a_1}{P}) = (\frac{a_2}{P})$.

Property 2 $(\frac{ab}{P}) = (\frac{a}{P})(\frac{b}{P})$.

Property 3 $(\frac{1}{P}) = 1$.

Property 4 $(\frac{ab^2}{P}) = (\frac{a}{P})$.

Property 5 $(\frac{-1}{P}) = (-1)^{\frac{P-1}{2}}$.

Property 5 needs proof!
Let $P = p_1p_2 \dots p_n$, where the p_i are primes; then by one of the results of Sect. 10.4,

$$\begin{aligned}\left(\frac{-1}{P}\right) &= \left(\frac{-1}{p_1}\right)\left(\frac{-1}{p_2}\right)\cdots\left(\frac{-1}{p_n}\right)\\ &= (-1)^{\frac{p_1-1}{2}}(-1)^{\frac{p_2-1}{2}}\dots(-1)^{\frac{p_n-1}{2}}\\ &= (-1)^{\frac{p_1-1}{2}+\frac{p_2-1}{2}+\dots+\frac{p_n-1}{2}}.\end{aligned}$$

Now an arithmetical trick:

$$\begin{aligned}\frac{P-1}{2} &= \frac{p_1p_2\dots p_n - 1}{2}\\ &= \frac{1}{2}\left[\left(1+2\frac{p_1-1}{2}\right)\left(1+2\frac{p_2-1}{2}\right)\dots\right.\\ &\qquad \left.\left(1+2\frac{p_n-1}{2}\right)-1\right].\end{aligned}$$

On multiplying this out, we get

$$\frac{P-1}{2} = \frac{1}{2}\left[1+2\frac{p_1-1}{2}+\dots+2\frac{p_n-1}{2}+4N-1\right]$$

for some integer N, and therefore

$$\frac{P-1}{2} = \frac{p_1-1}{2}+\frac{p_2-1}{2}+\dots+\frac{p_n-1}{2} = \frac{P-1}{2}-2N,$$

which gives us

$$\left(\frac{-1}{P}\right) = (-1)^{\frac{P-1}{2}-2N} = (-1)^{\frac{P-1}{2}},$$

as required.

Property 6 The next property, which extends Property 5 of the Legendre symbol, is proved by using a similar trick: this time we write

$$\frac{P^2-1}{8} = \frac{p_1^2p_2^2\cdots p_n^2-1}{8}$$

in the form

$$\frac{1}{8}\left[\left(1+8\frac{p_1^2-1}{8}\right)\left(1+8\frac{p_2^2-1}{8}\right)\cdots\right.$$
$$\left.\left(1+8\frac{p_n^2-1}{8}\right)-1\right]$$
$$=\frac{p_1^2-1}{8}+\frac{p_2^2-1}{8}+\cdots+\frac{p_n^2-1}{8}+N$$

for some integer N, as, once again, one can see by multiplying out. In fact, when multiplying out, one will find that N is even, as every term $\frac{p_i^2-1}{2}$ is even. Putting this into the expression for the Jacobi symbol we get

$$\left(\frac{2}{P}\right)=(-1)^{(p_1^2-1)/8+(p_2^2-1)/8+\cdots+(p_n^2-1)/8}$$
$$=(-1)^{(P^2-1)/8-2N'}$$
$$=(-1)^{(P^2-1)/8}.$$

At this stage the reader will be happily aware that the properties of the Legendre symbol appear to carry over to the Jacobi symbol in an almost unchanged form (ignoring the fact, of course, that a Jacobi symbol $(\frac{a}{P})=1$ does not mean that a is actually a quadratic residue modulo P). The reader may therefore ask whether the theorem of which Gauss was most proud, the quadratic reciprocity theorem, carries over to composite P as well. Neither Gauss nor this reader will be disappointed:

Property 7 If P and Q are relatively prime odd integers, then

$$\left(\frac{P}{Q}\right)\left(\frac{Q}{P}\right)=(-1)^{\frac{P-1}{2}\times\frac{Q-1}{2}}.$$

Proof Let $P=p_1p_2\ldots p_r$ and $Q=q_1q_2\ldots q_s$, where the p_i and q_j are primes. Although the p_i (or the q_j) need not be distinct, no p_i can occur as a q_j, because P and Q are relatively prime.

Now we have:

$$\left(\frac{Q}{P}\right)=\prod_i\left(\frac{Q}{p_i}\right)=\prod_i\prod_j\left(\frac{q_j}{p_i}\right)$$
$$=(-1)^{\sum_i\sum_j(p_i-1)/2\times(q_j-1)/2}\prod_i\prod_j\left(\frac{p_i}{q_j}\right)$$
$$=(-1)^{\sum_i\sum_j(p_i-1)/2\times(q_j-1)/2}\left(\frac{P}{Q}\right)$$
$$=(-1)^{\sum_i(p_i-1)/2)\times\sum_j(q_j-1)/2}\left(\frac{P}{Q}\right).$$

But in the proof of Property 5 we saw that $(-1)^{\sum_i (p_i-1)/2} = (-1)^{(P-1)/2}$ and similarly for Q, so that we get the required result.

Examples

1.

$$\begin{aligned}
\left(\frac{2371}{449}\right) &= \left(\frac{5 \times 449 + 126}{449}\right) \\
&= \left(\frac{126}{449}\right) \text{ by Property 1} \\
&= \left(\frac{2}{449}\right)\left(\frac{63}{449}\right) = \left(\frac{63}{449}\right) \text{ by Property 6} \\
&= \left(\frac{449}{63}\right) \text{ by Property 7} \\
&= \left(\frac{8}{63}\right) \text{ by Property 1} \\
&= \left(\frac{2^2}{63}\right)\left(\frac{2}{63}\right) = \left(\frac{2}{63}\right) \text{ by Property 3} \\
&= 1 \text{ by Property 6.}
\end{aligned}$$

2. From this example and Property 7, we also get immediately that $(\frac{449}{2371}) = 1$.

As an aside: since both 449 and 2371 are primes, we could have found the Legendre symbols using the formula $\left(\frac{a}{p}\right) = a^{\frac{p-1}{2}} \bmod p$. Using the square-and-multiply method, this would have taken about $\log_2(p)$ steps. As is clear from these examples, the complexity of finding the Legendre symbols if we use the properties of the Jacobi symbol is similar to that of the Euclidean algorithm, which, again, is about $\log_2(p)$. However, the arithmetic of our new method is considerably simpler, since we never actually needed to perform any exponentiations.

Exercises

1. Find $(\frac{123}{9973})$, $(\frac{9973}{123})$ and $(\frac{1583}{1603})$.
2. Show that $(\frac{257}{391}) = 1$. How many solutions does the congruence $x^2 \equiv 257 \bmod 391$ have? $391 = 17 \times 23$, by the way.
3. Show that $(\frac{360}{391}) = 1$. How many solutions does the congruence $x^2 \equiv 360 \bmod 391$ have?
4. If $P = p_1 p_2 \ldots p_r$ and $\left(\frac{a}{p_i}\right) = 1$ for all i, how many solutions are there for the congruence $x^2 \equiv a \bmod P$?
5. Let $n = pq$, where $p \equiv q \equiv 3 \bmod 4$.

   (a) Show that n cannot be written in the form $x^2 + y^2$ where x and y are integers. *Hint:* Consider $(\frac{x}{y})^2 \bmod p$.

(b) Define $M = \{x \in \mathbb{Z} | \left(\frac{x}{n}\right) = 1, 0 \leq x < \frac{n}{2}\}$. Show that the function ψ defined by

$$\psi(x) = \begin{cases} x^2 & \text{if } x^2 < \frac{n}{2}, \\ n - x^2 & \text{otherwise,} \end{cases}$$

is a permutation of M. (This is the fundamental function in the Modified Rabin Pseudorandom Number Generator.)

10.7 Solving Quadratic Congruences

We have now seen that it is easy to find Jacobi symbols and therefore also the Legendre symbol of any a modulo a prime p. So we know whether or not the congruence $x^2 \equiv a \bmod p$ has a solution. But how do we find that solution?

It can be easy, but it isn't always. Let a be a quadratic residue modulo p. Depending on p, we distinguish three cases:

1. Solving $x^2 \equiv a \bmod p$ is easy if $p \equiv 3 \bmod 4$, i.e. if $p = 4k + 3$. For we have then that $a^{(p-1)/2} \equiv x^{p-1} \equiv 1$, so that $a^{(p+1)/2} \equiv a$, i.e. $a^{2k+2} \equiv a$, so the required square root is $x = a^{k+1}$.

   Example The solution of the congruence $x^2 \equiv 11 \bmod 19$ is $x = \pm 11^5 = \pm 7 \bmod 19$.

2. For primes congruent to 1 modulo 4, the situation is more difficult:
   If p is of the form $p = 8k + 5$, the solution of the congruence $x^2 \equiv a \bmod p$ is either

$$x \equiv \pm a^{k+1} \bmod p$$

   or

$$x \equiv \pm 2^{2k+1} a^{k+1} \bmod p.$$

   If the first solution turns out to be incorrect, the second one will work.
3. But worst of all is the remaining case, where $p = 8k + 1$ for some k. In this case one has to search for the solution: an algorithm by Tonelli in 1891 and rediscovered by Shanks, and, not entirely surprisingly, called the Tonelli–Shanks algorithm, does the job. We shall not pursue this further.

Exercises Solve the following congruences:

1. $x^2 \equiv 8 \bmod 23$.
2. $x^2 + 3x \equiv 1 \bmod 23$.
3. $x^2 \equiv 30 \bmod 37$.

10.8 The Quadratic Residuosity Problem

We reiterate that the results of Sects. 10.5 and 10.6 have made it clear that computing the Jacobi symbol $(\frac{m}{n})$ for any m and any odd n is relatively simple. It is also obvious that if m is a square modulo n, then its Jacobi symbol $(\frac{m}{n})$ will be +1. But, as we have already remarked earlier, the converse is clearly not true.

The *Quadratic Residuosity Problem*[4] is the following: Given two integers x and n, determine whether x is a quadratic residue modulo n.

If one knows the factorisation of n, the problem is easily solved: determine whether $(\frac{x}{p}) = 1$ for all the prime factors p of n. If that is true, then x is a quadratic residue modulo n. (Each one of the quadratic congruences $z^2 \equiv x \bmod p$ can, in principle, if not in fact, be solved and a solution (or, rather, all the solutions) to $z^2 \equiv x \bmod n$ can be constructed using the Chinese Remainder Theorem.) But what is one to do when one doesn't know the factors of n?

Whether the quadratic residue problem is equivalent to the factoring problem, i.e. whether an algorithm which gives correct answers to all questions "Is x a quadratic residue modulo n?" could be used in answering the questions "What are the factors of n?" is unknown. But, like the factoring problem, the quadratic residue problem has some interesting applications in public key cryptography, which we shall discuss next.

Exercise Let $n = pq$, where p and q are distinct odd primes. We know that $\mathbb{Z}_n^*$, the set $\{x|0 < x < n \text{ and } \gcd(x, n) = 1\}$ forms a group under multiplication modulo n. Let us call this group G. Which of the following subsets are subgroups of G?

1. The set of all x with $\left(\frac{x}{n}\right) = 1$.
2. The set QR of all quadratic residues modulo n.
3. The set PS of all pseudosquares modulo n.
4. The set NQR of all quadratic non-residues modulo n.

Show that the cardinality of the set QR is $\frac{(p+1)(q+1)}{4} - 1$.

10.9 Applications of Quadratic Residues

10.9.1 Rabin Encryption

In our discussion of RSA we noted that if an attacker knows the factorisation $n = pq$, then the system is broken, in the sense that the attacker can decrypt any data encrypted under Alice's public key (or alternatively forge Alice's signature whenever she uses her private key for signing). However, it has never been proved that factoring n is the only way of breaking RSA, i.e. while factoring n is a sufficient condition for breaking RSA, it has not been shown to be a *necessary* condition as well (although it is widely thought to be).

[4]"Residuosity" must be one of the ugliest words in the English language, but this author declines all responsibility for creating it.

Rabin's encryption scheme, also based on the intractability of factoring, is *provably secure*[5] in the sense in which cryptologists use that term: It is, as we shall show, as difficult to break as factoring the modulus.

In Rabin's scheme, the private key again consists of a pair (p, q) of primes, whose product is n. The public key consists of the pair (B, n), where B is an integer, $0 \leq B < n$.

To encrypt the data m, compute $c = E(m) = m(m + B) \bmod n$.

To decrypt, the recipient (let's assume it's Bob this time), has to solve the congruence

$$x^2 + Bx \equiv c \bmod n.$$

Bob, who knows p and q, does this by solving the two congruences

$$u^2 + Bu \equiv c \bmod p,$$

$$v^2 + Bv \equiv c \bmod q.$$

Each of these congruences has two solutions. The complete solution then consists of the four possible values of

$$x = au + bv \bmod n,$$

where a, b satisfy

$$a \equiv 1 \bmod p, \qquad a \equiv 0 \bmod q,$$

$$b \equiv 0 \bmod p, \qquad b \equiv 1 \bmod q.$$

This yields, in general, four possible values for the plaintext. It is up to Bob to determine which one of these was the intended message. If the message contains enough redundancy (for example, if it is an English text message) this should not be a problem.

Note that this technique is essentially that of putting the solutions of the two congruences (modulo p and modulo q) together by using the Chinese Remainder Theorem, to find solutions modulo n. But finding a and b once, and storing them for future use, provides a practical speed-up; the stored values can be used as long as the modulus n remains unchanged.

---

[5]This is the technical term used, but it may easily be misinterpreted: the scheme itself is not proved to be secure. What is proved is only that the difficulty of breaking the scheme is the same as the difficulty of some well known mathematical problem, which, with advancing knowledge, may turn out not to be all that difficult after all!

This terminology has its critics, see for example the paper by N. Koblitz and A. Menezes, *Another look at "provable security,"* J. Cryptology **20** (2007), pp. 3–37 and the reactions it provoked. You may also find interesting the paper by Neal Koblitz: *The uneasy relationship between mathematics and cryptography*, Notices of the Amer. Math. Soc., **54** (2007), pp. 972–979 and the reactions in the letters to the editor of the AMS notices.

In choosing the modulus n, both prime factors p and q are chosen as congruent to 3 modulo 4, in order to make the computation of the square roots tractable, as we saw in the previous section.

Clearly, if an adversary knows the factorisation of n, she can do the same as Bob. We now claim that breaking the system also implies that the adversary can factor n. This follows from the fact that being able to solve congruences of the form $x^2 + Bx \equiv c \bmod n$ implies the ability to find square roots modulo n. So the attacker has an algorithm that on input a returns a square root, modulo n. The attacker therefore picks a random m, and submits $m^2 \bmod n$ to the algorithm. The algorithm returns one of four possible values: $m, -m, \mu m, -\mu m$, where μ is defined by

$$\mu \equiv 1 \bmod p,$$

$$\mu \equiv -1 \bmod q.$$

If it returns one of the first two, the attacker tries again with a different value for m. But with a probability of 50 %, it will return μm or $-\mu m$. If the former, then she can use the fact that m and μm are congruent modulo p, so that she can compute $\gcd(m, \mu m) = p$. Otherwise she will compute q. Either way, she factorises n.

The fact that the ability of factoring $n = pq$ is equivalent to the ability of finding square roots modulo n, and in particular that finding two square roots x_0, x_1 of a modulo n, with $x_0 \not\equiv \pm x_1$, immediately yields a factor of n, is vitally important. It is exploited in all the applications that follow in this section.

Exercises

1. Find all the roots of $x^2 \equiv 71 \bmod 77$.
2. Using the fact that $454^2 \equiv 6893^2 \bmod 10823$, find the factors of 10823.
3. Find all the roots of $x^2 + 7x \equiv 187 \bmod 221$. $(221 = 13 \times 17.)$

10.9.2 Goldwasser–Micali Encryption

This was the first *probabilistic* public key encryption system, in that a message will yield different ciphertexts each time it is encrypted.[6]

---

[6]Classical RSA, as should be evident, did not satisfy this requirement. An encryption system may be defined as *semantically secure* if, given the encryption of one of two messages m_0 and m_1, chosen by the adversary, he/she cannot determine (with a probability of success different from $\frac{1}{2}$) which plaintext was encrypted. In the asymmetric case, semantic security clearly implies probabilistic (since the adversary can find the encryption of any non-randomised message), but the converse is not true in general.

For example, using the notation of Sect. 4.4.1 for ElGamal encryption and $\mathbb{Z}_p^*$ as our group, the encryption of a message m is the pair $< c_1, c_2 >$ where $c_1 = g^k$ and $c_2 = my_A^k$. Using our knowledge of the Legendre symbol, and the fact that g cannot be a square, since it is a generator of the group, we see that $(\frac{g}{p}) = -1$. This gives us the parity of the randomly selected k, as we can easily find $\frac{c_1}{p}$. We

It is also, as we shall see, totally impractical, since in encrypting a single bit a ciphertext of at least 1024 bits in length will be created if the system is to be secure. As in Rabin encryption, the security relies on the difficulty of factoring products of two large primes.

Again as in Rabin's scheme, there is a public key of the form (y, n), where y is a quadratic *non*-residue modulo n, with Jacobi symbol equal to 1, and $n = pq$, where the pair of primes (p, q) constitute the secret key.

To encrypt a bit b, the sender selects a random integer x and transmits $c = x^2y^b \bmod n$. The receiver decrypts as follows:

$$b' = \begin{cases} 0 \text{ if } c \text{ is a quadratic residue mod } n, \\ 1 \text{ otherwise.} \end{cases}$$

Clearly the message sent will always have the Jacobi symbol +1, but if $b = 1$, c is a quadratic non-residue, whereas otherwise it is a proper residue. Only with knowledge of the factorisation of n will anyone be able to compute whether $(\frac{c}{p}) = 1$ (in which case also $(\frac{c}{q}) = 1$ and the transmitted $b = 0$), or not.

Exercise Alice's public key for Goldwasser–Micali encryption is $(221, 124)$. She receives the message '43' from Bob. What bit value did Bob communicate to Alice?

10.9.3 Coin Flipping by Telephone

This fascinating protocol, due to Manuel Blum in 1983, was the first in which the theory of quadratic residues was applied to obtain the *commitment* (about which more later) of one of the parties to some data.

The idea of the protocol is described by Blum as follows:

> Alice and Bob want to flip a coin by telephone. (They have just divorced, live in different cities, want to decide who gets the car.) Bob would not like to tell Alice HEADS and hear Alice (at the other end of the line) say "Here goes . . . I'm flipping the coin . . . You lost!"

The protocol runs as follows:

1. Alice generates two large primes, p and q, both congruent to 3 modulo 4, and computes their product $n = pq$, which she sends to Bob.
2. Bob picks a number a in the range $[2, n - 1]$, computes $z = a^2 \bmod n$, and sends this to Alice.
3. Alice, who knows the factorization of n, computes the four square roots of z: these are $\pm a$ and $\pm \mu a$, where μ satisfies the two congruences

$$\mu \equiv 1 \bmod p,$$
$$\mu \equiv -1 \bmod q.$$

also compute $(\frac{c_2}{p})$. But then we can find the parity of $\frac{m}{p}$, so, given a ciphertext, we can determine which of m_0 and m_1 was encrypted if these are a quadratic residue and non-residue respectively. Thus ElGamal encryption in this simple form is, although probabilistic, not semantically secure.

By the choice of p and q the congruences are easy to solve. Note that a and μa have opposite Jacobi symbols modulo n, since

$$\left(\frac{\mu}{n}\right) = \left(\frac{1}{p}\right)\left(\frac{-1}{q}\right) = 1 \times (-1) = -1.$$

4. Alice chooses one of her solutions, say b, and sends its Jacobi symbol $\left(\frac{b}{n}\right)$ to Bob.
5. Bob has computed $\left(\frac{a}{n}\right)$. If $\left(\frac{b}{n}\right) = \left(\frac{a}{n}\right)$, Bob concedes that Alice has won. Otherwise, Bob informs Alice that she has lost AND sends his value a to Alice, who then verifies that a is indeed one of the four solutions. If this verification fails, or if Bob refuses to tell Alice what value a he had chosen, Alice concludes that Bob was attempting to cheat.[7]

A few observations may be in order:

Firstly, Alice cannot cheat: she finds four roots of z, but has no means of knowing which is the correct one; as far as she is concerned, any one is as likely as any other.

Bob, on the other hand, cannot cheat either: he is committed to his value of a, and if it turns out that Alice has won, he cannot send μa to Alice, for the simple reason that he cannot compute μ. For this he would have needed to know the factorisation of n, which (by assumption) he cannot do.

Note also that, having chosen a, Bob has no difficulty in computing $\left(\frac{a}{n}\right)$. The properties we listed for the Jacobi symbol allow him to do so, even though he does not know p and q.

Finally, note that this protocol cannot be repeated: If Alice and Bob next want to flip a coin,[8] Alice has to choose a new value of n.

Exercise Explain why Alice should not continue to use the same value of n.

10.9.4 Oblivious Transfer

In the simplest form of an oblivious transfer protocol, Alice sends a message m to Bob. Either Bob receives it or else he gains no knowledge whatsoever about it, *without Alice knowing whether Bob received m*. (That is where the "oblivious" comes in.) The version we discuss (the original Rabin version) depends, like coin flipping by telephone, on quadratic residuosity. It runs as follows:

1. Alice generates large primes p, q, both congruent to 3 modulo 4, and their product $n = pq$, and chooses a public exponent e, and transmits to Bob the triple $(n, e, m^e \bmod n)$.
2. Bob randomly picks a positive integer $x < n$, computes $z = x^2 \bmod n$ and transmits z to Alice.
3. Alice computes the four square roots of z, and sends one of them, y, to Bob.

[7] And presumably goes back to the lawyer who was handling the divorce. It must be noted here that Alice and Bob, known as the first couple of Cryptology, appear to have gotten together later, since there are numerous later protocols in which they happily communicate.

[8] The loser gets custody of the teenage children.

If $y = \pm x$, Bob has gained no information whatsoever. But if $y \neq \pm x$, then Bob can find the factors of n, and hence compute the inverse of e modulo $\phi(n)$ and thereby find m.

Note that the probability of Bob being able to compute m is exactly $\frac{1}{2}$, and that Alice, who does not know what Bob chose as his x, has no means of knowing whether Bob got the message or not.

A more complicated form of oblivious transfer is the 1 out of 2 version: Here Alice has two messages m_0 and m_1, and the protocol ensures that Bob receives exactly one of them, again without Alice knowing which one.[9] We describe a version of the protocol which (again) is RSA-based. As before, Alice computes an RSA modulus, and secret and public keys d and e respectively, and sends n and e to Bob. Then

1. Alice sends two integers x_0 and x_1 to Bob.
2. Bob selects $b \in \{0, 1\}$, as well as a randomly selected "blinding" integer k, and sends to Alice the value $v = k^e + x_b \bmod n$.
3. Alice computes the two values $K_b = (v - x_b)^d \bmod n$. One of these will be equal to k, but the other will be entirely random; she does not know which and therefore sends both $m_0 + K_0$ and $m_1 + K_1$ to Bob.
4. Bob knows the value of b, so he knows which of the two messages he can "unblind" by subtracting k.

10.9.5 Exchanging Secrets with Oblivious Transfer

In the coin flipping protocol, only one secret is involved, which Alice reveals to Bob (after committing herself to it). What happens if Alice and Bob each have a secret, which each will reveal to the other if and only if the other reveals his/hers? The trouble is that cryptographic protocols always have a "first A, then B, then A, ..." structure so that it would seem that at the end one of the parties has an opportunity for a double cross: after Bob (say) receives Alice's secret, he abandons the protocol and never divulges his. What one would like is "I'll show you mine, if you'll show me yours *simultaneously*".

Rabin came up with a fairly complicated protocol, which guarantees that with a probability of 75 % Alice and Bob obtain each other's secrets, and with probability of 25 % neither receives anything. The protocol is based on the mechanism used in the oblivious transfer protocol. In the description that follows, the secrets that Alice and Bob are single bits,[10] denoted by S_A and S_B respectively.

The protocol runs as follows:

1. Alice chooses two large primes p and q and sends the one-time key $n_A = pq$ to Bob in a signed message. Bob similarly creates $n_B = p'q'$ and sends it to Alice in a signed message.

[9] A spy has two secrets that he is willing to sell to us. But, as we suspect him of being a double agent, we don't want him to know what matter we are really interested in.

[10] This seems silly, but we are only showing what can be done in principle.

2. Bob chooses randomly $x \leq n_A$, computes $c = x^2 \bmod n_A$ and sends Alice the signed message

   "$E_{n_B}(x)$ is the encryption under my public key n_B of my chosen number x, whose square modulo n_A is c."

3. Alice computes a square root x_1 of c and sends Bob the signed message:

   "x_1 is a square root of c modulo n_A."

4. Bob computes the g.c.d. of $x - x_1$ and n_A. With probability $\frac{1}{2}$ this is one of p and q, else it is 1. So with probability $\frac{1}{2}$ Bob can factor n_A, but Alice does **not** know whether this is the case. Now define

$$\nu_B = \begin{cases} 0 \text{ if } \gcd(x - x_1, n_A) \neq 1, \\ 1 \text{ otherwise.} \end{cases}$$

5. Repeat all the previous steps with the roles of Alice and Bob interchanged, and obtain ν_A in the same way.
6. Alice finds $\epsilon_A = S_A \oplus \nu_A$ and sends it to Bob in a signed message. Bob does the similar thing with $\epsilon_B = S_B \oplus \nu_B$.
7. Alice sends S_A to Bob, encrypted as $E_{n_A}(S_A)$ in a suitable way[11] with an encryption mechanism such that Bob can decrypt if and only if he knows the factorisation of n_A. Bob similarly sends $E_{n_B}(S_B)$.

At the end of the protocol, if Bob can factor n_A, he decrypts to find S_A. He now informs Alice whether or not he knows her secret. He also knows whether or not Alice can factor n_B, but if she can't, Alice now knows the value of ν_B and therefore also of S_B (since Bob has sent her the value of $S_B \oplus \nu_B$).

The case where Alice can factor n_B is similar.

The only case where the protocol fails occurs when neither Alice nor Bob can do the factoring.

One final point: what happens if Bob knows how to factor n_A and gets S_A, but refuses to tell Alice? Alice will conclude (a) that Bob is a cheat and more important (b) that he has no further need of her and therefore must have succeeded in factoring n_A. So he might as well have told her.

10.9.6 Commitment

The coin flipping by telephone protocol is an (early) example of a commitment protocol, in which one party (Alice) commits herself to some value (frequently, as in the coin flipping

[11] Some previously agreed upon padding is probably required, and if Alice is using RSA, she must have sent the public key to Bob in step 1, and encrypt under the private key.

scenario, just a single bit). Such protocols consist of two components:

1. The **Commit** phase: In which Bob commits himself to a value, but Alice has no information whatsoever about it; and
2. The **Unveil** phase: In which he sends the value itself as well as enough information to prove that was indeed the value he committed himself to.
3. At the end of the protocol Alice will know the value and be convinced.

Another way of looking at this is to separate the two important aspects as

1. **Hiding:** In which Bob hides the data in such a way that Alice cannot obtain any information;
2. **Binding:** But also in such a way that Bob cannot later change the data.

Whether the security of the hiding is in the absolute (i.e. information-theoretic) sense or only in a computational sense is an open question. The nature of the scheme will depend on the circumstances in which it will be used.

Bit commitment schemes are the fundamental building block of several other tasks which need cryptology. We mention only two:

- **Multiparty Computation:** Several parties want to compute a non-secret result based on their inputs which they want to keep secret. Elections, and the computation of demographic statistics (e.g. average salary of a group of acquaintances) are examples.
- **Zero Knowledge Proofs:** An entity wishes to prove knowledge of a secret (e.g. the private key associated with a published public key or of a six line proof of Fermat's last Theorem) without divulging the secret. We have seen an example of a zero knowledge proof of possession in Sect. 4.4.4.

There are simpler ways of ensuring commitment than the one used in coin flipping, at least if some other mechanism, such as a hash function, is available. For instance, if Bob needs to commit to some value x, without revealing it, he can commit himself by publishing $H(x||r)$, where H is a one-way hash function and r is a random value. In the reveal phase he reveals the values of x and r and Alice can verify by also computing $H(x||r)$. This might suffice in the case of, for example, bidding for a contract. As long as H is second-preimage secure (i.e. it is infeasible to find x' and r' such that $H(x'||r') = H(x||r)$) and $H(x||r)$ does not leak any information about x or r, this works fine.

Another way of constructing a commitment scheme is to exploit the discrete logarithm problem. Let g be the generator of a cyclic group of prime order p. To hide an integer x modulo p, Bob simply publishes $g^x \bmod p$. In the unveil stage, he reveals x, and Alice can do the obvious verification. Clearly Bob is bound by the published value, and cannot change it. Several variations on this theme have appeared in the literature.

The beauty of the original coin flipping protocol lies, however, in the obviousness of how Alice's action affects the outcome of the protocol: it is a clear case of a win-lose situation.

10.9.7 Blum–Blum–Shub Pseudo-Random Bit Generation

We have, in these notes, not said much about generating pseudorandom numbers or sequences. Only when we dealt with Linear Feedback Shift Registers (LFSRs) have we touched upon the subject, which is vitally important in the practice of cryptography. We stressed the fact that LFSRs are cryptologically unacceptably weak. It is therefore appropriate to end this chapter, and the book, with a very strong pseudo-random bit generator, due to L. Blum, M. Blum and M. Shub, and therefore logically, though perhaps not euphoniously, known as the Blum–Blum–Shub generator.

Let $n = pq$, where p and q are distinct primes, with $p \equiv q \equiv 3 \bmod 4$. The modulus n can be public, but the random seed x_0 must be kept secret ($1 < x_0 < n$). We then define

$$x_{i+1} = x_i^2 \bmod n$$

and the output bit sequence is $b_0, b_1, b_2, \ldots$, where b_i ($i = 0, 1, 2, \ldots$) is the least significant bit of x_i.

The beauty of this construction is that the operation of squaring is a permutation of the set QR of quadratic residues, as is very easy to see. Moreover, it is a one-way permutation, because of the difficulty of finding square roots modulo a number of the specified form. It has been shown that this generator is secure in the sense that an attacker, observing the output but not knowing the value of x_i (nor the factors of n), has no better than a 50 % probability of guessing the next output bit. In fact, we can use the generator more efficiently, by using the log log (n) least significant bits of x_i^2. It has been shown that it will still be secure.

The method is not very practical, even with this speed-up, as the squaring operation will be far too slow for most operations.[12] So we end this chapter with an application, which is, sadly, not of great practical interest. But then, not all mathematics needs to be useful as long as it's elegant and beautiful in itself.

---

[12]The same is obviously true for the Modified Rabin generator of Exercise 5 in Sect. 10.6.

11 Where Do We Go from Here?

The purpose of the book which you have just completed was to highlight the most basic of the mathematical, and in particular the algebraic, aspects of modern cryptography. In the process we have covered quite a lot of ground, but even so we have barely scratched its surface. So in these concluding remarks, we shall indicate where you may dig deeper and also refer you to matters which have been left out of discussion altogether, some because the algebraic content is negligible or uninteresting, others simply because we wanted to keep the size of the book within reasonable bounds.

In these final notes I shall suggest, a little idiosyncratically, some further sources of useful material and list some topics which are essential or topical in current Cryptology.

11.1 Further Reading

There may be some who read this book as an introduction to Algebra; for them the way ahead is simple, and their only difficulty will be in choosing the right book for further reading. From the pure algebraist's point of view we have hardly dealt with his/her subject, as we discussed only matters which directly concerned cryptology.

There are many good books on Algebra; my personal favourites are *Abstract Algebra* by Dummit and Foote (published by Wiley) and Hungerford's *Algebra* (Springer). But there are many others, equally good. You will probably like *A Computational Introduction to Number Theory and Algebra* by Victor Shoup, available from the author, under a Creative Commons licence, at http://shoup.net/ntb/. One of Victor Shoup's main interests is the design and analysis of cryptographic protocols.

In this book we have treated the Number Theory as a derivative of Algebra, which is only partially true. In fact Number Theory is a much older subdiscipline of Mathematics than the kind of Algebra we dealt with, and the vitally important *Analytic* Number Theory has little, if anything, in common with what we discussed. Again, there are many texts which deal with (non-analytic) elementary number theory. I personally like Niven, Zuckermann and Montgomery's *An Introduction to the Theory of Numbers* (Wiley).

A.R. Meijer, *Algebra for Cryptologists*, Springer Undergraduate Texts in Mathematics and Technology,
DOI 10.1007/978-3-319-30396-3_11

Although now rather dated, *Cryptology and Computational Number Theory*, edited by Carl Pomerance and published by the American Mathematical Society in 1990, provides interesting reading. I also recommend Koblitz's *A Course in Number Theory and Cryptography* (Springer), which is about the same age. Koblitz's *Algebraic Aspects of Cryptography* (Springer) covers, at a faster pace than this book, some of the same Algebra, and then discusses elliptic and hyperelliptic curve systems as well as various "non-standard" public key systems. It also deals, in an earlier chapter, with complexity matters.

None of the foregoing pay much attention to secret key cryptography, on which we have spent some time and effort. As I hope to have convinced you, this is unfortunate, since symmetric encryption provides for most of the world's data security, whether the data is in transit or in storage—at least once the difficulties involved in managing the required keys have been solved.

As for general reference works, it is hard to beat the *Handbook of Applied Cryptography*, edited by Menezes, van Oorschot and Vanstone (CRC Press, 1996), in spite of its age. The *Encyclopedia of Cryptography and Security*, edited by van Tilborg (Springer, 2005), is also extremely useful. Sadly, the two volume second edition (2011), edited by van Tilborg and Jajodia, which contains much the same cryptological material as the first, and also much more material on computer security, is also much more expensive and lacks an index, so that, for example, the user cannot know in advance that *resilient* Boolean functions are dealt with in the article on *Correlation immune and resilient functions*.

When we come down to more specific topics:

- On the subject of **finite fields**, one must start with Lidl and Niederreiter's *Introduction to Finite Fields and their Applications* (Cambridge University Press, 1994), described as the textbook version of their *Finite Fields* (volume 20 of Cambridge's *Encyclopedia of Mathematics and its Applications*) of which the second edition appeared in 2008. If you want to go even deeper than that, there is the *Handbook of Finite Fields* (Chapman and Hall/ CRC Press, 2013) edited by Mullin and Panario.
- When it comes to **Boolean functions**, the chapter by Claude Carlet on *Boolean functions for cryptography and error correcting codes* in *Boolean Models and Methods in Mathematics, Computer Science, and Engineering*, edited by Yves Crama and by Peter L. Hammer (Cambridge University Press, 2010) will probably more than satisfy your needs.
- For a good background on **block ciphers**, I highly recommend *The Block Cipher Companion* by Knudsen and Robshaw (Springer, 2011). This covers the Data Encryption Standard, AES and several other block ciphers, as well as the various modes of use and their cryptanalysis.

  Daemen and Rijmen's *The Design of Rijndael* (Springer, 2002) is essential reading *inter alia* for understanding the authors' "wide trail strategy" as a block cipher design principle.

  Algebraic Aspects of the Advanced Encryption Standard by Cid, Murphy and Robshaw (Springer, 2006) gives an indication of exploiting the algebraic properties of Rijndael/AES analytically.

  For further information about the AES selection process, see the NIST report at csrc.nist.gov/archive/aes/round2/r2report.pdf. Specifications and notes on the design principles of the finalists are readily available on the Internet.

- The literature on **stream ciphers** is more limited. Andreas Klein's *Stream Ciphers* (Springer, 2013) is a very good text and reference. *New Stream Ciphers and their Design*, edited by Robshaw and Billet (Springer LNCS 4986, 2008) is a useful collection of the documentation provided for the eStream finalists when submitted (See Sect. 7.9). *Analysis and Design of Stream Ciphers* by Rueppel (Springer, 1986) is by now out of date.

11.2 Further Topics

We have, in this book, concentrated on the algebra applicable to cryptography, and as a consequence left out of our discussion many matters which are of vital importance to the cryptologist. The most serious of the resulting omissions is the fact that we barely mentioned cryptanalysis.

We also never touched upon matters concerning the complexity of any algorithms, which is not only of theoretical but also of practical import. Discussion of such issues would have been necessary if we had, for example, wanted to consider appropriate key lengths for block ciphers or acceptable security parameters in general. As algebraists we have in fact disregarded all implementation issues, with the exception of a very brief reference to "lightweight" cryptography in our discussion of MDS matrices, where the algebra was relevant.

But there are some matters which are important both from their algebraic content and from the cryptological point of view, and to which we dedicate a few words, in the hope that the reader of this book will have the inclination and the opportunity to pursue them further.

11.2.1 Elliptic Curves

Elliptic curve cryptography in our treatment got a mere mention. They first gained the interest of cryptographers with H.W. Lenstra Jr.'s 1987 paper *Factoring integers with elliptic curves.*[1] However, their significance stretches well beyond this: their use in cryptography was first proposed by Miller and Koblitz (independently) in 1986 and 1987. An excellent introduction to this subject is L.C. Washington's *Elliptic curves: Number theory and cryptography* (Chapman and Hall/CRC Press, Boca Raton, 2003).

At high security levels, elliptic curve authenticated key agreement (ECMQV, named after Menezes, Qu and Vanstone) and elliptic curve signatures (ECDSA) analogous to the Digital Signature Algorithm (Sect. 4.4.3) are more efficient than the integer based versions.

Important in this connection is the concept of **pairings**. Let G_1 and G_2 be additively written groups of prime order q (not necessarily distinct) and let G_T be a finite multiplicatively

[1]Ann. Math. **126** (1987), pp. 649–673.

written group. A *pairing* is a function $e : G_1 \times G_2 \longrightarrow G_T$ satisfying the following

1. *Bilinearity:* $e(aP, bQ) = (e(P, Q))^{ab}$ for all $a, b \in GF(q)\backslash\{0\}$ and all $P \in G_1, Q \in G_2$;
2. *Non-degeneracy:* $e(P, Q) \neq 1$ if neither P nor Q is $\mathcal{O}$, the "point at infinity";
3. There exists an efficient algorithm for computing e.

Such functions were first used in cryptanalysis: If the group G_T is more tractable to analysis, it can be used to solve problems involving G_1; this was the case with analysis by Menezes, Okamato and Vanstone showing the vulnerability of the so-called supersingular curves when used for Diffie–Hellman purposes.[2] On the positive side, pairings have been used in *identity based encryption schemes*, public key schemes in which a party's public key is a simple function of publicly available information about him or her, such as the e-mail address.[3] We refer to Chap. 6 of the book by Washington for both cryptographic and cryptanalytical examples.

11.2.2 Lattices

As we noted very early in this book (in Sect. 3.3, to be precise) the term "lattice" has two possible meanings in Mathematics. The more important one is actually *not* that of Sect. 3.3, but its alternative: lattice as defined as a discrete subgroup of $\mathbb{R}^n$ (or of $\mathbb{Z}^n$), i.e. a regular arrangement of points in n-dimensional space. Their study, in Number Theory, goes back all the way to Gauss, but in modern cryptography they first appeared in breaking the public key system proposed by Merkle and Hellmann.[4]

An important problem in lattice theory was that of *lattice reduction*, concerning the finding of "interesting" bases for a lattice such as, for example, a basis consisting of short and (more or less) orthogonal vectors. The LLL algorithm, due to Lenstra, Lenstra and Lovász, for such reduction was used by these authors in factoring polynomials, but it was also crucial in attacking knapsack type cryptosystems and some other cryptographic constructions, including RSA with $e = 3$ as public exponent.[5]

But, somewhat surprisingly, lattices are useful in a positive way in cryptography too. Several proposals have been made for cryptosystems based on lattice-theoretical problems.

---

[2]Menezes, A.J., Okamoto, T. and Vanstone, S.A.: *Reducing elliptic curve logarithms to logarithms in a finite field* IEEE Trans. Info. Th. **39**, pp. 1639–1646.

[3]There are also *attribute based* encryption schemes, in which case the public key is a function of the attributes of the relevant party, e.g. his/her rights and privileges: only parties who satisfy the relevant policy can decrypt.

[4]Merkle, R.C. and Hellman, M.E.: *Hiding information in trapdoor knapsacks*, IEEE Trans. Info. Th. **IT-24**, pp. 525–530. See also Odlyzko, A.M.: *The rise and fall of knapsack cryptosystems* in *Cryptology and Computational Number Theory*, edited by Carl Pomerance, Am. Math. Soc., 1990.

[5]Joux, A. and Stern, J.: *Lattice reduction: a toolbox for the cryptanalyst*, In his book, *Algorithmic Cryptanalysis* (CRC Press, Boca Raton, 2009) Joux devotes an entire chapter to lattice reduction techniques.

A good overview, but now rather dated, on lattice based techniques is that of Ngyuen, P.Q. and Stern, J.: *The two faces of lattices in cryptology*, Proc. CaLC '01, Cryptography and Lattice Conference, LNCS 2148, Springer 2001, pp. 148–180.

The best know of these are the NTRU algorithms. These, for encryption and for signing, are based on the *closest vector problem* which is related to, but harder than, the *shortest vector problem*: Given a lattice $\mathcal{L}$ and a point $\mathbf{x}$, find the point $\mathbf{y} \in \mathcal{L}$ nearest to $\mathbf{x}$. This problem is known to be NP-hard. But more important may be the fact that NTRU is resistant against Shor's integer factorisation algorithm, which when run on a quantum computer would break RSA and elliptic curve signature schemes.

11.2.3 Homomorphic Cryptography

One of the most fascinating recent developments in cryptography has been the work of Craig Gentry. A cryptosystem is called homomorphic if it allows computations to be carried out on the ciphertext. For example, in the case of RSA, if $E(m) = c$ then $c^a = (m^e)^a = (m^a)^e = E(m^a)$ (all of this modulo n, of course), so RSA is at least partially homomorphic in that exponentiation carries through encryption. This shows that RSA encryption is *malleable* in that anyone (possibly an adversary) can modify an acceptable ciphertext into another acceptable ciphertext, related to the first. Usually malleability is an undesirable property and to be avoided.

However, there may be cases where the disadvantages are outweighed by the advantages. It might be very convenient if a complicated computation could be carried out "in the cloud" instead of on one's own inadequate system: the initial data is encrypted and transmitted to an entity with greater computational resources, the computations are performed, and the results returned, after which decryption is performed to give the answer to the original question. As a start, one would need an encryption scheme which maps sums onto sums and products onto products. Gentry[6] has shown that this can, in principle, be achieved without sacrificing security.

Smart and Vecauteren published an improved version of fully homomorphic encryption later in 2009.[7]

A popular description of Gentry's scheme was given in the September 2012 issue of *American Scientist*.[8] While the techniques are not yet usable in any practical sense, much work has been done since Gentry's first publication. And, if one starts thinking about it, the

[6]Gentry, C. *A fully homomorphic encryption scheme*. Ph.D. dissertation. Stanford University, 2009. Available at http://crypto.stanford.edu/craig.

[7]Smart, N.P. and Vercauteren, F.: *Fully homomorphic encryption with relatively small key and ciphertext sizes*, IACR ePrint Archive 2009/571.

Gentry's technique depends on the use of *ideal lattices*, which are not to be confused with lattices of ideals, which might be defined as we did for subgroups in Sect. 3.3. Instead an ideal lattice is a lattice (of the kind referred to above) which is itself isomorphic to an ideal in a ring of the form $\mathbb{Z}[x]/ < f(x) >$ where $f(x)$ is an irreducible polynomial.

[8]Hayes, B.: *Alice and Bob in Cyberspace*, American Scientist, **100** (2012), pp. 362–367.

question arises: Is it really such a good idea to depend on "the cloud" for handling one's sensitive data?[9]

11.2.4 Post-Quantum Cryptography

Another cryptosystem known to be resistant against quantum computing is that of McLeice[10]: this is also based on an NP-hard problem, namely that of decoding a general linear code. It dates from 1978, but has never achieved popularity because of the immense length of the public key required and the expansion of the text in the process of encryption. Should quantum computers become a reality, these disadvantages will presumably have to be accepted!

And quantum computers may be far into the future, or maybe not. The National Security Agency is taking the matter seriously: In August 2015 they announced:

> IAD [the NSA's Information Assurance Directorate] will initiate a transition to quantum resistant algorithms in the not too distant future. ...... [W]e have determined to start planning and communicating early about the upcoming transition to quantum resistant algorithms. Our ultimate goal is to provide cost effective security against a potential quantum computer. ... Unfortunately, the growth of elliptic curve use has bumped up against the fact of continued progress in the research on quantum computing, which has made it clear that elliptic curve cryptography is not the long term solution many once hoped it would be.[11]

Interestingly, Bernstein[12] offers some reassurance, or at least tells us not to panic. He observes that "there is no justification for the leap from 'quantum computers destroy RSA and DSA and ECDSA' to 'quantum computers destroy cryptography'." He lists several cryptosystems whose security will not be affected by the advent of quantum computers (apart from possibly a need for greater key lengths) including the two we have mentioned, but also symmetric cryptography in general (with AES/*Rijndael* as his chosen example).

Other matters that we have not discussed, in spite of their cryptological importance, are hash functions, and cryptographically secure random bit generators. Neither of these topics is very interesting from the point of view of the algebraist; a mathematically oriented statistician would be more appreciative.

Finally we reiterate that we have in this book ignored a host of **practical matters** of lesser algebraic/mathematical content. These include implementation issues (and a lot can go wrong there!), key management, security parameters such as key lengths, as well as most aspects

[9]Read Philip Rogaway's article *The Moral Character of Cryptographic Work*, IACR ePrint Archive, 2015/1162 for a view on this, and on many other important matters which are in danger of being ignored by the cryptological community.

[10]McEliece, R.J.: *A public-key cryptosystem based on algebraic coding theory*, DSN Progress Report, Jet Propulsion Lab, California Inst. Techn.

[11]https://www.nsa.gov/ia/programs/suiteb_cryptography/ Accessed 16 December 2015.

[12]Bernstein, D.J., Buchmann, J. and Dahmen, E. (eds.): *Post-Quantum Cryptography*, Springer, 2009.

of algorithm complexity. Also omitted from our treatment are all issues concerning protocol design (and it is surprising how difficult that is in practice!).[13]

11.3 Even More Further Reading

If you decide to become a cryptologist in the real sense, you will find that your work is not over yet.

In the course of an article which provoked much comment, entitled *The uneasy relationship between Mathematics and Cryptology*[14] Neal Koblitz observes a "clash of research cultures between math and cryptography". He asserts

> Cryptography has been heavily influenced by the disciplinary culture of computer science, which is quite different from that of mathematics. Some of the explanation for the divergence between the two fields might be a matter of time scale. Mathematicians, who are part of a rich tradition going back thousands of years, perceive the passing of time as an elephant does. In the grand scheme of things it is of little consequence whether their big paper appears this year or next. Computer science and cryptography, on the other hand, are influenced by the corporate world of high technology, with its frenetic rush to be the first to bring some new gadget to market. Cryptographers, thus, see time passing as a hummingbird does. Top researchers expect that practically every conference should include one or more quickie papers by them or their students.

Whether one agrees with this somewhat cynical assessment or not, it is certainly true that one cannot keep track of current trends in Cryptology without keeping a careful watch on what appears at conferences, especially those organised by the International Association for Cryptological Research (IACR) (Crypto, Eurocrypt, Asiacrypt) or in cooperation with it. The proceedings of these conferences are published in Springer Verlag's Lecture Notes in Computer Science, but are also made available on the IACR's website after a few years.

In addition, the IACR maintains the Cryptology ePrint Archive, also accessible from their website, which "provides rapid access to recent research in cryptology". The growth of this medium of distributing research results is shown by the fact that whereas in 2000 there were 69 submissions, this has grown to more than 1100 in 2015. While these submissions are unrefereed, they are often the first versions of later conference or journal papers, and thus give an accurate view of current research trends. But don't be discouraged if, looking at these papers, you don't even understand the titles of many of them. That happens to most of us.

Best wishes!

[13]And if you want to see in how many ways things can go wrong in protocol design, read Boyd and Mathurias's *Protocols for Key Authentication and Key Establishment*, Springer 2003.

[14]Notices of the American Mathematical Society **54** (2007), pp. 972–979.

Appendix A

Probability

In this Appendix we give a brief summary of some facts from Probability Theory which are useful in Cryptology, concentrating on those probability distributions which are of greatest importance to us.

A.1 Definitions

A.1.1 Introduction

Consider an experiment which has a finite number of possible outcomes. The set of all outcomes is called the *sample space*. Some of the outcomes may be considered successes. An event is a subset of the set of all outcomes.

For our purposes we start with the classical definition of the probability of an event as the ratio of successes to the number of total outcomes possible in an equiprobable sample space. Our life is made simpler by the fact that we only need to concern ourselves with discrete sample spaces.

For example, let Ω be the "space" of outcomes of the throw of a single (fair and six-sided) die, then the probability of throwing a 3 is $\Pr(X = 3) = \frac{1}{6}$, since exactly one of the six possible outcomes, all of which are equally likely, is successful. Similarly $\Pr(X \text{ is even}) = \frac{3}{6} = 0.5$.

A more elegant and more acceptable definition would run as follows: Let Ω be a finite (or countable) set, then a *probability mass function* is a function p with the properties

1. $p(x) \in [0, 1]$ for all $x \in \Omega$;
2. $\sum_{x \in \Omega} p(x) = 1$.

Then an event E is a subset of Ω and the probability of E is defined as

$$\Pr(E) = \sum_{x \in E} p(x).$$

A.R. Meijer, *Algebra for Cryptologists*, Springer Undergraduate Texts in Mathematics and Technology,
DOI 10.1007/978-3-319-30396-3

Given events E_1 and E_2, $\Pr(E_1 \cup E_2)$ denotes the probability that E_1 or E_2 occurs or possibly both, and, similarly $\Pr(E_1 \cap E_2)$ the probability of both occurring.

The following hold:

1. For any event E, $\Pr(E) \in [0, 1]$;
2. $\Pr(\phi) = 0$, $\Pr(\Omega) = 1$;
3. $\Pr(\Omega \backslash E) = 1 - \Pr(E)$;
4. If $E_1 \subseteq E_2 \subseteq \Omega$, then $\Pr(E_1) \leq \Pr(E_2)$;
5. $\Pr(E_1 \cup E_2) + \Pr(E_1 \cap E_2) = \Pr(E_1) + \Pr(E_2)$ for any $E_1, E_2 \subseteq \Omega$;
6. In particular, if $E_1 \cap E_2 = \phi$, i.e. if E_1 and E_2 are mutually exclusive, then $\Pr(E_1 \cup E_2) = \Pr(E_1) + \Pr(E_2)$.

The continuous analogue of the discrete probability mass function is that of the *probability density function*: a function f_X such that

$$f_X(x) \geq 0 \text{ for all } x,$$

$$\int_{-\infty}^{\infty} f_X(x)dx = 1,$$

defines a probability distribution X. Probability density functions correspond to the properties of mass functions. But the correspondence must be treated with some caution: interpreting $f(x)$ as the probability of "event" x does not make sense. What does make sense are statements like the following:

$$\Pr_X(x \leq a) = \int_{-\infty}^{a} f_X(x)dx,$$

$$\Pr_X(a \leq x \leq b) = \int_{-\infty}^{b} f_X(x)dx - \int_{-\infty}^{a} f_X(x)dx.$$

A.1.2 Conditional Probability

Let E_1 and E_2 be two events, with $\Pr(E_2) \neq 0$. Then the *conditional probability* of E_1, given E_2, is defined as

$$\Pr(E_1|E_2) = \frac{\Pr(E_1 \cap E_2)}{\Pr(E_2)}.$$

This is a measure of the likelihood that event E_1 will occur, given that E_2 has occurred. Note that $\Pr(E_1|E_2) \geq \Pr(E_1 \cap E_2)$, with equality iff $\Pr(E_2) = 1$.

For example, the probability that the sum of the outcomes X and Y of two dice throws is 11 is $\frac{2}{36} = \frac{1}{18}$. But if we already knew the outcome x ($x \geq 5$) of the first throw, then this probability would increase to $\Pr(X + Y = 11|X = x) = \Pr(Y = 11 - x) = \frac{1}{6}$. If the outcome x of the first throw is less than 5, then $\Pr(X + Y|x) = 0$.

The following statement is easily proved: Let $\{B_i\}$, $i \in \{0, 1, 2, \ldots, n-1\}$, be a set of mutually exclusive events (i.e. $B_i \cap B_j = \phi$ if $i \neq j$) and let A be any event. Then

$$\Pr(A) = \sum_{i=0}^{n-1} \Pr(A|B_i) \Pr(B_i),$$

i.e. $\Pr(A)$ is the weighted average of the conditional probabilities.

A.1.3 Independence of Events

Two events E_1 and E_2 are called *independent* if $\Pr(E_1|E_2) = \Pr(E_1)$, or, in other words, if the occurrence of E_2 does not influence the likelihood (or the unlikelihood) of E_1 occurring. It is obvious that independence of E_1 and E_2 is equivalent to $\Pr(E_1 \cap E_2) = \Pr(E_1) \cdot \Pr(E_2)$, and that $\Pr(E_1|E_2) = \Pr(E_1)$ implies that also $\Pr(E_2|E_1) = \Pr(E_2)$.

We could have added this as item 7 on our list of properties of the function Pr.

In many applications one considers events which are independent and identically distributed, frequently simply denoted as i.i.d. For example, we may consider sequences of zeros and ones, where each term in the sequence is independent of all the others, and where in each case the probability of the term being 0 is exactly 0.5. This would, for example, correspond to the outcome of a sequence of (fair) coin tosses. This is notably different from a Markov sequence, where the probabilities of the nth outcome depend on the preceding outcomes.

A.1.4 Mode, Mean and Variance

The expected value $E(X)$ of a discrete random variable X is defined as

$$E(X) = \sum_x x \cdot \Pr(x).$$

In the continuous case with probability density function $f(x)$ this becomes

$$E(X) = \int_{-\infty}^{\infty} xf(x)dx.$$

The *mean* μ and the *variance* σ^2 of these distributions are defined as follows:

$$\mu = E(X),$$
$$\sigma^2 = E(X^2) - (E(X))^2 = E((X - \mu)^2).$$

σ (= the square root of the variance) is called the *standard deviation* of the distribution.

The *mode* of a distribution is the value of x for which $\Pr(x)$ or $f(x)$ is a maximum, if such a value exists.

A.2 Discrete Probability Distributions

A.2.1 The Uniform Distribution

This is the case where all outcomes (N of them, say) are equally likely, i.e. the probability mass function f is given by

$$f(x) = \frac{1}{N}.$$

For example, in a random byte generator (labelling the bytes numerically), one would have $\Pr(x = i) = f(i) = \frac{1}{256}$ for all $i \in \{0, 1, \ldots, 255\}$.

A.2.2 The Binomial Distribution

A binomial distribution with parameters p and n represents the distribution of successes in a sequence of n successive experiments, statistically independent of each other, with a probability p of a successful outcome. It is easy to see that the probability of exactly k successful outcomes in n experiments is $\binom{n}{k}p^k(1-p)^{n-k}$.

The function $f(k, n) = \binom{n}{k}p^k(1-p)^{n-k}$ is the corresponding probability mass function.

It is not hard to verify that the mean of the distribution is $\mu = np$, and the variance $\sigma^2 = np(1-p)$. The standard deviation is $\sigma = \sqrt{np(1-p)}$.

A.2.3 The Bernoulli Distribution

The Bernoulli distribution is the special case $n = 1$ of the binomial distribution, with outcomes from the set $\Omega = \{0, 1\}$. The obvious example is the single toss of a coin, which can come up with heads (1) or tails (0). If the probability of heads is p, then the probability mass function is simply

$$f(x) = \begin{cases} p & \text{if } x = 1, \\ 1-p & \text{if } x = 0, \end{cases}$$

or $f(x) = p^x \times (1-p)^{1-x}$.

The expected value is p, and the variance is $p(1-p)$.

A.2.4 The Poisson Distribution

The Poisson distribution models events occurring in a fixed interval of time (or segment of space). It represents the probability of a given number of events occurring in such an interval if these events occur with a known average rate λ and independently of the time since the last event.

For example, from a "good" random byte generator, one would expect the byte `0xff` to occur, on average, twice in a set of 512 bytes generated.

The Poisson probability mass function $\Pr(x = k)$, with this value of $\lambda = 2$, will then give the probability of `0xff` occurring k times in a given segment of 512 bytes of output.

This probability mass function is given by

$$\Pr(x = k) = \frac{\lambda^k e^{-\lambda}}{k!}$$

as a solution of the iterative definition:

$$f(0) = e^{-\lambda},$$

$$f(k+1) = \frac{\lambda}{k+1} f(k) \quad k = 0, 1, 2, \ldots$$

Note that this is a one-sided distribution, since fewer than 0 occurrences is clearly impossible. Note also that, unlike the previous distributions (and the subsequent ones) this distribution depends on only a single parameter. The Poisson distribution also has the remarkable property that its mean equals its variance: $\mu = \lambda$. The mode of the distribution is $\lfloor \lambda \rfloor$.

Moreover,

$$\text{if } x > \lambda, \text{ then } \Pr(X \geq x) \leq \frac{e^{-\lambda}(e\lambda)^x}{x},$$

$$\text{if } x < \lambda, \text{ then } \Pr(X \leq x) \leq \frac{e^{-\lambda}(e\lambda)^x}{x}.$$

Returning to the example with which we started: the probability of a block of 512 randomly generated bytes contain no or one copy of `0xff` is $\Pr(X \leq 1) \leq \frac{e^{-\lambda} * e * \lambda}{1} \approx 0.4060$. The probability of such a block containing exactly two copies of `0xff` is $\Pr(X = 2) = \frac{2^2 * e^{-2}}{2!} \approx 0.271$.

A.2.5 The Geometric Distribution

There are actually two forms of the geometric distribution:

1. The first is the probability distribution of the number of unsuccessful experiments before a success occurs.
2. The second gives the probability distribution of the first successful experiment being the kth one.

Thus the first is defined on the set $\{0, 1, 2, \ldots\}$ and the second on the set $\{1, 2, 3, \ldots\}$. If the probability of success is p, then the probability mass functions are, respectively

1. $f(k, p) = (1-p)^k p$,
2. $f(k, p) = (1-p)^{k-1} p$,

or, cumulatively:

1. $\sum_{i=0}^{k} f(i, p) = 1 - (1-p)^{k+1}$,
2. $\sum_{i=1}^{k} f(i, p) = 1 - (1-p)^k$.

The expected value for $E(X)$ if X has the first distribution is

$$E(X) = \sum_{k=1}^{\infty} (1-p)^k pk = p(1-p) \sum_{i=0}^{\infty} k(1-p)^{k-1}.$$

The reader may recall the following trick from Calculus: for $x \in (-1, 1)$, $\sum_{k=0}^{\infty} kx^{k-1} = \frac{d}{dx}\left(\sum_{k=1}^{\infty} x^k\right) = \frac{d}{dx}\frac{x}{1-x} = \frac{1}{(1-x)^2}$ which, with $x = 1-p$ gives us

$$E(X) = p(1-p)\frac{1}{[1-(1-p)]^2} = \frac{p(1-p)}{p^2} = \frac{1-p}{p}.$$

In the same way, the expected value for the second distribution can be shown to be $\frac{1}{p}$. In both cases the variance is $\frac{1-p}{p^2}$.

A.3 Continuous Probability Distributions

A.3.1 The Uniform Distribution

If the range of possible outcomes is the interval $[a, b]$, then the probability density function is

$$f(x) = \frac{1}{b-a} \quad \text{for } x \in [a, b]$$

(and $f(x) = 0$ outside the interval, obviously). This leads to the obvious conclusion that

$$\Pr(x \leq k) = \begin{cases} 0 & \text{if } k < a, \\ \frac{k-a}{b-a} & \text{if } k \in [a,b], \\ 1 & \text{if } k > b. \end{cases}$$

It is easy to show that $\mu = \frac{a+b}{2}$ and $\sigma^2 = \frac{(b-a)^2}{12}$.

A.3.2 The Normal Distribution

For large values of n, the binomial distribution $B(n, p)$ of a random variable X can be approximated by the (continuous) normal distribution $N(\mu, \sigma^2)$, which is defined by the probability density function

$$f(x|\mu, \sigma) = \frac{1}{\sigma\sqrt{2\pi}} e^{-\frac{(x-\mu)^2}{2\sigma^2}},$$

where as usual μ is the mean of the variable and σ the standard deviation, i.e. σ is the square root of the *variance* σ^2.

By making the transformation $Z = (X - \mu)/\sigma$, one obtains the random variable Z with normal distribution $N(0, 1)$ with mean equal to 0, and standard deviation equal to 1.

The values of Z for $N(0, 1)$ are readily available in tabular form. The table below provides an extract; it is worth noting that, by the symmetry of the normal distribution, $\Pr(Z < -\alpha) = \Pr(Z > \alpha)$ for any $\alpha \in [0, 1)$. So, for example, $\Pr(|Z| > 3.0) = \Pr(Z < -3.0) + \Pr(Z > 3.0) = 2 \times 0.00135 = 0.00270$.

The probability that Z exceeds	is
0.0	0.50000
0.5	0.30854
1.0	0.15866
1.5	0.06681
2.0	0.02275
2.5	0.00621
3.0	0.00135
3.5	0.00023
4.0	0.00003

From the tables one can conclude that

$$\begin{gathered}38\,\%\text{ of the values of } X \text{ lie between } \mu - 0.5\sigma \\ \text{and } \mu + 0.5\sigma, \\ 68\,\%\text{ lie between } \mu - \sigma \text{ and } \mu + \sigma, \\ 95\,\%\text{ lie between } \mu - 2\sigma \text{ and } \mu + 2\sigma, \\ 99.73\,\%\text{ lie between } \mu - 3\sigma \text{ and } \mu + 3\sigma,\end{gathered}$$

or, in other words, the probability that X lies more than 3σ away from the mean is approximately 0.27 %.

Example The table tells us that

$$\Pr(Z > 3.5) \approx 0.0005.$$

So if, in the test of a pseudorandom bit generator (where 0s and 1s are expected to be equally likely as outputs) we find 5193 zeros in 10000 output bits, we should be somewhat wary: In this case $\mu = 5000$, $\sigma = \sqrt{10^4 \times \frac{1}{2} \times \frac{1}{2}} = 50$, and $(5193{-}5000)/\sigma = 193/\sigma = 193/50 > 3.5$. The result of the experiment would therefore be very improbable if the 0 and 1 outputs of the generator were indeed equally likely. Some more tests of the generator are certainly indicated. (We hasten to point out that such a frequency analysis, even if a pseudorandom bit generator passes it, is by no means the only test that should be performed. But it would certainly be the first test one tries.)

A.3.3 The Exponential Distribution

The probability density function for this distribution is given by

$$f(x,\lambda) = \begin{cases} 0 & \text{if } x < 0, \\ \lambda \cdot e^{-\lambda x} & \text{if } x \geq 0. \end{cases}$$

λ is called the *rate parameter*. For this distribution the mean is $\mu = \frac{1}{\lambda}$ and the variance $\sigma^2 = \frac{1}{\lambda^2}$.

A.4 The Central Limit Theorem

Suppose that X_n is a sequence of random variables, such that, for any $\epsilon > 0$, and for some α,

$$\Pr(|X_n - \alpha| > \epsilon) \xrightarrow[\infty]{n} 0,$$

then the sequence is said to *converge in probability to* α.

The law of large numbers states that if $X_1, X_2, \ldots, X_i \ldots$ is a sequence of random variables each having mean value μ and variance σ^2, and

$$Z_n = \frac{1}{n}\sum_{i=1}^{n} X_i$$

(i.e. Z_n is the average of the first n values) then the sequence Z_n converges in probability to μ. (This corresponds to our intuitive belief that if we flip a fair coin a very large number of times, then the number of heads should be close to half that number.)

Now the Central Limit Theorem states that the distribution of $\sqrt{n}(Z_n - \mu)$, as n gets larger, approximates the normal distribution with mean 0 and variance σ^2. This is (more or less) regardless of the original distribution of the X_i.

Example If a card is drawn from a well shuffled pack, the probability of drawing a king is $p = \frac{1}{13}$, so if this is done repeatedly (with replacement and reshuffling) the number of times cards are drawn until a king is drawn is described by a geometric distribution with $\mu = \frac{1}{p} = 13$ and $\sigma^2 = \frac{1-p}{p^2} = 12 \times 13 = 156$.

Performing this experiment 100 times, with outcomes X_i, we get that $10(Z_{100} - 13)$ behaves approximately like the normal distribution $N(0, \sqrt{156}) = N(0, 12.5)$ where $Z_{100} = \frac{\sum_{i=1}^{100} X_i}{100}$. Thus with probability 68 % Z_{100} will lie between 1.5 and 25.5.

A.5 The Chi-Squared Distribution

The χ^2 distribution is probably the most widely used probability distribution in inferential statistics, such as hypothesis testing or in construction of confidence intervals.

It is defined as follows: The χ^2 distribution with k degrees of freedom is the distribution of the sum of the squares of k independent variables, each with a standard normal distribution (i.e. the normal distribution with $\mu = 0$ and $\sigma = 1$).

Suppose therefore that $\{X_i | i = 1, 2, \ldots k\}$ is a set of k independent random variables with standard normal distributions.

Then

$$Q = \sum_{i=1}^{k} X_i^2$$

has a χ^2 distribution with (by definition) $k - 1$ *degrees of freedom.*[1]

[1] $k - 1$ rather than k, essentially because once Q is fixed there is a constraint on the X_i, or, in other words, then $X_1, \ldots, X_{k-1}$ can vary, but once they have been chosen, there remains no choice for X_k.

If X is a random variable with normal distribution $N(\mu, \sigma)$, with mean μ and standard deviation σ, then $\frac{X-\mu}{\sigma}$ has a standard normal distribution. We could therefore take (with the obvious notation)

$$Q = \sum_{i=1}^{k} \left(\frac{X_i - \mu_i}{\sigma_i} \right)^2.$$

The following properties of this χ^2-distribution can be proved:

1. Mean = k;
2. Standard deviation = $\sqrt{2k}$;
3. Mode = $k - 2$ if $k > 2$;
4. Median = $k(1 - \frac{2}{9k})^2$.

The central limit theorem implies that, the χ^2 distribution being the sum of k independent random variables, it converges to a normal distribution for large k. Convergence is actually quite slow.

Example The monobit test for the output of a random bit generator determines whether the frequencies of 0s and 1s in its output are approximately the same. If, for a sequence of length n, the number of 0s occurring is n_0 and n_1 is the number of 1s, then the distribution of

$$X = \frac{(n_1 - n_0)^2}{n}$$

approximates a χ^2 distribution with one degree of freedom.

In the example of Sect. 3.2 above, $X = \frac{193^2}{10000} \approx 3.72$. A table for the χ^2 distribution shows that $\Pr(X \geq 3.72) \approx 0.05$, which suggests that further tests might be advisable before accepting this generator.

A.6 The Birthday "Paradox"

Note the quotation marks around the word "paradox": the fact that we discuss in this section is not a paradox in the classical sense, but more of a surprise when one is confronted with it the first time. The name derives from the fact that when as few as 23 randomly selected people gather at a party, there is a probability of more than 50 % that two of them celebrate their birthdays on the same day of the year. (This does not, of course, mean that when you walk in on the party that one of them has the same birthday as you: that probability is just $\frac{23}{365}$, if we ignore leap years.)

We phrase the question more generally: If k values are assigned randomly and independently from a set of n, each value with the same probability, what is the probability that the same value is assigned twice? Let us call such a repetition a *collision*.

The probability of the second assignment differing from the first is $1 - \frac{1}{n}$, then the probability that the third differs from both the first two is $(1 - \frac{1}{n})(1 - \frac{2}{n})$, and so on. The probability that the first k are all distinct will be

$$q(n,k) = \prod_{i=1}^{k-1} \left(1 - \frac{i}{n}\right),$$

so that, taking (natural) logarithms, we have

$$\log q(n,k) = \sum_{i=1}^{k-1} \log\left(1 - \frac{i}{k}\right).$$

Now, for small x, $\log(1 + x) \approx x$, $\log(1 - x) \approx -x$, so that

$$\log q(n,k) \approx \sum_{i=1}^{k-1} -\frac{i}{n} = -\frac{(k-1)k}{2n}.$$

Hence $q(n,k) \approx e^{\frac{(k-1)k}{2n}}$ and the probability $p(n,k)$ of a collision is

$$p(n,k) = 1 - q(n,k) = 1 - e^{-\frac{(k-1)k}{2n}}.$$

Putting $p(n,k) = \epsilon$, and solving for k we find, after some simplification, that

$$k \approx \sqrt{2n \log\left(\frac{1}{1-\epsilon}\right)}.$$

For $\epsilon = 0.5$, this reduces to $k \approx 1.177\sqrt{n}$. (If $n = 365$, this works out at 22.49, confirming our introductory statement.)

This result is of some importance in cryptography, where various attacks are based on finding collisions. For example, a digital signature is obtained through an operation on an n-bit "digest" of the data that is certified. If n is too small, a collision between two digests may be found, so that a signature on one document may falsely be used on another. Such an attack is commonly referred to as a *birthday attack*.

A.7 Bayes' Theorem

From our definition of conditional probability, it is easy to verify that

$$\Pr(E_2) \cdot \Pr(E_1|E_2) = \Pr(E_1 \cap E_2) = \Pr(E_1) \cdot \Pr(E_2|E_1)$$

or

$$\Pr(E_1|E_2) = \frac{\Pr(E_1) \cdot \Pr(E_2|E_1)}{\Pr(E_2)}.$$

This is known as Bayes' theorem, and has great applicability. It allows one to convert, for example on the basis of experimental results, to improve estimates for the conditional probability $\Pr(E_2|E_1)$ based on improved estimates for $\Pr(E_1|E_2)$.

From communication theory, for example, we have the following application:

Example Let m_1 be a message and m_2 be a bitstring (say of the same length as m_1) which is not a valid message. Let E_1 be the event that m_1 was transmitted and E_2 the event that m_2 is received. Then, by Bayes' theorem

$$\Pr(E_1|E_2) = \frac{\Pr(E_1) \cdot Pr(E_2|E_1)}{\Pr(E_2)}.$$

On receiving m_2, therefore, the receiver decodes this as that message m_1 for which $\Pr(E_1) \cdot \Pr(E_2|E_1)$ assumes its maximum value.

This is not the same as decoding as the nearest codeword (in the Hamming distance sense), as in the case of the error-correcting codes of Sect. 9.5, since it takes the probability distribution of the possible transmitted words m_1 into account.

Index

A.R. Meijer, *Algebra for Cryptologists*, Springer Undergraduate Texts in Mathematics and Technology,
DOI 10.1007/978-3-319-30396-3

Printed in the United States
By Bookmasters